OXIDATIVE STRESS: CLINICAL AND BIOMEDICAL IMPLICATIONS

OXIDATIVE STRESS: CLINICAL AND BIOMEDICAL IMPLICATIONS

BASHIR M. MATATA
AND
MAQSOOD M. ELAHI
EDITORS

Nova Biomedical Books
New York

For permission to use material from this book please contact us:
Telephone 631-231-7269; Fax 631-231-8175
Web Site: http://www.novapublishers.com

Library of Congress Cataloging-in-Publication Data

Oxidative stress : clinical and biomedical implications / Bashir M. Matata, Maqsood M. Elahi (editors).

p. ; cm.

Includes bibliographical references and index.

ISBN-13: 978-1-60021-888-0 (hardcover)

ISBN-10: 1-60021-888-1 (hardcover)

1. Oxidative stress. I. Matata, Bashir M. II. Elahi, Maqsood M.

[DNLM: 1. Oxidative Stress--physiology. 2. Inflammation--physiopathology. 3. Nitric Oxide--physiology. 4. Reactive Oxygen Species--adverse effects. QZ 180 O976 2008]

RB170.O9615 2008

616.3'9--dc22 2007030044

Published by Nova Science Publishers, Inc. ✢ New York

Contents

Preface

There is an emerging concept that nitric oxide (NO), its derivatives and reactive oxygen intermediates (ROS) are the major determinants of the involvement of oxidant stress that contributes to the control of physiological cell signaling processes, blood flow regulation and pathological hypoxia. There is recent published evidence for additional functions of these oxygen metabolites in cardiac gene expression; functions that include the modulation of the cytokine response of lymphocytes and the regulation of immune cell apoptosis as well as immuno-deviating effects. In vitro and in vivo experimental models investigating the biological nature of ROS have shown that there is a concentration dependent effect of NO and that differences in cell sensitivity to NO may have an impact upon the degree of immune or inflammatory state. Disturbances in NO bio-availability leads to loss of cardioprotective actions within biological systems and in some cases may even increase oxidative stress and disease progression. Hence, considerable interest in identifying the mechanisms involved in their generation and implications have spurted in the hope that inhibition strategies can be developed either bio-materially or pharmacologically.

This book brings together the emerging biological concepts of oxidative stress and its connotation in the pathogenesis of various inflammatory processes in biological systems.

In: Oxidative Stress: Clinical and Biomedical Implications ISBN: 978-1-60021-888-0
Editors: B. M. Matata, M. M. Elahi, pp. 1-22

Chapter I

Ethnic Diversity and Oxidative Stress in Health and Disease

***Kamlesh Asotra**[*1] **and Virender K. Rehan**[2]*
[1]Tobacco-Related Disease Research Program
University of California Office of President
Oakland, CA 94612, USA
[2]Department of Pediatrics, Harbor-UCLA Medical Center,
Los Angeles Biomedical Research Institute at Harbor-UCLA,
David Geffen School of Medicine at UCLA,
1124 West Carson St., RB1, Torrance, CA 90502, USA

Abstract

Global human populations of more than 6 billion people of diverse ethnic and racial groups, distributed over distinct geographical regions, are increasingly exposed to similar environmental and life style factors in the 21st century. In addition, dietary intake of high levels of fats, proteins and glucose and tobacco smoking also contribute to increased oxidative stress and subsequently, chronic inflammation and myriad diseases. This review discusses a select number of major diseases and pathological conditions that are causally related to reactive oxygen species including cardiovascular disease, cerebrovascular disease and stroke, insulin resistance and diabetes, neurological disorders and dementias, chronic lung diseases and macular degeneration, among other. Glucose-caused oxidative stress and the use of insulin infusion to reverse hyperglycemia in diabetic patients following myocardial infarction can reduce mortality by 50%, are significant recent advances in the clinical use of insulin as an antioxidant. Because of high susceptibility to oxidative stress in premature babies, myriad neonatal and postnatal conditions develop such as bronchopulmonary dysplasia, retinopathy of prematurity, intra/periventricular hemorrhage, and necrotizing enterocolitis. It is increasingly clear

* Address for Correspondence: Kamlesh Asotra, M.Phil., Ph.D. Tobacco-related Disease Research Program, University of California Office of President, 300 Lakeside Drive, 6th Floor, Oakland, CA 94612, USA. Tel: (510) 287-3366. Fax: (510) 835-4740. E-mail: kamlesh.asotra@ucop.edu

that ethnicity is an important factor in the prevalence and severity of different diseases in general, although the precise mechanistic aspects of this association are not yet clear and warrant further research. Future research needs to propel the oxidant-antioxidant research in the context of human disease among diverse ethnic groups, as well as novel metabolomic and confirmatory imaging technologies for early diagnosis of oxidative stress-caused diseases are highlighted.

Introduction

During evolution of human species, the most significant developments have occurred over the past 100,000 years, with emergence of *Homo sapiens* in Africa and its distribution to and further diversification as distinct racial and ethnic groups on several continents of the world. There are hundreds of ethnic human groups distributed all over the world, and until the last 2-3 centuries, these populations have been relatively confined within well demarcated geographical locales and nations. With the advent of global industrialization and agricultural revolution coupled with changing patterns of intercontinental immigration, and innovative developments in modern science, medicine and technology, new public health problems have emerged globally that both deserve attention and need efficacious remedies.

Chronic oxidative stress caused in the human body by environmental factors, primarily air pollutants that originate from automobile emissions, agricultural pesticide aerosols, sporadic forest fires, and tobacco smoking, are linked with ever increasing incidence and severity of various diseases in diverse human populations. Epidemiological and mechanistic biomedical studies are currently the focus of extensive research on gene-environment interactions that lead to initiation, progression and exacerbation of several diseases and attempt to obtain better molecular understanding of disease processes and subsequently for devising improved diagnoses, treatment, and prevention of these diseases that exact a substantial toll on public health and economic burden. This article presents a brief review of the current knowledge and the need for further research on both resistance and susceptibility of diverse ethnic populations to oxidative stress in normal health and major diseases from the perspective of disease prevention and treatments using anti-oxidants and improved therapeutic approaches.

Oxidants, Antioxidants and Oxidative Stress

An excellent and critical account of these terms synonymous with free radical biology has been recently reviewed (Azzi, Davies and Kelly, 2004). While a constant supply of oxygen is a critical requirement for cardiac and brain function, the role of oxygen and physiological processes associated with oxygen can be either beneficial and life sustaining or cause dysfunction and death of both heart and brain (Giardano, 2005). At resting pulse rate, the heart consumes 8-15 ml oxygen per min per 100 g tissue, and more than 70 ml oxygen per min per 100 g tissue during vigorous exercise, which is significantly higher that the 3 ml oxygen consumed per min per 100 g by the brain (West, 1991; Braunwald, 2001). Oxidants, formed at low concentrations during aerobic metabolism and at elevated levels under pathological states, are produced as superoxide anion radical, hydrogen peroxide and the hydroxyl radical, respectively, as intermediary steps in reduction of one, two or three electrons of molecular oxygen to water (Sies, 1997). The half-lives of the major reactive oxygen species (ROS) differ widely, from the short-lived hydroxyl radical that catalyzes oxidative reactions of target substrates at the site of generation to the peroxyl radical with half-life in the range of seconds and diffusible to distant target sites, necessitating appropriately diverse approaches for mounting defense responses against ROS (Sies, 1993). An antioxidant is defined as *any substance that, when present at low concentrations compared with that of an oxidizable substrate, significantly delays or inhibits oxidation of that substrate* (Halliwell and Gutteridge (1989). Antioxidants may be enzymatic or non-enzymatic in nature, and match oxidants in diversity. Oxidative stress is defined as *an imbalance between oxidants and antioxidants in favor of the oxidants, potentially leading to damage* (Sies, 1997). The human diet, used in diverse ethnic populations, contains both oxidants and antioxidants (Ames, 1983). For example, intake of large amounts of glucose (75 g), lipid and protein leads to increased oxidative stress via elevated levels of oxygen "free" radical ($O_2^{\cdot}$), the reactive oxygen species, and subsequent activation of redox-sensitive proinflammatory molecules including NF-κB, activator protein-1, early growth response factor-1 and hypoxia-inducible factor-α (Mohanty et al., 2000, 2002), but intake of moderately high calorie meal of fruit and fiber neither produces oxidative stress nor inflammation (Mohanty et al., 2004). This observation would suggest that human populations with high consumption of sweets (glucose and sucrose) would be susceptible to greater oxidative stress, and diseases associated with oxidative damage. A significant proportion of South Asians, especially people from the Indian subcontinent, who have a sweet tooth, do indeed present with a variety of diseases such as coronary artery disease, metabolic syndrome, and diabetes, all of which may be ascribed to greater oxidative stress on account of high dietary glucose content. The potential anti-oxidant role of dietary polyphenols in the regulation of inflammation and redox signaling has been recently reviewed by Rahman et al. (2006).

Ethnicity in Health and Disease

There is considerable information available on variations in both the prevalence and severity of different diseases in general and cardiovascular disease in particular among various ethnic populations from different parts of the world. However, the underlying causes of these variations are not fully understood. It is likely that some of these variations are due to the differences in genetic background, diet, environmental exposures, stress, socioeconomic status, education, access to health care, social discrimination and other factors alone or in combination (Sankar et al., 2004; Olden and White, 2005). However, the effect of each of these variables in individual disease or health-related outcomes is unclear. For example, despite vagueness about the strictness with which dietary patterns are described in various studies, major epidemiologic studies from the USA and Europe show that vegetarians live longer than omnivores from the same regions of the world (White, 1997). However, at present, it is not well established whether vegetarians live longer because of the vegetarian diet itself or because of some other characteristics and habits that are associated with longer lives in vegetarians (Fonnebo, 1994). Similarly, a number of studies show that people from the South Asian background have a greater prevalence and propensity for cardiovascular disease and diabetes. Although higher degrees of dyslipidemia including higher triglycerides and lower HDL-cholesterol levels, higher fasting glycemia and impaired glucose tolerance, higher prothrombotic factors such as fibrinogen and PAI-1, elevated C-reactive protein concentrations, and lower homocysteine levels along with various life style factors have been suggested for these associations, the exact underlying cause(s) for these associations remain unclear (Mukhopadhyay and Sattar, 2005). Whether increased oxidative stress due to genetic variations plays a critical role in these associations remains to be clarified (Wong et al., 2003). To accurately tease out the relative contributions of the genetic and environmental factors, and their interactions, rigorously designed, large-scale studies in multiple populations in which equally rigorously genetic and environmental data are collected, are needed. Further, to validate quantitative conclusions about genes, environment and their interactions in health and disease for multiple groups, long-term, longitudinal prospective cohort studies with appropriate interventions are needed (Collins, 2004).

Oxidative Stress and Disease

With the advent of industrialization, the human populations over the last century have been incessantly exposed to industrial pollutants and those arising from the automobile exhaust aerosols. Yet another and perhaps, the greatest threat associated with modern day life style choices stems from the use of tobacco introduced to the world merely 500 years ago by traders in North America. This single mixture of toxicant antigens in tobacco smoke inhaled by over 1 billion smokers and exposure of an estimated 1-2 billion non-smokers to secondhand smoke the world over, causally culminates in myriad diseases because the human immune system has not had sufficient time to evolve defense mechanisms to handle tobacco smoke-induced, chronic inflammatory responses involving the oxidative stress. Most tobacco-related diseases including cardiovascular disease, chronic lung diseases, type 2 diabetes, selected cancers and dementias are being increasingly recognized to have their origins in oxidative stress-related imbalance and the insufficient immune responses.

Tobacco and Oxidative Stress

Tobacco is the second major cause of death in the world, and is responsible for 10% of all deaths, killing approximately 5 million people each year. Tobacco accounted for nearly 3 million deaths in 1990, increasing its toll to 4 million deaths in 1998, and it is estimated that by 2030, it will cause 10 million deaths annually worldwide (Brudtland, 1999). It is mind-boggling that this mayhem, unleashed on humanity with the introduction of commercial tobacco to the world by early settlers and traders in North America some 5 centuries ago has become a global public health disaster. Each of these deaths, one every 6.5 seconds, is caused by a variety of tobacco-related diseases predominantly cancer, heart disease, pulmonary disease and cerebrovascular disease culminating in stroke.

Tobacco use accelerates development of both cardiovascular and cerebrovascular diseases by promoting atherosclerosis – a condition of arterial wall hardening due to plaque build-up that causes diminished blood flow – in the arteries that supply blood to the heart muscle or to the brain, respectively. The relationship between smoking and atherosclerosis was observed in the early 1900's by Buerger in young male smokers and an association between cerebrovascular disease and extracranial atherosclerosis was reported by Gowers as early as 1875. Since then, a number of risk factors for cerebrovascular disease culminating in incidents of stroke have been identified (see below).

Cardiovascular Disease

Within the context of cardiovascular disease, lipid peroxides produced by tobacco smoke rapidly inhibit plasma lecithin cholesteryl acetyl transferase (LCAT), a structural component of the high density lipoprotein (HDL) particle (Bielicki and Forte, 1999; Bielicki et al., 1995, 2001). Every puff of tobacco smoke that a smoker inhales while smoking a cigarette, and repeatedly when smoking 10 or more cigarettes a day, obviously results in damaged HDL with impaired capability to re-absorb any deposited lipids from the lumen of coronary arteries on the one hand, and as a consequence of feedback mechanism produces a net decrease in the levels of "good" HDL cholesterol and simultaneously increased levels of LDL – the "bad cholesterol". Oxidation of LDL in the sub-endothelial space of the artery wall may be a major source of oxidative damage to the arterial wall, initiates activation and recruitment of macrophages from circulating blood into the arterial wall where macrophages transform into fat laden "foam cells" and in turn trigger downstream local inflammatory responses and release of interleukins and TNF-alpha to perpetuate a "chronic inflammation" within the arterial wall causing atherosclerosis and at the fringes of coronary plaques which can rupture to cause a clinical and often fatal event, usually via myocardial infarction and sudden cardiac death (Morel et al., 1984; Steinbrecher et al., 1984; Steinberg et al, 1989; Berliner et al., 1990; Witzum and Steinberg, 1991; Ross, 1993; Navab et al., 1988, 1991, 2004, 2005). In advanced stages of atherosclerosis, the expression of matrix metalloproteinases increases in atherosclerotic plaque due to local chronic inflammation (Uzui et al., 2002). There is also evidence for transdifferentiation of arterial smooth muscle cells into osteoblast-like cells and that of subendothelial macrophages into osteoclast-like cells (Doherty et al., 2002). An imbalance in the dynamic equilibrium of osteoblastic and osteoclastic functions within the arterial wall cells, especially among diabetics with atherosclerosis, also results in arterial calcification (Doherty et al., 2003).

In addition to the widely used statins, primarily for lowering LDL levels by inhibiting HMG CoA reductase activity and thereby averting the formation of oxidized LDL with atherogenic properties, raising HDL levels also blunts the damaging effects of LDL. The underlying mechanism for HDL increase is via inhibition of the cholesterol ester transfer protein (CTEP). Niacin intake usually increases HDL by 25% to 35% when used at the highest doses. Several emerging strategies are being currently developed for increasing HDL (Forrester and Shah, 2006). Although initial clinical trial results with a pharmaceutical inhibitor of CTEP were promising, the clinical trial was halted in early 2007 due to unexpectedly high deaths due to unknown reasons during later stages of clinical trial. Lowering dietary saturated fats is known to reduce the oxidative susceptibility of LDL in healthy men and women (Yu-Poth et al., 2000). There is some evidence that LDL in Asian vegetarians is less oxidizable than that from omnivores (Lu et al., 2000). One promising therapeutic approach to treat cardiovascular disease in humans is to exploit the thiol-bearing synthetic peptides that retain the antioxidant activity of Apo A-I_{Milano} (Jia et al., 2002). Experimental evidence in animal models and patients with advanced atherosclerosis shows that intravascular infusion of microgram quantities of Apo A-I_{Milano}–derived peptides for 10-15 weeks results in significant disappearance of luminal plaques and deposited fat (Shah et al, 2001; Nicholls et al, 2006). Intramural delivery of recombinant Apo A-I_{Milano}-Phospholipid complex, ETC-216, is reported to inhibit in-stent restenosis in porcine coronary arteries (Kaul et al., 2003), which may find use in clinical setting in the future.

Coronary heart disease (CHD) is the second leading cause of mortality in the United States, with 80% of people who die of CHD are 65 years of older, regardless of gender or ethnicity (Ferdinand, 2006). The average age at the time of first myocardial infarction among Americans is 65.8 years in men and 70.4 years in women (Thom et al., 2006). CHD death rates per 100,000 are higher among blacks than whites (Thom et al., 2006), and although age-adjusted mortality due to cardiovascular disease is 19% greater for black men than white men, black women are one-third more likely to die of CHD than white women (Minino et al., 2002). Also, the occurrence of multiple risk factors for CHD in blacks is 50% higher than the whites (3rd Report of NCEP, 2002). Clearly, the prevalence and influence of CHD risk factors differ by ethnicity (Liao et al., 1999; Jones et al., 2002). Unfortunately, blacks and other racial minorities have been underrepresented in statin safety and efficacy trials, making their benefit assessment difficult for these ethnic groups. Health disparities in cardiovascular health in the United States reveal a strong link between poverty and incidence of reported heart disease, coronary heart disease, hypertension and stroke (Mensah et al., 2005). Racial minorities including South Asians, African-Carribeans in the United Kingdom also have high prevalence of CHD as in African Americans and Mexican Americans, due to recent changes in life style and "Westernization" (Forouhi and Sattar, 2006). In India, there is higher prevalence of CHD in urban areas than in rural areas (Gupta, 1996), and immigrant South Asians present greater levels of CHD risk factors than their siblings in India (Bhatnagar et al., 1995), demonstrating the relationship between acculturation of Western lifestyle and CHD risk.

Cerebrovascular Disease and Stroke

The precise mechanisms by which tobacco smoking causes cerebrovascular disease culminating in stroke are poorly understood and there is an immediate need to carry out further research into causes and therapeutic management of stroke in diverse populations as well as need for additional research funding for this hitherto under-funded area (Asotra, 2004). It is generally believed that tobacco smoke causes oxidation of LDL, and this formation of oxidized LDL is one of the earliest events that initiate the process of plaque formation in systemic and carotid arteries. Tobacco smoke results in compromised protective function of HDL, lowered HDL levels, and elevated LDL/HDL ratio in chronic smokers, increasing their risk for atherosclerotic disease in the arteries. Exposure of endothelial cells to cigarette smoke condensate increases the levels of inflammatory cytokines suggesting a complex pro-inflammatory response to tobacco smoke constituents (Nordskog et al., 2003).

Stroke, the third leading cause of death, is caused by the interruption of blood supply carrying oxygen and nutrients to the brain due to rupture or blockage of an artery to the brain. About 700,000 individuals suffer from stroke annually in the U.S.A., and at any given time there are nearly 3 million stroke survivors in this country. On average, someone in the United States suffers a stroke every 53 seconds, and every 3.3 minutes someone dies of one (American Stroke Association, 2004). Two-thirds of stroke survivors suffer moderate or severe impairment. An immediate consequence of stroke is brain injury in the region deprived of blood flow, resulting in loss of consciousness and unilateral paralysis, affecting speech and vision.

The most common variety of stroke called an ischemic stroke is caused by blockage of arterial blood flow due to a blood clot or due to fatty deposits in the blood vessel wall (plaque). This type accounts for over 85% of all cases of stroke. There are two clinical forms of ischemic stroke: (a) embolic stroke, caused by a blood clot formed somewhere in the body (usually the heart) that cannot pass through the narrow bore of an artery, thereby blocking blood circulation to the brain, and (b) thrombotic stroke, the more common variety of ischemic stroke, occurs when blood flow is reduced due to blockage of one or more arteries (Caplan, 2000). Large vessel occlusive disease, commonly due to atherosclerotic involvement of the common and internal carotid arteries, can cause carotid ischemic stroke. Small vessel occlusive disease is thought to be non-atherosclerotic narrowing of small end-arteries in the brain. Cardiogenic stroke occurs due to blood clots arising from the heart, typically as a result of atrial fibrillation or ischemic (i.e., atherosclerotic) heart disease (Fisher, 1982; Adams et al., 1993). Strokes caused by bleeding within the brain – called intracerebral (within the brain) and subarachanoid (between the skull and the brain) hemorrhagic strokes – have a higher risk of fatality and cause greater functional impairment among survivors than the ischemic strokes. Stroke is *not* one disease but rather a heterogeneous group of disorders reflecting differences in pathological mechanisms (Hassan and Markus, 1995; Alberts, 1999). The mortality of ischemic stroke is 15–30% within the first 30 days. Hemorrhagic stroke has a graver prognosis, with a 30-day mortality rate of 40–80%. Treatment and rehabilitation of stroke patients costs nearly $53.6 billion annually in the United States (American Heart Association, 2004).

A precursor and predictor of stroke, called mini-stroke or transient ischemic attack (TIA), is a neurologic deficit lasting less than 24 hours and is attributed to focal cerebral or retinal ischemia (Johnston, 2002). Although many cases of TIA never come to medical attention, it is estimated that more than 5 million Americans are diagnosed with a TIA annually. The causes of true TIAs, such as atrial fibrillation, carotid artery disease, and large- and small-artery disease in the brain, are identical to those of stroke. Cigarette smoking is reported to be a major preventable risk factor for possibly 200,000 to as many as 500,000 cases of TIA that come to medical attention each year (Johnston, 2002).

The Framingham Heart Study was the first to assess the relation of smoking to the type of stroke, number of cigarettes smoked and the beneficial effect of smoking cessation (Wolf et al., 1988). It concluded that smoking was a significant independent factor contributing to the risk of stroke in general, and to brain infarction specifically. The risk of stroke has been directly related to the number of cigarettes smoked. Heavy smokers (>40 cigarettes/day) have a relative risk of stroke 2-4 times greater than non-smokers (Wolf et al., 1988; Wannmathee et al., 1995), and twice that of people who smoke less than 10 cigarettes/day. The large cohort study of 22,071 US male physicians showed that individuals smoking >20 cigarettes/day had nearly three times the risk of total non-fatal stroke and 50% higher chance of fatal stroke than non-smokers Robbins et al.(1994). However, other studies have shown that within 5 years after stopping, the risk for stroke can be reduced to that of non-smokers (Wolf et al., 1988).

The female-to-male mortality ratio differs for stroke subtypes by ethnicity and age in the United States (Gillum 1995 a, b; 1999; Ayala et al., 2001, 2002). Although the fatality rates for hemorrhagic and ischemic stroke are similar for men and women (The CROSS Group, 2000), more women than men are admitted to nursing home for stroke disabilities (Leibsom et al., 1998). Prospective studies have suggested an increased risk for total hemorrhagic stroke, intracereberal stroke and subarachanoid stroke among men and women who are current smokers; the risk increases with the number of cigarettes smoked (Kurth et al., 2003 a, b). Generally, women have a lower risk of stroke than men at ages less than 65 years, but women have a higher risk than men at ages of 65 years and above.

The deaths attributed to intracerebral hemorrhagic strokes among Blacks, American Indian/Alaskan Natives, Asian Pacific Islanders, and Hispanic men and women are higher than those among White men and women. The proportion of death due to subarachanoid strokes was greater among American Indian/Alaska Natives, Asian/Pacific Islanders and Hispanic women than Black or White women. (Ayala et al., 2001, 2002). An analysis of sex-specific age-standardized death rates for ischemic stroke (n = 507, 256), intracerebral hemorrhagic stroke (n = 98,709) and subarachanoid hemorrhagic stroke (n = 27,334) among Whites, Blacks, American Indian/Alaska Natives, Asians/Pacific Islanders, and Hispanics revealed that overall 81.7% women and 77.7% men succumbed to ischemic stroke, 13.7% women and 18.5% men succumbed to intracerebral hemorrhagic stroke, and 4.6% women and 3.8% men died of subarachanoid hemorrhagic stroke (Ayala et al., 2002).

There has been a lot of interest in learning if stroke has some genetic determinants. Although genetic determinants of the common forms of stroke are largely unknown, some mutations in specific genes causing rare forms of stroke have been reported (Joutel et al., 1996) including brain hemorrhage (Palsdottir et al., 1988). The first main locus associated with stroke, called STRK1, was mapped to chromosome 5q12 (Gretarsdottir et al., 2002). Recently, two genes have been identified in the pathogenesis of stroke: a gene called PDE4D, encoding the enzyme phosphodiesterase 4D, is reported to be important in ischemic stroke (Gretarsdottir et al., 2003), and another gene encoding 5-lipoxygenase activating protein confers risk of myocardial infarction and stroke (Helgadottir et al., 2004).

Insulin Resistance, Diabetes and Anti-Oxidant Role of Insulin

Reactive oxygen species have a causal role in multiple forms of insulin resistance (Houstis et al., 2006). Insulin resistance is a salient feature of metabolic syndrome first described by Reaven (1988). It is becoming increasing clear that development of insulin resistance due primarily to oxidative stress and elevated plasma free fatty acids (Reaven, 1988), invariably accelerates the onset of type 2 diabetes, especially among chronic smokers by 2 to 4 years compared with non-smokers (Farin et al., 2007), and also plays a role in the development of atherosclerosis (Ingelsson et al., 2005). Development of insulin resistance occurs due to obesity as well, and obese people are insulin resistant to varying degrees (Boden, 1997; Bray, 2004). Insulin resistance also occurs clinically during pregnancy, sepsis, carcinomatous cachexia, and frequent periods of starvation or in people who observe fasts. TNF-alpha and glucocorticoid dexamthasone are causally related to development of insulin resistance both in vitro and in vivo. Insulin secreting pancreatic beta cells lack glutathione, and would therefore, be predictably extremely susceptible to damage due to oxidative stress.

Breakthrough discovery of the anti-inflammatory effect of insulin and the pro-inflammatory properties of glucose by Paresh Dandona and colleagues in the last 5 years have provided a rationale for the treatment of hyperglycemia in several acute clinical conditions (Dandona et al., 2005, 2007). Importantly, potent anti-oxidant effects of insulin, when infused and tightly monitored to restore normoglycemia in hyperglycemic patients within an hour after myocardial infraction in the intensive care unit, resulted in a 50% reduction in mortality, with several other beneficial effects including reduced incidence in renal failure, septicemia and ICU neuropathy (Van den Berghe et al., 2001). This finding may one day translate in preventing cardiac muscle damage due to impending cardiomyocyte death within the infarct region of normoglycemic patients following myocardial infarction.

Neurological Disorders and Dementias

Neurological disorders such as Alzheimer's disease characterized by loss of neurons in brain regions involved in learning and memory (DeKosky, 2001; Mattson, 2006), Parkinson's disease and Huntington's correa that result in motor dysfunction and uncontrolled limb movement and muscle spasms ( ), and multiple sclerosis have been causally associated with oxidative stress (Cutler et al., 2004; Alimonti et al., 2006). There are two principle factors that are believed to be involved in neuronal dysfunction and neurodegeneration in Alzheimer's disease (AD): (i) increased oxidative stress that causes enzymatic production of ceramides from sphingomyelinase action on sphingomyelin (Billis et al., 1998) and (ii) increased production of neurotoxic variants of amyloid peptide beta (Mattson, 1997). Ceramides have also been implicated in ischemic stroke (Yu et al., 2000) and Parkinson's disease (Brugg et al., 1996) that also involve neuronal apoptosis (Alessenko, 2000). The oxidative stress-induced ceramides are known to induce the death of cultured hippocampal (Goodman and Mattson, 1996), cortical (Willaime et al., 2001)), mesencephalic (Brugg et al., 1996) and motor (Irie and Hirabayashi, 1998) neurons.

Chronic Lung Diseases

Chronic obstructive pulmonary disease (COPD) is the fourth leading cause of death in the United States and mortality continues to increase particularly among African Americans (Dransfield and Bailey, 2006). In adults and older children, it is well established that ethnic group is an important determinant of normal lung function, with both lung volumes and forced expiratory flows tending to be lower in black than white subjects based on values predicted from standing height (American Thoracic Society, 1991; Pool and Goodenough, 1989; Dransfield et al., 206) Although these discrepancies have been largely attributed to ethnic differences in trunk:leg ratio, there is evidence that genetic, environmental, and dietary factors may be important in the etiology of chronic lung diseases such as COPD and asthma. Of the environmental factors, exposure to cigarette smoking is the most important risk factor for COPD-nearly 90% of COPD patients are smokers. However, only 15% of cigarette smokers develop clinically significant COPD and there are clear differences in the prevalence of COPD among different ethnic groups (Bascom, 1991; Marcus et al., 1988). One study found that although African Americans have lower overall exposure to tobacco smoke, they are more susceptible to developing smoking-related illnesses (Chatila et al., 2006). Polymorphisms for genes encoding for proteinases, anti-proteinases, anti-oxidants, xenobiotic metabolising enzymes, and inflammatory mediators have been suggested to account for much of the variation in predisposition to COPD among different ethnic groups (Hagub et al., 2004a, b). However, much of the data have been inconclusive and contradictory and segregation analysis has demonstrated that the genetic components determining forced expiratory volume in 1 second (a marker of the severity of COPD) consist of multiple genes, each of which has a small influence, rather than a single Mendelian gene (Giverlber et al., 1998).

Among other factors, differences in dietary patterns between different ethnic groups might also account for differences in predisposition to COPD. For example, it has been reported that eating mostly meat, refined starches, and sodium might increase the likelihood of developing COPD. Individuals whose diets are rich in meat, refined starches and sodium were 1.43 times more likely to report new onset of persistent coughs with phlegm than those who consume a diet high in fruit and soy (Butler et al., 2006).

An association between ethnicity and risk of allergen sensitization has also been described. In one study, among children with mild to severe asthma, those of either Puerto Rican or African-American descent, had a higher rate of sensitization than did the white children (Celedon et al., 2004). While the Puerto Rican and African-American children were more likely to be sensitized to certain indoor allergens (such as dust mites and cockroaches) than were the white children, they were also more likely to be sensitized to outdoor allergens. Therefore, there is no doubt that there is an association between the ethnicity and the development of COPD and asthma; however, the biological basis of this association remains unknown.

Macular Degeneration

Age-related macular degeneration (AMD) is the most common cause of irreversible visual loss in the developed world. It is a complex disorder with multiple phenotypes, which are ultimately associated with vision loss in individuals with late stage disease. A recent study, involving six thousand one hundred seventy-six 45- to 85-year-old subjects, selected from 6 United States communities, found AMD prevalence to be 2.4% in Blacks, 4.2% in Hispanics, 4.6% in Chinese, and 5.4% in whites ($P<0.001$ for any differences among groups). The highest prevalence of any AMD occurred in those 75 to 84 years old, varying from 7.4% in blacks to 15.8% in whites and Chinese ($P = 0.03$). Differences in age, gender, pupil size, body mass index, smoking, alcohol drinking history, diabetes, and hypertension status did not explain the variability among the 4 racial/ethnic groups (Klein et al., 2006). Although both the prevalence and the phenotypic spectrum of AMD varies widely among persons of different ethnicities, the underlying reasons for these differences are not known. Some studies have demonstrated that the variations in single nucleotide polymorphisms in complement factor H (CFH) gene frequency might be associated with an increased risk for AMD (Chen et al., 2006; Despriet et al., 2006). Complement activation initiates a cascade of events, which possibly via free oxygen radicals ultimately leads to macular AMD. Complement factor H is an important inhibitor of the complement pathway. It preferentially binds and inactivates complement component 3b, and prevents the production of C3 convertase and progression of cascade that ultimately leads to AMD. Individuals homozygous for CFH Y402H polymorphism have a 48% risk of developing AMD by the age 95 years while this risk does not exceed 22% for non-carriers, thereby putting the carriers at greater than 50% risk compared to general population. However, other potential modifiers and genetic factors might play important roles since one recent study suggested that other yet unidentified genetic factors might be important in pathogenesis of AMD that might mitigate the effects of certain CHF polymorphisms such as the c.1204T>C, p.Tyr402His polymorphism, another CHF polymorphism has been linked to AMD (Grassi et al., 2006).

Neonatal and Postnatal Responses

There are plenty of animal data suggesting developmental increases in expression of antioxidant enzymes during fetal development. Importantly, during embryonic development in vertebrates, glycolytic and glycogenolytic metabolism are utilized preferentially, and the oxidative and lipolytic metabolic pathways evolve rapidly only postnataly (Asotra and Malhotra, 1978) such that certain enzymes like glucose 6-phopshatase may serve as a metabolic switch from glycolytic to oxidative metabolism (Asotra, 1984; Asotra, 1986). These findings have special relevance to premature babies, who, in the recent past, have been subjected to high oxygen environment with unexpected oxidative damage effects. However with the exception of limited data supporting a developmental increase in the expression of lung catalase activity, there are very few human studies that have directly examined the expression of antioxidant enzymes during the human development. In fact, some studies have failed to demonstrate any late fetal surges in antioxidant enzyme activities in human tissues (McElroy et al., 1992). However, on the other hand, there is no doubt that significant lower antioxidant enzymes have been observed in many studies in premature infants and their impaired ability to adequately up-regulate antioxidant enzymes in response to oxidant stresses, making them highly susceptible to oxidant injury (Saugstad, 2005). Therefore, premature infants are almost certainly developmentally unprepared for extrauterine life in an oxygen-rich environment and exhibit a unique sensitivity to oxidant injury.

There is compelling experimental and clinical evidence for the involvement of oxidant and antioxidant imbalance in at least four main disorders of prematurity 1) chronic lung disease or Bronchopulmonary Dysplasia (BPD); 2) retinopathy of prematurity (ROP); 3) intra/periventricular hemorrhage (IVH/PVH); and 4) necrotizing enterocolitis (NEC). Further, greater the prematurity, higher is the risk. Due to underdeveloped lungs, extremely premature infants often require respiratory support, either just with supplemental oxygen alone or in conjunction with ventilatory support. The mechanistic basis for free radical involvement in these disorders is that free oxidant radicals are formed too rapidly to be detoxified by the limited antioxidant defenses of the premature infant mechanisms with the resultant tissue-specific damage. In the case of chronic lung disease, the evidence currently favors excess exposure to hyperoxia as the primary reason for the increased free radical production, whereas in retinopathy of prematurity, necrotizing enterocolitis, and intraventricular hemorrhage, it is likely that periods of ischemia/hypoxia followed by periods of reperfusion/reoxygenation is the more likely mechanism for the excess free radical formation (Gitto et al., 2002).

Chronic lung disease of prematurity, or BPD, although multifactorial in its pathogenesis, is strongly linked to the exposure of the premature infant to supraphysiologic concentrations of oxygen (Rehan, 2006; Rehan et al., 2006). In support of this, a number of studies have reported high concentrations of metabolites reflecting increased peroxidation products such as pentane, ethane, protein carbonyl, o-tyrosine, allantoin and F2-isoprostanes, as well as low levels of glutathione and sulfhydryl/total protein ratio, also reflecting increased oxidative load, in premature infants at risk of or developing chronic lung disease. Surfactant production and function are also altered by both hyperoxia and reactive oxygen species per se, making the lungs of premature infants more vulnerable to injury (Schock et al., 2001; Saugstad, 2003). Furthermore, a number of animal and human studies have suggested a protective benefit of antioxidant strategies, further supporting the critical role played by oxidant lung injury in the pathogenesis of BPD (Davis, 2002). However, BPD may not be all iatrogenic since twinning, family and population studies have recently implicated genetic factors in the evolution of BPD. The candidate genes examined for their potential role in BPD have included surfactant apoprotein and inflammatory genes. However, differentiating the effects of environmental and environmental-genetic interactions from isolated genetic etiologies has been problematic (Parton et al., 2006).

Retinopathy of prematurity is a vaso-proliferative disorder of the premature retina. It is of variable severity and often heals spontaneously with no sequelae. However, it can progress to retinal detachment or severe retinal scar formation and is one of the major causes of visual impairment and blindness in infants. The more premature the infant and the more complicated the initial neonatal clinical course, more likely is the disease (Lily et al., 2003). Like BPD, the pathogenesis of ROP is also multifactorial and exposure to excessive reactive oxygen species is potentially the major contributing factor in causing the initial retinal injury (Saugstad, 2005).

In a recent study that examined the prenatal risk factors for severe ROP among very preterm infants, prematurity was the dominant risk factor, with infants with gestation age of <25 weeks having 20 times greater odds of severe ROP than infants with gestational age of 28 weeks (Darlow et al., 2005). Birth weight for gestational age also had a "dose-response" effect; the more growth-restricted infants had greater risk, with infants below the 3rd percentile of weight for gestational age having 4 times greater odds of severe ROP than those between the 25th and 75th percentiles. Male gender was also a significant risk factor (odds ratio: 1.73; 95% confidence interval: 1.25–2.40). Although attempts have been made previously to show an association between genetic factors and susceptibility to ROP, it was only in a recent study, in which after controlling for known and unknown nongenetic factors, genetic factors were calculated to account for 70% of the variance in susceptibility to ROP (Matthew et al., 2006).

With the improvement in neonatal care in general, although the incidence of IVH/PVH has declined in recent years, it still remains the most important cause of brain injury in premature infants (Inder and Volpe, 2000). It most frequently occurs in infants born before 32 weeks' gestation or less than 1500 g birth weight and virtually all IVH/PVH occurs within the first five postnatal days of life. Although its pathogenesis is mostly unknown and is likely to be multifactorial, hemodynamic instability, impaired autoregulation, and free radical-mediated cellular injury are probably the key contributing factors (Volpe, 2001). The role of free radical-induced injury in the pathogenesis of IVH/PVH is also supported by experimental and clinical studies demonstrating the beneficial effects of either free radical scavengers such as superoxide dismutase or agents that reduce free radical production such as indomethacin or ethamsylate (Fowlie, 2002).

Recently, polymorphisms in TNF alpha , IL-1 beta , IL-4, IL-6 and IL-10 as well as mutations in coagulation proteins have been investigated as potential candidate genes to modify risk and or severity of perinatal brain IVH/PVH (Baier, 2006).

Necrotizing enterocolitis is the most common gastrointestinal emergency in the newborn infant and it accounts for significant morbidity and mortality in infants admitted to neonatal intensive care unit. Its incidence is greatest in infants born before 32 weeks' gestation or less than 1500 g birth weight, occurring in 2 to 10 percent of such infants, and decreases with increasing gestational age (Lin and Stowll, 2006). Like, BPD, IVH/PVH and ROP, the pathogenesis of NEC is unknown, but it is caused by multiple factors in a susceptible host, which include but are not limited to prematurity, milk feedings, circulatory instability, infection, impaired mucosal defense and medications (Lin and Stowll, 2006). Several lines of evidence support the role of oxidant injury as the key contributing factor. For example, intestine is rich in xanthine oxidase, enabling brisk production of reactive oxygen species during reperfusion following ischemia and in experimental models of NEC pretreatment with superoxide dismutase, vitamin E (both potent antioxidants), or deferoxamine (an iron chelator), at least to some extent ameliorates intestinal damage (O'Donovan et al., 2004).

In summary, premature infants are particularly susceptible to oxidant damage, and free oxygen radical-mediated injury has been implicated as a key mechanism for a wide variety of disorders of the premature infants that contribute to significant morbidity and mortality. In recent years, although significant advances have been made in understanding the tissue-, cell-, and disease-specific molecular pathways involved in free oxidant radical damage in the premature infant, there are still several challenges to be solved before these advances could be translated into clinical practice (Das, 2004). In particular, complete understanding of the molecular mechanisms involved in oxidant radical damage, therapeutic efficacy, safety, route of administration, pharmacokinetics, and pharmacodynamics of a host of potential antioxidant agents in specific diseases need to be determined.

Future Research Needs

As reviewed in this chapter, increased oxidative stress has been implicated in a number of conditions including atherosclerosis, cardiac and cerebrovascular diseases, various cancers, diabetes, chronic lung diseases, ageing, and a large number of serious neonatal and postnatal conditions. The link between increased oxidative stress and atherosclerosis, ageing, and various neonatal and postnatal conditions has been most vigorously pursued. However, despite clear associations between these conditions and increased oxidative stress, no prospectively randomized clinical study has shown a clear benefit of supplemental antioxidant therapy. Further studies to determine specific signal transduction mechanisms used by free radicals for each particular condition will elucidate important disease-specific molecular targets for future therapeutic interventions. Further, which particular oxidant marker is to be monitored for each specific condition is not established. To monitor the beneficial effects of antioxidants or antioxidant-eliciting therapeutic agents, monitoring of disease-specific oxidant markers should also be standardized and should be the focus of future studies. In addition, since previously used antioxidants have not worked, either these need to be reformulated, or newer antioxidants need to be identified. This all has to occur in conjunction with the efforts to eliminate tobacco exposure worldwide, the biggest preventable factor for the increase in oxidative stress faced by us today. Hopefully, further clinical studies on the anti-oxidant properties of and benefits from insulin infusion may demonstrate its broader usefulness in acute clinical conditions. Novel, innovative confirmatory diagnostic tests, such as DNP and hyperpolarization nuclear magnetic resonance imaging, currently being developed to visualize the damaged heart muscle, following primary myocardial infarction, stem from ceased mitochondrial function within the affected regions and rely primarily on mechanisms underlying oxidative metabolism within cardiomyocyte mitochondria. Similar oxidative stress based imaging technologies should help confirm early diagnoses of a variety of cancers and chronic lung diseases.

Acknowledgments

This work was supported in part by grants from the National Institutes of Health (HL075405 and HL55268) and the California Tobacco-Related Disease Research Program (14RT-0073 and 15IT-0250) to VKR, and with funds from the research account of the Tobacco Tax and Health Protection Act of 1988 (Proposition 99) to KA.

References

Abbasi, F., Chu, J.W., Lamendola, C., McLaughlin, T., Hayden, J., Reaven, G.M., and Reaven, P.D. Discrimination between obesity and isulin resistance in the relationship with adiponectin. *Diabetes* 2004; 53(3): 585-590.

Adams, H.P., Jr. et al. Classification of subtypes of acute ischemic stroke. Definitions for use in a multicenter clinical trial. TOAST. Trial of Org 10172 in Acute Stroke Treatment. *Stroke 1993;* 24: 35-41.

Alberts, M.J. *Genetics of Cerebrovascular Disease* 1999; Futura, New York.

Alessenko, A.V. The role of sphingomyelin cycle metabolites in transduction of signals of cell proliferation, differentiation and death. *Membr. Cell Biol.* 2000 13: 303-320.

Alimonti, A., Ristori, G., Giubilei, F., Stazi, M.A., Pino, A., Visconti, A., Brescianini, S., Monti, M.S., Fortre, G., Stanzione, P., Bocca, B., Bomboi, G., D'Ippolito, C., Annibali, V., Salvetti, M. and Sancesario, G. Serum chemical elements and oxidative stress in

Alzheimer's disease, Parkinson's disease and multiple sclerosis. *Neurotoxicol.* 2006; doi.10.1016/j.neuro.2006.12.001.

Ames, B.N. Dietary carcinogens and anticarcinogens. *Science* 221: 1256-1263, 1983.

American Stroke Association. www.strokeassociation.org

American Thoracic Society. Lung function testing: selection of reference values and interpretative strategies. *Am. Rev. Respir. Dis.* 1991; 144: 1202-1218.

Asotra, K.. Does Glucose 6-Phosphatase Regulate Metabolic Transformation in Developing Skeletal Muscle? *Biomedica Biochimica Acta* 1984; 43:309-319..

Asotra, K. Glucose 6-phosphatase activity in normal and denervated developing chick Gastrocnemii: Reappraisal of glycogenolytic and glycolytic metabolism in skeletal muscle, *Experimental Pathology* 1986; 29:103-112.

Asotra, K. Tobacco-Caused Cerebrovascular Disease: Urgent Need for Increased Research Funding. *Burning Issues* 2004; 7(1): 8-9, 13-15.

Asotra, K. and Malhotra, R.K. Growth metabolism of diaphragm in Swiss albino mouse. *Experientia* 1978; 34:75-77.

Ayala, C., Greenlund, K.J., Croft, J.B., Keenan, N.L., Donehoo, R.S., Giles, W.H., Kittner, S.J. and Marks, J.S. Racial/ethnic disparities in mortality by stroke subtype in the United States, 1995-1998. *Am. J. Epidemiol.* 2001; 154: 1057-1063.

Ayala, C., Croft, J.B., Greenlund, K.J., Keenan, N.L., Donehoo, R.S., Malacher, A.M., and Mensah, G.A. Sex differences in US mortality rates for stroke and stroke subtypes by race/ethnicity and age, 1995-1998. *Stroke* 2002; 33: 1197-1201.

Azzi, A., Davies, K.J., and Kelly, F. Free radical biology – terminology and critical thinking. *FEBS Letters,* 558: 3-6, 2004.

Baier RJ.Genetics of perinatal brain injury in the preterm infant. *Front Biosci.* 2006; 11:1371-1387.

Bascom, R. Differential susceptibility to tobacco smoke: possible mechanisms. *Pharmacogenetics* 1991; 1:102–6).

Berliner, J.A., Territo, M.C., Sevanian, A., Ramin, S., Kim, J.A., Bamshad, B., Esterson, M. and Fogelman, A.M. Minimally modified low density lipoprotein stimulates monocyte endothelial interactions. *J. Clin. Invest.* 1990; 85: 1260-1266.

Bhatnagar, D., Anand, I.S., Durrington, P.N. et al. Coronary risk factors in people from the Indian subcontinent living in West London and their siblings in India. *Lancet* 1995; 345: 405-409.

Bielicki, J.K., Forte, T.M., and McCall, M.R.. Gas-phase cigarette smoke inhibits plasma lecithin-cholesterol acyltransferase activity by modification of the enzyme's free thiols. *Biochim Biophys Acta.* 1995; 24;1258(1):35-40.

Bielicki, J.K. and Forte, T.M. Evidence that lipid peroxides inhibit plasma lecithin:cholesterol acyltranferase activity. *J. Lipid. Res.* 1999; 40: 948-954.

Bielicki, J.K., Knoff, L.J., Tribble, D.L. and Forte, T.M. Relative sensitivities of plasma lecithin:cholesterol acyltransferase, platelet-activiating factor acyhydrolase, and paraoxonase to in vitro gas-phase cigarette smoke exposure. *Atherosclerosis 2001;* 155: 71-87.

Billis, W., Fuks, A. and Kolesnick, R. Signaling in and regulation of ionizing radiation-induced apoptosis in endothelial cells. *Recent Prog. Horm. Res.* 1998; 53: 85-92.

Bizzarro, M.J., Hussain, N., Jonsson, B., Feng, R. et al. Genetic susceptibility to retinopathy of prematurity. *Pediatrics* 2006; 118: 1858-1863.

Braunwald, E. Coronary blood flow and myocardial ischemia. *Heart Disease: A Textbook of Cardiovascular Medicine*. W. B. Saunders Company. Philadelphia, PA, USA, 2001: 1161-1183.

Brundtland, G.M. Address to the WHO International Conference on Tobacco and Health, Kobe City, Japan, 15th November, 1999. http://www.who.int/director-general/speeches/1999/english/19991115_kobe.html.

Brugg, B., Michael, P.P., Agid, Y. and Ruberg, Y. Ceramide induces apoptosis in cultured mesencephalic neurons. *J. Neurochem.* 1996; 66(2):733-9.

Buerger, L. Thromboangiitis obliterans: a study of the vascular lesions leading to presenile spontaneous gangrene. *Am. J. Med. Sci.* 1908, 208:567.

Butler, L.M., Koh, W.P., Lee, H.P., Tseng, M. et al. Prospective study of dietary patterns and persistent cough with phlegm among Chinese Singaporeans. *Am. J. Respir. Crit. Care Med.* 2006;173:264-70.

Cadenas, E. Mitochondrial free radical production and cell signaling. *Molecular Aspects of Medicine* 2004; 25: 17-26.

Caplan, L.R. *Caplan's Stroke: A Clinical Approach 2000;* Butterworth-Heinemann, Boston.

Celedón, J.C., Sredl, D., Weiss, S.T. et al. Ethnicity and skin test reactivity to aeroallergens among asthmatic children in Connecticut. *Chest.* 2004;125:85-92.

Chatila, W.M., Hoffman, E.A., Gaughan, J., Robinswood, G.B. et al, Advanced emphysema in African-American and white patients: do differences exist? *Chest.* 2006;130(1):108-18.

Chen, L.J., Liu, D.T., Tam, P.O., Chan, W.M., et al. Association of complement factor H polymorphisms with exudative age-related macular degeneration. *Mol. Vis.* 2006 ; 12:1536-42.

Collin, F.S. The case for a U.S. prospective cohort study of genes and environment. *Nature* 2004; 429:475–477.

Culter, R.G., Kelly, J., Storie, K., Pedersen, W.A., Tammara, A., Hatanpaa, K., Troncoso, J.C., and Mattson, M.P. Involevment of oxidative stress-induced abnormalities in ceramide and cholesterol metabolism in brain aging and Alzheimer's disease. *Proc. Natl. Acad. Sci. USA* 2004; 101: 2070-2075.

Dransfield, M.T. and Bailey, W.C. COPD: racial disparities in susceptibility, treatment, and outcomes. *Clin Chest Med.* 2006; 27:463-471.

Dandona, P., Mohanty, P., Chaudhury, A., Garg, R. and Aljada, A. Insulin infusion in acute illness. *J. Clin. Invest.* 2005; 115: 2069-2072.

Dandodna, P., Chaudhary A., Ghanim, H. and Mohanty, P. Proinflammatory effects of glucose and anti-inflammatory effect of insulin: Relevance to cardiovascular disease. *Am. J. Cardiol.* 2007; 99 (Suppl): 15B-26B.

Dandona, P., Aljada, A. and Bandyopadhyaya, A. The potential therapeutic role of insulin in acute myocardial infarction in patients admitted to intensive care and in those with unspecified hyperglycemia. *Diabetes Care* 2003; 26(2): 516-519.

Darlow, B.A., Hutchinson, J.L., Henderson-Smart, D.J., Donoghue, D.A., Simpson, J.M., and Evans, N.J.. Prenatal risk factors for severe retinopathy of prematurity among very preterm infants of the Australian and New Zealand neonatal network. *Pediatrics* 2005;115:990-996.

Das, K.C. Redox control of premature birth and newborn biology. *Antioxidants and Redox Signaling.* 2004, 6 (1): 105-107.

Davis, J.M. Role of oxidant injury in the pathogenesis of neonatal lung disease. *Acta Paediatr. Suppl.* 2002;437:23-25.

DeKosky, S.T. Epidemiology and pathophysiology of Alzheimer's disease. *Clin. Cornerston 2001; e* 3: 15-26.

Despriet, D.D., Klaver, C.C., Witteman, J.C., Bergen, A.A. et al. Complement factor H polymorphism, complement activators, and risk of age-related macular degeneration. *JAMA.* 2006;296:301-309.

Doherty, T.M., Asotra, K., Fitzpatrick, L.A., Dunstan, C.R., Qiao, J-H., Wilkin, D.J., Detrano, R.C., Shah, P.K. and Rajavashisth, T.B. Calcification in Atherosclerosis: Bone Biology and Chronic Inflammation at the Arterial Crossroads.,*Proc. Natl. Acad. Sci. USA* (2003); 100(20):11201-11206.

Dransfield, M.T., Davis, J.J., Gerald, L.B., Bailey, W.C. Racial and gender differences in susceptibility to tobacco smoke among patients with chronic obstructive pulmonary disease. *Respir Med.* 2006; 100(6):1110-6).

Farin, H.M.F., Abbasi, F., Kim, S.M., Lamendola, C., McLaughlin, T. and Reaven, G.M. The relationship between insulin resistance and dyslipidemia in cigarette smokers. *Diabetes, Obesity and Metabolism* 2007; 9: 65-69.

Ferdinand, K.C. Ethinc, gender and age-related differences in the treatment of dyslipidemia. Am J. Manag Care 2006; 12: S400-S404.

Fisher, C.M. Lacunar strokes and infarcts: A review. *Neurology* 1982; 32: 871-876.

Fonnebo, V. The healthy Seventh-Day Adventist lifestyle: what is the Norwegian experience? *Am. J. Clin .Nutr.* 1994; 59(5 Suppl):1124S-1129S.

Forrester, J.S. and Shah, P.K. Emerging strategies for increasing high-density lipoprotein. *Am. J. Cardiol.* 2006; 98: 1542-1549.

Forouhi, N.G., and Sattar, N. CVD risk factors and ethnicity-A homogenous relationship? *Athersocelrosis Supplements* 2006; 7: 11-19.

Fowlie, P.W. Prophylactic intravenous indomethacin for preventing mortality and morbidity in preterm infants. Cochrane Review. In: *The Cochrane Library* 2002; issue 4. Oxford, U.K.:Update Software.

Garg, R., Chaudhuri, A., Munschuer, F. And Dandona, P. Hyperglycemia, insulin and acute ischemic stroke: A mechanistic justification for a trial of insulin infusion therapy. *Stroke* 2006; 37: 267-273.

Giordano, F.J. Oxygen, oxidative stress, hypoxia and heart failure. *J. Clin. Invest.* 2005; 115: 500-508.

Gillum, R.F. Epidemiology of stroke in Native Americans. *Stroke 1995a*; 26: 514-521.

Gillum, R.F. Epidemiology of stroke in Hispanic Americans. *Stroke* 1995b; 26: 1707-1712.

Gillum, R.F. Stroke mortality in blacks: disturbing trends. *Stroke* 1999;30: 1711-1715.

Gitto, E., Reiter, R.J., Karbownik, M., Tan, D., Gitto, P., Barberi, S., and Barberi, I.. Causes of Oxidative Stress in the Pre- and Perinatal Period. *Biol. Neonate* 2002; 81:146-157.

Givelber, R.J., Couropmitree, N.N., Gottlieb, D.J., Evans, J.C. et al. Segregation analysis of pulmonary function among families in the Framingham study. *Am. J. Respir. Crit Care Med* 1998;157:1445–51.

Goodman, Y. and Mattson, M.P. Ceramide protects hippocampal neurons against excitotoxic and oxidative insults, and amyloid beta-peptide toxicity. *J. Neurochem*. 1996; 6: 869-872.

Gowers, W.R. On a case of simultaneous embolism of central retinal and middle cerebral arteries. *Lancet 1875*; ii:794.

Grassi, M.A., Fingert, J.H., Scheetz, T.E., Roos, B.R. et al. *Hum Mutat*. 2006;27(9):921-5.Ethnic variation in AMD-associated complement factor H polymorphism p.Tyr402His.

Gretarsdottir, S., et al.. Localization of a susceptibility gene for common forms of stroke to 5q12. *Am. J. Hum. Genet*. 2002; 70:593-603.

Gretarsdottir, S., et al. The gene encoding phosphodiesterase 4D confers risk of ischemic stroke. *Nat Genet*. 2003; 35:131-138.

Gupta, R. Lifestyle risk factors and coronary heart disease prevalence in Indian men. *J. Assoc. Phys. India* 1996; 44: 689-693.

Halliwell, B. and Gutteridge, J.M.C. *Free Radicals in Biology and Medicine 1989;* 2nd Edition, Clarendon Press, Oxford, UK.

Hassan, A., and Markus, M.J. Genetics and ischemic stroke. *Brain* 1995; 123: 1784-1812.

Heart Disease and Stroke Statistics – 2004 Update. American Heart Association, pp.13-16. (www.americanheart.org).

Hegab, A.E., Sakamoto, T., Uchida, Y. and Nomura, A. *CLCA1* gene polymorphisms in chronic obstructive pulmonary disease. *Journal of Medical Genetics* 2004; 41:e27.

Hegab, A.E., Sakamoto, T., Saitoh, W., Massoud, H.H. et al. Polymorphisms of IL4, IL13, and ADRB2 Genes in COPD. *Chest* 2004; 126: 1832 - 1839.

Helgadottir, A., et al. The gene encoding 5-lipoxygenase activating protein confers risk of myocardial infarction and stroke. *Nat. Genet*. 2004; 36:233-239.

Houstis, N., Rosen, E.D. and Lander, E.S. Reactive oxygen species have a causal role in multiple forms of insulin resistance. *Nature* 2006; 440: 944-948.

Inder, T.E. and Volpe, J.J. Mechanisms of perinatal brain injury. *Semin. Neonatol*. 2000; 5:3.

Irie, F. and Hiyabayashi, Y. Application of exogenous ceramide to cultured rat spinal motoneurons promotes survival or death by regulation of apoptosis depending on its concentrations. *J. Neurosci. Res*. 1998; 54(4): 475-85.

Jia, Z., Natarajan, P., Forte, T.M., and Bielicki, J.K. Thiol-bearing synthetic peptides retain the antioxidant activity of Apolipoprotein A-I $_{\text{Milano}}$. *Biochem Biophys Res Commun* 2002; 297: 206-213.

Johnston, S.C. Transient ischemic attack. *N. Engl. J. Med*. 2002; 347 (21): 1687-692.

Joutel, A. et al. *Notch3* mutations in CADASIL, a hereditary adult-onset causing stroke and dementia. *Nature* 1996; 383: 707-710.

Kangavari, S., Matetzky, S., Shah, P.K., Yano, J., Chyu, K-Y, Fishbein, M.C. and Cercek, B. Smoking increases inflammation and metalloproteinase expression in human carotid atherosclerotic plaques. *J. Cardiovasc. Pharmacol*. Therapeut 2004; 9(4): 291-298.

Kaul, S., Rukshin, V., Santos, R., Azarbal, B., Bisgaier, C.L., Johansson, J., Tsang, V.T., Chyu, K-Y., Cercek, B., Mirocha, J. and Shah, P.K. Intramural delivery ofrecombinant apolipoprotein A-I Milano/Phospholipid Complex (ETC-216) inhibits in-stent stenosis in porcine coronary arteries. *Circulation* 2003; 107: 2551-2554.

Klein, R., Klein, B., Knudtson, M., Wong, T. et al. .Prevalence of Age-Related Macular Degeneration in 4 Racial/Ethnic Groups in the Multi-ethnic Study of Atherosclerosis. *Ophthalmology* 2006;113: 373-380.

Kurth, T., Kase, C.S, Berger, K., Schaeffner, E.S., Buring, J.E., and Gaziano, M. Smoking and the risk of hemorrhagic stroke in men. *Stroke* 2003a; 34: 1151-1155.

Kurth, T., Kase, C.S, Berger, K., Gaziano, M., Cook, N.S., and Buring, J.E.. Smoking and risk of hemorrhagic stroke in women. *Stroke* 2003b; 34: 2792-2795.

Leibsom, C.L., Ranson, J.E., Brown, R.D., O'Fallon, W.M., Hass, S.L., and Whisnant, J.P. Stroke-attributable nursing home use. *Neurology* 1998; 51: 163-168.

Li, X., Chyu K-Y., Faria Neto J.R., Yano, J., Nathwani, N., Ferreira, C., Dimayuga, P.C., Cercek, B., Kaul, S. and Shah, P.K. Differential effects of apolipoprotein A-I-mimetic peptide on evolving and established atherosclerosis in Apolipoprotein E-null mice. *Circulation* 2004;10: 1701-1705.

Lily C. Chow, Kenneth W. Wright, and Augusto SolaCan Changes in Clinical Practice Decrease the Incidence of Severe Retinopathy of Prematurity in Very Low Birth Weight Infants? *Pediatrics*, 2003; 111: 339 – 345.

Lin PW, Stoll BJ. Necrotizing enterocolitis. *Lancet* 2006;368:1271-1283.

Lu, S-C., Wu, W-H., Le, C-A., Chou, H-F., Lee, H-R., and Huang, P-C. LDL of Taiwanese vegetarians are less oxidizable than those of omnivores. *J. Nutr.* 2000; 130: 1591-1596.

Marcus, E.B., Buist, A.S., Curb, J.D., MacLean, C.J. et al. Correlates of FEV1 and prevalence of pulmonary conditions in Japanese-American men. *Am. Rev. Respir. Dis.* 1988; 138:1398–404.

Mattson, M.P. Neuronal life-and-death signaling, apoptosis, and neurodegenerative disorders. *Antioxid Redocx Signal* 2006; **8:** 1997-2006.

Mattson, M.P. Cellular actions of beta-amyloid precursor protein and its soluble and fibrillogenic derivatives. *Physiol. Rev.* 1997; 77**:** 1081-1132.

McElroy, M.C., Postle, A.D. and Kelly, F.J. Catalase, superoxide dismutase and glutathione peroxidase activities of lung and liver during human development. *Biochim. Biophys. Acta* 1992; 1117: 153–158.

Mensah, G.A., Mokdad, A.H., Ford, E.S., Greenlund, K.J. and Croft, J.B. State of disparities in cardiovascular health in the United States. *Circulation* 2005; 111: 1233-1241.

Minimo, A.M., Arias, E., Kochanek, K.D. et al. Death: final data for 2000. National Vital Statistics Reports. Hyattsville, MD: *National Center for Health Statistics*. 2002; 20(15).

Mohanty, P., Hamouda, W., Garg, R., ALjada, A., Ghanim, H. and Dandona, P. Glucose challenge stimulates reactive oxygen species (ROS) generation by leukocytes. *J. Clin. Endocrinol. Metab*. 2000; 85: 2970-2973.

Mohanty, P., Ghanim, H., Hamouda, W., ALjada, A., Garg, R. and Dandona, P. Both lipid and protein intakes stimulate increased generation of reactive oxygen species by polymorphonuclear leukocytes and mononuclear cells. *Am. J. Clin. Nutr*. 2002; 75: 767-772.

Mohanty, P., Daoud, N., Ghanim, H., Ravishankar, S., Szudzik, E., ALjada, A. and Dandona, P. Absence of oxidative stress and inflammation following the intake of a 900 kcalorie meal rich in fruit and fiber. *Diabetes* 2004; 53: A405.

Mukhopadhyay, B. and Sattar, F.M. Diabetes and cardiac disease in South Asians. *Br. J. Diabetes Vasc. Dis*. 2005; 5: 253-259.

Navab, M., Hough, G.P., Stevenson, L.W., Drinkwater, D.C., Laks, H., and Fogelman, A.M. Monocyte migration into the subendothelial space of a coculture of adult human aortic endothelial and smooth muscle cells. *J. Clin. Invest*. 1988; 82: 1853-1963.

Navab, M., Imes, S.S., Hama, S.Y., Hogh, G.P., Ross, L.A., Bork, R.W., Valente, A.J., Berliner, J.A., Drinkwater, D.C., Laks, H. and Fogelman, A.M. Monocyte transmigration induced by modification of low density lipoprotein in cocultures of human aortic wall cells is due to induction of monocyte chemotactic protein 1 synthesis and is abolished by high density lipoprotein. *J. Clin. Invest*. 1991; 88: 2039-2046.

Navab, M., Ananthramaiah, G.M., Reddy, S.T., Van Lenten, B.J., Ansell, B.J., Fonarow, G.C., Vahabzadeh, K., Hama, S., Hough, G., Kamranpour, N., Berliner, J.A., Lusis, A.J. and Fogelman, A.M. The oxidation hypothesis of Atherogenesis: the role of oxidized phospholipids and HDL. *J. Lipid. Res*.2004; 45: 993-1007.

Navab, M., Anantharamaiah, G.M. and Fogelman, A.M. An Apolipoprotein A-I mimetic works best in the presence of apolipoprotein A-I. *Circ. Res*. 2005; 97: 1085-1086..

Nicholls, S.J., Tuzcu, E.M., Sipahi, I., Schoenhagen, P., Crowe, T., Kapadia, S. and Nissen, S.E. Relationship between atheroma regression and changein lumen size after infusion of Apolipoprotein A-I Milano. *J. Amer. College Cardiol*. 2006; 47: 992-997.

Nordskog, B.K., Blixt, A.D., Morgan, W.T., Fields, W.R., and Hellmann, G.M. Matrix-degrading and pro-inflammatory changes in human vascular endothelial cells exposed to cigarette smoke condensate. *Cardiovasc Toxicol*. 2003; 3(2):101-117.

O'Donovan DJ, Fernandes CJ. Free radical and diseases in premature infants. *Antioxidants and redox signaling* 2004;6:169-176.

Olden, K. and White, S.L. Health-related disparities: influence of environmental factors. *Med. Clin. North Am*. 2005; 89:721-738.

Palsdottir, A. et al. Mutation in cystatin C gene causes hereditary brain hemorrhage. *Lancet* 1988; 2: 603-604.

Parton LA, Strassberg SS, Qian D, Galvin-Parton PA, Cristea IA. The genetic basis for bronchopulmonary dysplasia. *Front Biosci*. 2006;11:1854-1860.

Pool, J. B., and Greenough, A. Ethnic variation in respiratory function in young children. *Respir. Med*. 1989; 83: 123-125.

Rahman, I., Biswas, S.K. and Kirkham, P.A. Regulation of inflammation and redox signaling by dietary polyphenols. *Biochemical Pharmacology* 2006; doi:10.1016/j.bcp.2006.07.004.

Reaven, G.M. The metabolic syndrome: is this diagnosis necessary? *Am. J. Clin. Nutr*. 2006; 83: 1237-1247.

Rehan, V.K., Wang, Y., Patel, S., Santos, J. and Torday, J.S. Rosiglitazone, a peroxisome proliferator-activated receptor-gamma agonist, prevents hyperoxia-induced neonatal rat lung injury in vivo. *Pediatr. Pulmonol*. 2006; 41(6):558-69.

Rehan, V.K. Prevention of Bronchopulmonary Dysplasia: Finally, something that works. *Indian J. Pediatr* 2006; 73(11): 1027-1032.

Robbins, A.S., Mansfield, J.E., Lee, I., Satterfield, S., and Hennekens, C.H. Cigarette smoking and stroke in a cohort of US male physicians. *Ann. Intern. Med* 1994; 1210: 458-462.

Sankar, P., Cho, M.K., Condit, C.M., Hunt, L.M. et al. Genetic research and health disparities. *JAMA* 2004; 291:2985–2989.

Saugstad, O.D. Bronchopulmonary dysplasia-oxidative stress and antioxidants. *Semin. Neonatol.* 2003; 8:39-49.

Saugstad, O.D. Oxidative stress in the newborn--a 30-year perspective. *Biol. Neonate*. 2005; 88(3):228-236.

Schock, B.C., Sweet, D.G., Halliday, H.L., Young, I.S. and Ennis, M. Oxidative stress in lavage fluid of preterm infants at risk of chronic lung disease. *Am. J. Physiol. Lung Cell Mol. Physiol.* 2001; 281: L1386–L1391.

Shah, P.K., Yano, J., Reyes, O., Chyu, K-Y, Kaul, S., Bisgaier, C.L., Drake, S. and Cercek, B. 2001; 103: 3047-3050.

Siess, H. Strategies of antioxidant defense. *Eur. J. Biochem.* 1993; 215: 213-219.

Sies, H. Oxidative stress: oxidants and antioxidants. *Exp. Physiol.* 1997; 82: 291-295.

Steinberg, D., Parthasarthy, S., Carew, T.E., Khoo, J.C., and Witzum, J.L. Beyond cholesterol. Modifications of low density lipoprotein that increase its atherogenicity. *N. Engl. J. Med.* 1989; 320: 915-924.

Steinbrecher, U.P., Parthasarthy, S., Leake, D.S., Witzum, J.L. and Steinberg, D. Modification of low density lipoprotein by endothelial cells involves lipid peroxidation and degradation of low density lipoprotein phospholipids. *Proc. Natl. Acad. Sci. USA* 1984; 81: 3883-3887.

The CROSS Group. Epidemiology of aneurismal subarachnoid hemorrhage in Australia and New Zealand: incidence and case fatality from the Australian Cooperative Research on Subarachnoid Study (ACROSS). *Stroke* 2000; 31: 1843-1850.

Third Report of the National Cholesterol Education Program (NCEP) Expert Panel on Detection, Evaluation, and Treatment of High Blood Cholesterol in Adults (Adult Treatment Panel III) Final Report. *Circulation*. 2002; 106: 3143-3421.

Thom, T., Haase, N., Rosemund W. et al. Heart disease and stroke statistics-2006 Update: A report from the American Heart Association Statistics Committee and Stroke Statistics Subcommittee. *Circulation* 2006; 113: e85-e151.

Uzui, H., Harpf, A., Liu, M., Doherty, T.M., Shukla, A., Chai, N-N., Tripahi, P.V., Jovinge, S., Wilkin, D.J., Asotra, K., Shah, P.K. and Rajavashisth, T.B. Increased expression of membrane-type 3 matrix metalloproteinase in human atherosclerotic plaque: Role of activated macrophages and inflammatory cytokines. *Circulation* 2002; 106: 3024-3030.

Volpe, J.J. Intracranial hemorrhage: Germinal matrix-intraventricular hemorrhage. In: *Neurology of the Newborn*, 4th Ed, WB Saunders, Philadelphia, 2001. p.428.

Walter, P. Effects of vegetarian diets on aging and longevity. Nutr Rev 1997; 55:S61-S65.

Wannmathee, S.G., Shaper, A.G., Whicup, P.H., and Walker, M. Smoking cessation and the risk of stroke in middle-aged me. *JAM* 1997;, 274: 155-160.

West, J.B. Cardiac energetics and myocardial oxygen consumption. *Physiologic basis of medical practice.* Williams and Wilkins. Baltimore, MD, USA. 1991: 250-260.

Willaime, S., Vanhoutte, P., Caboche, J., Lemaigre-Dubreuil, Y., Mariani, J. and Brugg, B. Ceramide-induced apoptosis in cortical neurons is mediated by an increase in p38 phosphorylation and not by the decrease in ERK phosphorylation. *Eur. J. Neurosci.* 2001;13(11): 2037-46.

Witzum, J.L and Steinberg, D. Role of oxidized low density lipoprotein in atherogenesis. J. Clin. Invest. 1991; 88: 1785-1792.

Wolf, P.A., D'Agostino, R.B., Kannel, W.B., Bonita, R., and Belanger, A.J. (1988). Cigarette smoking as a risk factor for stroke. The Framingham Study. *JAMA* 1988; 259: 1025-1029.

Wong, W.M., Hawe, E., Li, L. K., Miller, G. J. et al. Apolipoprotein AIV gene variant S347 is associated with increased risk of coronary heart disease and lower plasma apolipoprotein AIV levels. *Circ. Res* 2003; 92: 969-975.

Yu, Z., Nikolova-Karakashian, M., Zhou, D., Cheng, G., Schuchman, E.H., and Mattson, M.P. Pivotal role for acidic sphingomyelinase in cerebral ischemia-induced ceramide and cytokine production, and neuronal apoptosis. *J. Mol. Neurosci*. 2000; 15: 85-98.

Yu-Poth, S., Etherton, T.D., Redyy. C.C., Pearson, T.A., Reed, R., Zhao, G., Jonnlagadda, S., Wan, Y. and Kris-Etherton, P.M. Lowering dietary saturated fat and total fat reduced the oxidative susceptibility of LDL in health men and women. *J. Nutr.* 2000; 130: 2228-2237.

In: Oxidative Stress: Clinical and Biomedical Implications ISBN: 978-1-60021-888-0
Editors: B. M. Matata, M. M. Elahi, pp. 23-38

Chapter II

Sources of Reactive Oxidants Species in Biology and Disease

Bashir M. Matata1 and Maqsood M. Elahi*[2]
[1]The Cardiothoracic Centre, Liverpool NHS Trust,
Liverpool, United Kingdom
[2]The Wessex Cardiothoracic Centre,
Southampton, United Kingdom

Abstract

During biological processes molecules undergo chemical changes involving reducing and oxidising reactions. A molecule with an unpaired electron can combine with a molecule capable of donating an electron. The donation of an electron is termed as oxidation whereas the gaining of an electron is called reduction. Reduction and oxidation can render the reduced molecule unstable and free to react with other molecules to cause damage to cellular and sub-cellular components such as membranes, proteins and DNA. The current chapter will systematically discuss the mechanisms of formation of reactive oxidant species in the form of oxygen-derived reactive oxygen species (ROS) and reactive nitrogen species (RNS, e.g. nitric oxide, NO). This will include a review of RNS and ROS generation by a system of tightly regulated enzymes, such as NO synthase (NOS), xanthine oxidases, the cyclooxygenases, NAD(P)H oxidase isoforms and metal-catalysed reactions. In addition, a discussion of physiological defences such as specialised enzymes and antioxidants that maintain reduction-oxidation (redox) balance. The chapter will also discuss how enzymes and antioxidants can become overwhelmed by the excessive production of reactive oxidants resulting in oxidative stress/nitrosative stress, a process that is an important mediator of damage to cells and a ubiquitous feature of human disease.

* Correspondence To: Bashir M Matata PhD. The Cardiothoracic Centre, Liverpool NHS Trust, Thomas Drive, Liverpool, L14 3PE, UK. Email: matata_bashir@hotmail.com

1. Introduction

1.1. Historical Perspective

The historical perspective of reactive oxidants has been discussed in more details in the recent manuscript by Valko et al 2007. In the excellent review by Valko the authors highlighted the fact that it was not until the publication of Gershman's free radical theory of oxygen toxicity in 1954 (Gerschman et al, 1954) did the causes of the poisonous properties of oxygen begun to be understood. In the same year, observations of a weak electron paramagnetic resonance (EPR) signal attributable to the presence of free radicals in a variety of lyophilised biological materials were reported by Commoner et al 1954. Denham Harman proposed the concept that free radicals played a role in the ageing process (Harman, 1956). This work gradually triggered intense research into the field of free radicals in biological systems. A second epoch of the research of free radicals in biological systems was explored in 1969 when McCord and Fridovich discovered the enzyme superoxide dismutase (SOD) and thus provided convincing evidence about the importance of free radicals in living systems (McCord and Fridovich, 1969). A third era of free radicals in biological systems dates from 1977 when Mittal and Murad 1977 provided evidence that the hydroxyl radical, •OH, stimulates activation of guanylate cyclase and formation of the "second messenger" cyclic guanosine monophosphate (cGMP). Since then, a large body of evidence has been accumulated that living systems have not only adapted to a coexistence with free radicals but have developed various mechanisms for the advantageous use of free radicals in various physiological functions.

Reactive oxygen species (ROS), as well as reactive nitrogen species (RNS), are products of normal cellular metabolism. ROS and RNS are well recognised for playing a dual role as both deleterious and beneficial species, since they can be either harmful or beneficial to living systems (Valko et al 2006). Beneficial effects of ROS occur at low/moderate concentrations and involve physiological roles in cellular responses to insults, as for example in defence against infectious agents and in the function of a number of cellular signalling systems. One further beneficial example of ROS at low/moderate concentrations is the induction of a mitogenic response. The harmful effect of free radicals causing potential biological damage is termed oxidative stress and nitrosative stress (Kovacic and Jacintho 2001; Ridnour et al 2005; Valko et al 2001). This occurs in biological systems when there is an overproduction of ROS/RNS on one side and a deficiency of enzymatic and non-enzymatic antioxidants on the other. In other words, oxidative stress results from the metabolic reactions that use oxygen and represents a disturbance in the equilibrium status of prooxidant/antioxidant reactions in living organisms. The excess ROS can damage cellular lipids, proteins, or DNA inhibiting their normal function. Because of this, oxidative stress has been implicated in a number of human diseases as well as in the ageing process. The delicate balance between beneficial and harmful effects of free radicals is a very important aspect of living organisms and is achieved by mechanisms called "redox regulation". The process of "redox regulation" protects living organisms from various oxidative stresses and maintains "redox homeostasis" by controlling the redox status in vivo (Dröge 2002). This chapter examines the available evidence for the mechanisms and sources of cellular oxidants in normal physiological functions as well as in the pathogenesis of various diseases, including cancer, diabetes mellitus, ischemia/reperfusion injury, inflammatory diseases, neurodegenerative disorders and ageing. The various protective pathways that may be provided by the antioxidant network against the deleterious action of free radicals are also discussed.

2. Reactive Oxygen Species (ROS)

Free radicals can be defined as molecules or molecular fragments containing one or more unpaired electrons in atomic or molecular orbitals (Halliwell and Gutteridge 1999). This unpaired electron(s) usually gives a considerable degree of reactivity to the free radical. Radicals derived from oxygen represent the most important class of radical species generated in living systems (Miller et al 1990). Molecular oxygen (dioxygen) has a unique electronic configuration and is itself a radical. The addition of one electron to dioxygen forms the superoxide anion radical ($O_2^{\bullet-}$) (Miller et al 1990).

2.1. Superoxide

The oxygen molecule is used in respiration to biologically oxidize carbon and hydrogen rich molecules in order to obtain chemical energy and heat essential for life. In this process, the oxygen molecule is reduced to water by a stepwise addition of 4 electrons giving rise to intermediate ROS.

$$O_2 + e^- \longrightarrow O_2^{\bullet-} \quad (1)$$

$$O_2^{\bullet-} + e^- + 2H^+ \rightarrow H_2O_2 \quad (2)$$

$$H_2O_2 + e^- + H^+ \rightarrow {}^{\bullet}OH + H_2O \quad (3)$$

$$OH + e^- + H^+ \rightarrow H_2O \quad (4)$$

In this way the hydroxyl radical, hydrogen peroxide and the superoxide radicals are produced. For instance, when opsonised particles are contacted by neutrophils a 'respiratory burst' occurs, with oxygen uptake and the release of superoxide radicals into the phagocytic vacuole. Superoxide anion, arising either through metabolic processes or following oxygen "activation" by physical irradiation, is considered the "primary" ROS, and can further interact with other molecules to generate "secondary" ROS, either directly or prevalently through enzyme- or metal-catalysed processes (Valko et al 2005). The production of superoxide occurs mostly within the mitochondria of a cell (Cadenas and Sies 1998) and the endoplasmic reticulum. The mitochondrial electron transport chain is the main source of ATP in the mammalian cell and thus is essential for life. During energy transduction, a small number of electrons "leak" to oxygen prematurely, forming the oxygen free radical superoxide, implicated in the pathophysiology of a variety of diseases (Kovacic et al 2005; Valko et al 2004). Within the mitochondria superoxide is produced from both Complexes I and III of the electron transport chain, and once in its anionic form it is too strongly charged to readily cross the inner mitochondrial membrane. Recently, it has been demonstrated that Complex I-dependent superoxide is exclusively released into the matrix and that NO detectable levels escape from intact mitochondria (Muller et al 2004). This finding fits well with the proposed site of electron leak at Complex I, namely the iron–sulphur clusters of the (matrix-protruding) hydrophilic arm. In addition, experiments on Complex III show direct extra-mitochondrial release of superoxide, but measurements of hydrogen peroxide production revealed that this could only account for <50% of the total electron leak even in mitochondria lacking Cu, Zn-SOD. The thinking is that the remaining 50% of the electron leak must be due to superoxide released into the matrix.

2.1.1. Phagocytic Cells

Phagocytic cells such as the neutrophils when exposed to a stimulus have the ability of recognising the foreign particle and undergoing a series of reactions called the respiratory burst (DeCoursey and Ligeti 2005). Nicotine adenine dinucleotide phosphate (NAD(P)H) oxidase is best characterised in neutrophils, where its production of $O_2^{\cdot -}$ generates the respiratory burst necessary for bacterial destruction. The enzyme complex consists of two membrane-bound components gp91phox and p22phox, which comprise cytochrome b558, the enzymatic centre of the complex. After activation, cytosolic components, involving p47phox, p67phox, p40phox and the small G coupled proteins, Rac and Rap1A, translocates to the membrane to form the active enzyme complex.

2.1.2. Non-Phagocytic Cells

NAD(P)H oxidases derived from non-phagocytic cells- produce superoxide at a fraction (1–10%) of the levels produced in neutrophils and are thought to function in intracellular signaling pathways. In the endothelial cells several other potential sources of $O_2^{\cdot -}$ are implicated in vascular physiology and pathophysiology, these include the mitochondrial electron transport chain, xanthine oxidase, cytochrome P-450 enzymes, uncoupled NOSs (NO synthases), the phagocytic myeloperoxidase system (for further information see [Li and Shah 2004; Fleming 2001; Duchen 2004; Channon and Guzik 2002; Landmesser et al 2003]). The mitochondrial electron transport chain can be a significant ROS source and increased mitochondrial ROS generation is implicated in diabetic vasculopathy and in ischemia/reperfusion. Interestingly, mitochondria are quite susceptible to oxidative damage which can result in enhanced mitochondrial ROS production [Duchen 2004]. Xanthine oxidase is expressed on the luminal surface of the endothelium in many organs and catalyses the conversion of hypoxanthine into urate in a process that generates $O_2^{\cdot -}$. The enzyme is normally present as xanthine dehydrogenase, which does not generate $O_2^{\cdot -}$, but is converted into xanthine oxidase either through oxidation or by proteolytic cleavage of a segment of xanthine dehydrogenase. The former mode of activation is notable in that it provides a mechanism for an increase in xanthine oxidase activity in response to oxidative stress, i.e. a mechanism that may potentially amplify oxidative stress originally arising from other sources. Increased xanthine oxidase- derived $O_2^{\cdot -}$ production may be involved in ischemia/reperfusion and in endothelial dysfunction in several diseases [12]. NO synthases normally generate NO, but can become 'uncoupled', usually in the context of deficiency of the essential NOS cofactor H_4B (tetrahydrobiopterin) [Cai and Harrison 2000]. When this occurs, eNOS (endothelial NOS) becomes a significant source of $O_2^{\bullet -}$. ROS generation by uncoupled eNOS is reportedly involved in the pathophysiology of diabetic vasculopathy, atherosclerosis, hypertension and hypercholesterolemia. Of interest, H_4B is highly susceptible to oxidative degradation by $O_2^{\bullet -}$or $ONOO^-$; thus, initial degradation of H_4B by ROS derived from another source can induce NOS uncoupling and amplification of oxidative stress [Channon and Guzik 2002; Landmesser et al 2003]. In experimental hypertension, it has been convincingly demonstrated that initial ROS generation by NADPH oxidase leads to eNOS uncoupling and amplification of ROS production [Landmesser et al 2003].

2.2. Hydrogen Peroxide

Any system producing superoxide would simultaneously also be expected to produce hydrogen peroxide, by the dismutation reaction:

$$O_2^{\bullet -} \xrightarrow{2H^+} 2H_2O_2 + O_2 \qquad (5)$$

This reaction is accelerated when catalyzed by the ubiquitous superoxide dismutase (SOD) enzymes, which removes superoxide from the solution at the expense of forming hydrogen peroxide. In the absence of transition metal ions, hydrogen peroxide is relatively stable and can diffuse across membranes in much the same way as water. Superoxide can react with hypochlorous acid (HOCl) to form hydroxyl radicals [Candeias et al 1993] and both hydrogen peroxide (H_2O_2) and HOCl can react with iron species to generate hydroxyl radicals [Candeias LP et al 1994]. Neutrophils in the presence of hydrogen peroxide, which is formed by the dismutation of superoxide, can also oxidize chloride ions into the powerful oxidant HOCl.

$$H_2O_2 + Cl^- \xrightarrow[\text{Myeloperoxidase}]{H^+} HOCl + H_2O \qquad (6)$$

Peroxisomes are known to produce H_2O_2, but not $O_2^{\bullet -}$, under physiologic conditions [Valko et al, 2004]. Peroxisomes are major sites of oxygen consumption in the cell and participate in several metabolic functions that use oxygen. Oxygen consumption in the peroxisomes leads to H_2O_2 production, which is then used to oxidize a variety of molecules. The organelle also contains catalase, which decomposes hydrogen peroxide and presumably prevents accumulation of this toxic compound. Thus, the peroxisome maintains a delicate balance with respect to the relative concentrations or activities of these enzymes to ensure NO net production of ROS. How the organelle maintains this equilibrium is unclear. When peroxisomes are damaged and their H_2O_2 consuming enzymes downregulated, H_2O_2 releases into the cytosol which significantly contributes to oxidative stress.

2.3. Hydroxyl Radicals

The hydroxyl radical, •OH, is the neutral form of the hydroxide ion. The hydroxyl radical has a high reactivity, making it a very dangerous radical with a very short *in vivo* half-life of approximately 10^{-9} s [Pastor et al 2000]. Thus when produced *in vivo* $^{\bullet}OH$ reacts close to its site of formation. The redox state of the cell is largely linked to an iron (and copper) redox couple and is maintained within strict physiological limits. It has been suggested that iron regulation ensures that there is NO free intracellular iron; however, *in vivo*, under stress conditions, an excess of superoxide releases "free iron" from iron-containing molecules. The release of iron by superoxide has been demonstrated for [4Fe-4S] cluster containing enzymes of the dehydratase-lyase family [Liochev and Fridovich 1994]. The released Fe2+ can participate in the Fenton reaction, generating highly reactive hydroxyl radical [Halliwell and Gutteridge 1999].

$$(Fe^{2+} + H_2O_2 \rightarrow Fe^{3+} + {}^{\bullet}OH + OH^-) \quad (7)$$

Thus under stress conditions, $O_2^{\bullet-}$ acts as an oxidant of [4Fe–4S] cluster-containing enzymes and facilitates $^{\bullet}OH$ production from H_2O_2 by making Fe^{2+} available for the Fenton reaction [Valko et al 2005; Leonard et al 2004]. The superoxide radical participates in the Haber–Weiss reaction (Equation 8) which combines a Fenton reaction (Equation 7) and the reduction of Fe^{3+} by superoxide, yielding Fe^{2+} and oxygen [Liochev and Fridovich 2002].

$$O_2^{\bullet-} + H_2O_2 \rightarrow O_2 + {}^{\bullet}OH + OH^- \quad (8)$$

$$Fe^{3+} + O_2^{\bullet-} \rightarrow Fe^{2+} + O_2 \quad (9)$$

The Fe–S cluster contains also iron responsive elements (IRE)-binding protein (IRE-BP). This Fe–S cluster has been implicated as the region of the protein that senses intracellular iron levels and accordingly modifies the ability of the IRE-BP to interact with iron-responsive elements (IREs). IRE-BP produced in iron-replete cells has aconitase activity [Han et al, 2005]. In mammalian cells, oxidants are able to convert cytosolic aconitase into active IRE-BP, which increases the "free iron" concentration intracellularly both by decreasing the biosynthesis of ferritin and increasing biosynthesis of transferrin receptors.

Hydroxyl radicals in free solution can attack most biological molecules at almost diffusion-controlled rates, but have little or NO specificity in Fenton chemistry. They are considerably site-specificity because the hydroxyl radical is formed close to where the iron is located. Excessive biological production of superoxide can generate all the ingredients necessary for the formation of hydroxyl radicals. The most realistic *in vivo* production of hydroxyl radical according to the Fenton reaction occurs in the presence of metal ions such as iron, copper, chromium, or cobalt. However, Rae and co-workers recently reported that the upper limit of so-called "free pools" of copper was far less than a single atom per cell [Rae et al 1999]. This finding casts serious doubt on the *in vivo* role of copper in Fenton-like generation of hydroxyl radical. Although Fenton chemistry is known to occur *in vitro*, its significance under physiological conditions is not clear, noting particularly the negligible availability of "free catalytic iron" due to its effective sequestration by the various metal-binding proteins [Kakhlon and Cabantchik, 2002]. However, organisms overloaded by iron (as in the conditions of hemochromatosis, β-thalassemia, hemodialysis) contain higher amounts of "free available iron" and this can have deleterious effects. "Free-iron" is transported into an intermediate, labile iron pool (LIP), which represents a steady state exchangeable and readily chelatable iron compartment [Kakhlon and Cabantchik 2002].

2.4. Peroxyl Radicals ($ROO^{\bullet}$)

Additional reactive radicals derived from oxygen that can be formed in living systems are peroxyl radicals ($ROO^{\bullet}$). Oxygen free radicals attack polyunsaturated fatty acids (PUFA) causing a radical chain reaction known as lipid peroxidation. The oxygen free radicals abstract a hydrogen atom from a methylene ($-CH_2-$) group. For example, if the hydroxyl radical reacts with the methylene group, the following reaction would occur:

$$-CH_2- + {}^{\bullet}OH \longrightarrow {}^{\bullet}CH- + H_2O \qquad (10)$$

Abstraction of the hydrogen atom leaves an unpaired electron on the carbon. These carbon radicals then react with oxygen to form peroxyl radicals, which are capable of abstracting hydrogen atoms from other lipid molecules. Thus a chain reaction is initiated. The simplest peroxyl radical is $HOO^{\bullet}$, which is the protonated form (conjugate acid; p*K*a ~ 4.8) of superoxide ($O_2^{\bullet-}$) and is usually termed either hydroperoxyl radical or perhydroxyl radical. Given this p*K*a value, only ~ 0.3% of any superoxide present in the cytosol of a typical cell is in the protonated form [De Grey 2002]. It has been demonstrated that hydroperoxyl radical initiates fatty acid peroxidation by two parallel pathways: fatty acid hydroperoxide (LOOH)-independent and LOOH-dependent [Aikens and Dix 1991]. The LOOH-dependent pathway of $HO_2^{\bullet-}$ initiated fatty acid peroxidation may be relevant to mechanisms of lipid peroxidation initiation *in vivo*. Xanthine oxidase (XO, EC 1.1.3.22) and xanthine dehydrogenase (XD, EC 1.1.1.204) are inter-convertible forms of the same enzyme, known as xanthine oxidoreductase (XOR) [Borges et al 2003]. In purine catabolism, XOR catalyzes the oxidative hydroxylation of hypoxanthine to xanthine and subsequently of xanthine to uric acid. Uric acid acts as a potent antioxidant and free radical scavenger. XOR has, therefore, important functions as a cellular defence enzyme against oxidative stress. With both XO and XD forms, but particularly with the XO form, numerous ROS and RNS are synthesized [Vorbach et al, 2003]. Thus, the synthesis of both an antioxidant (uric acid) and numerous free radicals (ROS and RNS) makes XOR an important protective regulator of the cellular redox potential.

3. Reactive Nitrogen Species (RNS)

NO is a small molecule that contains one unpaired electron on the antibonding $2\pi.y$ orbital and is, therefore, a radical. NO• is generated in biological tissues by specific nitric oxide synthases (NOSs), which metabolise arginine to citrulline with the formation of NO *via* a five electron oxidative reaction [Ghafourifar and Cadenas 2005]. NO is an abundant reactive radical that acts as an important oxidative biological signalling molecule in a large variety of diverse physiological processes, including neurotransmission, blood pressure regulation, defence mechanisms, smooth muscle relaxation and immune regulation [Bergendi et al 1999]. NO has a half-life of only a few seconds in an aqueous environment. NO has greater stability in an environment with a lower oxygen concentration (half-life >15 s). However, since it is soluble in both aqueous and lipid media, it readily diffuses through the cytoplasm and plasma membranes [Chiueh 1999]. NO has effects on neuronal transmission as well as on synaptic plasticity in the central nervous system. In the extracellular milieu NO reacts with oxygen and water to form nitrate and nitrite anions. NO readily binds certain transition metal ions; in fact many physiological effects of NO are exerted as a result of its initial binding to Fe^{2+}-Haem groups in the enzyme soluble guanylate cyclase (sGC) [Archer, 1993]. The product is represented here as *{Fe^{2+}–NO•}*, however, *{Fe^{3+}–NO^-}* is also commonly seen.

$$Fe^{2+}\{sGC\} + NO \longrightarrow Fe^{2+}\{sGC\}\text{–}NO \qquad (11)$$

Overproduction of reactive nitrogen species is called nitrosative stress [Klatt and Lamas 2000; Ridnour et al 2004]. This may occur when the generation of reactive nitrogen species in a system exceeds the system's ability to neutralise and eliminate them. Nitrosative stress may lead to nitrosylation reactions that can alter the structure of proteins and so inhibit their normal function. Cells of the immune system produce both the superoxide anion and nitric oxide during the oxidative burst triggered during inflammatory processes. Under these conditions, nitric oxide and the superoxide anion may react together to produce significant amounts of a much more oxidatively active molecule, peroxynitrite anion ($ONOO^-$), which is a potent oxidising agent that can cause DNA fragmentation and lipid oxidation [Carr et al 2000]:

$$NO + O_2^{\bullet -} \longrightarrow ONOO^- \qquad (12)$$

This reaction has one of the highest rate constants known for reactions of NO, $7.0\times10^9 M^{-1} s^{-1}$. Thus NO• toxicity is predominantly linked to its ability to combine with superoxide anions. Once formed, peroxynitrite becomes rapidly protonated in body fluids to peroxynitrous acid (ONOOH) and leads to ONOOH-dependent depletion of –SH- groups and other antioxidants, nitration and oxidation of lipids [Graham et al 1993]. In proteins, peroxynitrite reacts with tyrosine residues to form 3-nitrotyrosine [Beckmann and Koppenol

1996]. It has been shown that phosphorylation of tyrosine residues, are a key feature of intracellular signal transduction, which is blocked by nitration of tyrosine kinases [Berlett et al 1996; Gow et al 1996].

4. Antioxidants

Organisms have developed a series of defence mechanisms [Cadenas 1997] against free radicals from a variety of sources. Mechanisms against free radical-induced oxidative stress include: (i) preventative mechanisms, (ii) repair mechanisms, (iii) physical defences, and (iv) antioxidant defences. Enzymatic antioxidant defences include superoxide dismutase (SOD), glutathione peroxidase (GPx), and catalase (CAT). Non-enzymatic antioxidants are represented by ascorbic acid (Vitamin C), α-tocopherol (Vitamin E), glutathione (GSH), carotenoids, flavonoids, and other antioxidants. Under normal conditions, there is a balance between both the activities and the intracellular levels of these antioxidants. This balance is essential for the survival of organisms and their health.

Antioxidants are molecules which when present at low concentrations compared to those of the oxidizable substrate, significantly delay or inhibit oxidation of that substrate. Primary antioxidant defences are those, which prevent radical formation. The iron binding properties of transferrin and lactoferrin fulfill such roles in extracellular fluids, since iron correctly attached to their high affinity binding sites no longer catalyses radical formations. Secondary defences are those, which remove or inactivate formed ROS. In some cases they may be enzymes such as SOD, catalase and glutathione peroxidase, or low molecular mass molecules such as vitamin E, ascorbate and glutathione (GSH). Tertiary defences operate to remove and repair oxidatively-damaged molecules and are particularly important for DNA integrity; an example of this is the DNA repair enzyme O^6-methylguanine-DNA-methyltransferase [Halliwell and Gutteridge 1999], polypolymerase-1 [Pieper et al 1999] and p53 [Bennett 1999].

4.1. Cellular Antioxidant Defences

The major antioxidant and redox buffer of the cell is the tripeptide glutathione (GSH) [Masella et al 2005] and its oxidised form glutathione disulphide, GSSG. Glutathione is highly abundant in the cytosol (1–11 mM), nuclei (3–15 mM), and mitochondria (5–11 mM) and is the major soluble antioxidant in these cell compartments [Volka et al 2007]. Because GSH is synthesized in the cytosol by the sequential action of glutamate–cysteine ligase and glutathione synthetase, its mitochondrial presence requires inner membrane transport. Two mitochondrial electron-neutral antiport carrier proteins have been shown to have the capacity to transport GSH the dicarboxylate carrier protein and the 2-oxoglutarate carrier protein. Recently, it has been shown that externally added GSH is readily taken up by mitochondria, despite the ~8mM GSH present in the mitochondrial matrix [Shen et al 2005]. It therefore appears that GSH is taken up against a concentration gradient. GSH in the nucleus maintains the redox state of critical protein sulfhydryls that are necessary for DNA repair and expression. Oxidised glutathione is accumulated inside the cells and the ratio of GSH/GSSG is a good measure of oxidative stress of an organism [Nogueira et al 2004; Jones et al 2000]. Too high a concentration of GSSG may damage many enzymes oxidatively. The main protective roles of glutathione against oxidative stress are [Masella et al 2005]: (i) glutathione is a cofactor of several detoxifying enzymes against oxidative stress, e.g. glutathione peroxidase (GPx), glutathione transferase and others; (ii) GSH participates in amino acid transport through the plasma membrane; (iii) GSH scavenges hydroxyl radical and singlet oxygen directly, detoxifying hydrogen peroxide and lipid peroxides by the catalytic action of glutathione peroxidase; (iv) glutathione is able to regenerate the most important antioxidants, Vitamins C and E, back to their active forms; glutathione can reduce the tocopherol radical of Vitamin E directly, or indirectly, *via* reduction of semi-dehydroascorbate to ascorbate. The capacity of glutathione to regenerate the most important antioxidants is linked with the redox state of the glutathione disulphide-glutathione couple (GSSG/2GSH) (Pastore et al 2003). The various roles of enzymatic antioxidants (SOD, Catalase, glutathione peroxidase) and non-enzymatic antioxidants (Vitamin C, Vitamin E, carotenoids, lipoic acid and others) in the protection against oxidative stress can be found in a numerous reviews and original papers [Burton and Ingold 1984; Cameron and Pauling 1976; Carr and Frei 1999; Catani et al 2001; El-Agamey et al 2004; Hirota et al 1999; Kojo 2004; Landis and Tower 2005; Makropoulos et al 1996; Mates et al 1999; Miller et al 2005; Nakamura et al 1997; Packer and Suzuki 1993; Pryor 2000; Schrauzer 2006; Sharoni et al 2004; Smith et al 1997].

Oxygen metabolism occurs within cells, where antioxidants have evolved to deal speedily and specifically with reactive oxidants or reductants. These are likely to be protein-based catalysts i.e. enzymes. The SODs rapidly promote the dismutation of superoxide into hydrogen peroxide and oxygen at a rate considerably faster than it occurs uncatalysed.

$$O_2^{\cdot -} + H^+ \rightarrow H_2O_2 + O_2 \qquad [13]$$

Three forms of SOD are present in the body; copper-zinc SOD in the cytoplasm, manganese SOD in the mitochondria and extracellular SOD present in the extracellular matrix. Hydrogen peroxide, a product of the dismutation reaction, can be destroyed by 2 different enzymes, namely catalase and glutathione peroxidases (GSHPx), a selenium-containing enzyme requiring glutathione (GSH).

$$H_2O_2 \xrightarrow{\text{Catalase}} H_2O + O_2 \qquad [14]$$

$$H_2O_2 + GSH \xrightarrow{\text{GSHPx}} GSSG + H_2O \qquad [15]$$

During normal oxygen metabolism, enzymes such as these function as secondary anti-oxidants to eliminate toxic intermediates of oxygen inside the cell.

4.2. Membrane Antioxidant Defenses

Within the hydrophobic lipid interior of membranes, different types of lipophilic radicals are formed to those seen in the intracellular milieu. Lipophilic radicals require different types of antioxidants for their removal. Vitamin E (α-tocopherol), a fat soluble vitamin is a poor antioxidant outside a membrane bilayer but is extremely effective when structurally incorporated into it. Lipid soluble antioxidants are, therefore, extremely important in protecting membrane polyunsaturated fatty acids (PUFAs) from undergoing autocatalytic free radical chain reactions known as lipid peroxidation. The peroxidation of a PUFA leads to its destruction and the formation of a plethora of oxidation products such as lipid peroxides, carbonyls and carboxylic acids. Many of these are biologically reactive and may be used as signal molecules by the body.

4.3. Extracellular Antioxidant Defences

Normally enzymes such as the intracellular SODs, catalase and glutathione peroxidase are not present in extracellular fluids. Nevertheless extracellular fluids are often subjected to fluxes of superoxide, nitric oxide, hypochlorous acid and hydrogen peroxide by 'activated' phagocytic cells and some ROS can also arise by auto-oxidation reactions. Blood is an effective anti-oxidant 'buffering' system with both plasma and red blood cells offering their own specialized protective systems. Erythrocytes have an anion channel through which superoxide can enter the cell and be destroyed by copper-zinc SOD. Hydrogen peroxide is an uncharged molecule, which behaves much like water and can be destroyed by erythrocyte catalase and glutathione peroxidase. Unlike plasma, erythrocytes contain high levels of GSH. Plasma contains both primary and secondary anti-oxidants and is usually a powerful inhibitor of free radical reactions.

5. Reactive Oxidant Species and Disease

Excessive production of reactive species has been implicated in various pathological such as in cancer, cardiovascular disease, neurological disorders, diabetes mellitus, ischemia/reperfusion injury and ageing [Dalle-Donne et al 2006; Dhalla et al 2000; Jenner 2003; Sayre et al 2001]. These diseases fall into two groups [Valko et al 2007]; the first group involves diseases characterized by pro-oxidants shifting the thio/disulphide redox state and impairing glucose tolerance the so-called "mitochondrial oxidative stress" conditions (cancer and diabetes). The second group involves diseases characterized by "inflammatory oxidative conditions" and enhanced activity of either NAD(P)H oxidase (leading to atherosclerosis and chronic inflammation) or xanthine oxidase-induced formation of ROS (implicated in ischemia and reperfusion injury). The process of ageing is to a large extent due to the damaging consequences of free radical action such as lipid peroxidation, DNA damage, protein oxidation [Harmann 1956]. A more comprehensive review of this topic is covered by Cadenas and Davies 2000; Valko et al 2007; Elahi et al 2007; and Dalle-Donne et al 2006.

Conclusion

Decades of oxidative stress research in the biological system have concluded that there are two types of reactive oxidants (ROS and RNS), products of normal cellular metabolism. In addition, ROS/RNS act as secondary messengers for controlling various normal physiological functions of the organism. Therefore, the production of NO by NOS and superoxide by NAD(P)H is tightly regulated by hormones, cytokines, and other mechanisms. ROS and RNS also participate in various redox-regulatory mechanisms of cells in order to protect cells against oxidative stress and maintenance of cellular "redox homeostasis". Overproduction of ROS, most likely through excessive stimulation of NAD(P)H by cytokines or by the mitochondrial electron transport chain and xanthine oxidase results in oxidative stress. Oxidative stress is a deleterious process that can be an important mediator of damage to cell structures and consequently various disease states and ageing. This chapter has briefly described the types of ROS and RNS radicals and antioxidant molecules significant in physiological and pathological processes.

References

Aikens J, Dix TA. Perhydroxyl radical (HOO$^{\cdot}$) initiated lipid-peroxidation - The role of fatty-acid hydroperoxides. *J. Biol. Chem.* 1991; 266:15091-15098.

Archer S. Measurement of nitric oxide in biological models. *FASEB J.* 1993; 7:349-360.

Beckman JS, Koppenol WH. Nitric oxide, superoxide and peroxynitrite: the good, the bad and the ugly. *Am. J. Physiol.* 1996; 271:C1424-37.

Bennett MR. Mechanisms of p53-induced apoptosis. *Biochem. Pharmacol.* 1999; 58:1089-1095.

Bergendi I, Benes I, Durackova Z, Ferencik M. Chemistry, physiology and pathology of free radicals. *Life Sci.* 1999; 65:1865-1874.

Berlett BS, Friguet B, Yim MB, Chock PB, Stadtman ER. Peroxynitrite-mediated nitration of tyrosine residues in Escherichia coli glutamine-synthetase mimics adenylation: Relevance to signal transduction. *Proc. Natl. Acad. Sci. USA*. 1996; 93:1776-1780.

Borges F, Fernandes E, Roleira F. Progress towards the discovery of xanthine oxidase inhibitors. *Curr. Med. Chem*. 2002; 9:195-217.

Burton GW, Ingold KU. Beta-carotene - An unusual type of lipid antioxidant. *Science*. 1984; *224*, 569-573.

Cadenas E. Basic mechanisms of antioxidant activity. *Biofactors*.1997; 6:391-397.

Cadenas E, Davies KJ. Mitochondrial free radical generation, oxidative stress, and aging. *Free Radic. Biol. Med*. 2000; 29:222-230.

Cadenas E and Sies H. The lag phase. *Free Radic Res*.1998; 28:601-609.

Cai H and Harrison DG. Endothelial dysfunction in cardiovascular diseases: the role of oxidant stress. *Circ. Res*. 2000; 87, 840-844.

Cameron E, Pauling L. Supplemental ascorbate in supportive treatment of cancer – prolongation of survival times in terminal human cancer. *Proc. Natl. Acad. Sci. USA.* 1976(1); *73*:3685-3689.

Candeias LP, Patel KB, Stratford MRL, Wardman P. Free hydroxyl radicals are formed on reaction between the neutrophils-derived species superoxide anion and hypochlorous acid. *FEBS Lett*. 1993; 333:151-153.

Candeias LP, Stratford MRL, Wardman P. Formation of hydroxyl radicals on reaction of hypochlorous acid with ferrocyanide, a model iron (II) complex. *Free Radic Res*. 1994; 20(4):241-249.

Carr A, Frei B. Does Vitamin C act as a pro-oxidant under physiological conditions? *FASEB J*., 1999; *13*: 1007-1024.

Catani MV, Rossi A, Costanzo A, Sabatini S, Levrero M, Melino G, et al. Induction of gene expression *via* activator protein-1 in the ascorbate protection against UV-induced damage. *Biochem J*. 2001; 356:77-85.

Channon KM, Guzik TJ. Mechanisms of superoxide production in human blood vessels: relationship to endothelial dysfunction, clinical and genetic risk factors. *J. Physiol.* Pharmacol. 2002; 53, 515-524.

Chiueh CC. Neuroprotective properties of nitric oxide. *Ann. N.Y.Acad.Sci.*1999; 890:301-311.

Commoner B, Townsend J, Pake GE. Free radicals in biological materials. *Nature,* 1954; 174: 689-691.

Dalle-Donne I, Rossi R, Giustarini D, Milzani A. Biomarkers of oxidative damage in human disease. *Clin. Chem*. 2006; 52:601-623.

DeCoursey TE, Ligeti E. Regulation and termination of NADPH oxidase activity. *Cell Mol Life Sci*. 2005; 62:2173-2193.

De Grey ADNJ. $HO_2^{\cdot}$: The forgotten radical. *DNA Cell Biol*. 2002; 27:251-257.

Dhalla NS, Ternsah RM, Netticadan T. Role of oxidative stress in cardiovascular diseases. *J. Hypertens*. 2000; 18:655-673.

Dröge W. Free radicals in the physiological control of cell function. *Physiol. Rev*. 2002; 82, 47-95.

Duchen MR. Mitochondria in health and disease: perspectives on a new mitochondrial biology. *Mol. Aspects Med.* 2004; 25, 365-451.

El-Agamey A, Lowe GM, McGarvey DJ, Mortensen A, Phillip DM, Truscott,TG. Carotenoid radical chemistry and antioxidant/pro-oxidant properties. *Arch. Biochem. Biophys.* 2004; 430:37-48.

Elahi MM, Naseem KM, Matata BM. Nitric oxide in blood: The nitrosative-oxidative disequilibrium hypothesis on the pathogenesis of cardiovascular disease. *FEBS J.* 2007; 274:906-923.

Fleming I. Cytochrome P450 and vascular homeostasis. *Circ. Res.* 2001; 89, 753-762.

Graham A, Hogg N, Kalyanaraman B, O'Leary V, Darley-Usmar V, Moncada S. Peroxynitrite modification of low-density lipoprotein leads to recognition by the macrophage scavenger receptor. *FEBS Lett*. 1993; 330(2):181-185.

Gerschman R, Gilbert DL, Nye SW, Dwyer P, Fenn WO. Oxygen poisoning and x-irradiation - a mechanism in common. *Science*. 1954; 119:623-626.

Gow AJ, Duran D, Malcolm S, Ischiropoulos H. Effects of peroxynitrite-induced protein modifications on tyrosine phosphorylation and degradation. *FEBS Lett*. 1996; 385:63-66.

Halliwell B, Gutteridge JMC. *Free Radicals in Biology and Medicine*. 3rdEdition. Oxford: Oxford University Press. 1999.

Han D, Canali R, Garcia J, Aguilera R, Gallaher TK, Cadenas E. Sites and mechanisms of aconitase inactivation by peroxynitrite: modulation by citrate and glutathione. *Biochemistry*. 2005; 44:11986-11996.

Harman D. Aging – A theory based on free-radical and radiation-chemistry. *J. Gerontol.* 1956; 11:298-300.

Hirota K, Murata M, Sachi Y, Nakamura H, Takeuchi J, Mori K, et al. Distinct roles of thioredoxin in the cytoplasm and in the nucleus - A two-step mechanism of redox regulation of transcription factor NF-kappa B. *J. Biol. Chem.*, 1999; *274*: 27891– 27897.

Jenner P. Oxidative stress in Parkinson's disease. *Ann Neurol*. 53: S26-S36.

Jones DP, Carlson JL, Mody VC, Cai JY, Lynn MJ, Sternberg P. Redox state of glutathione in human plasma. *Free Radic Biol Med*. 2000; 28:625-635.

Kahklon O, Cabantchik ZI. The labile iron pool: characterisation, measurement and participation in cellular processes. *Free Radic Biol Med*. 2002; 33:1037-1046.

Klatt P, Lamas S. Regulation of protein function by S. glutathiolation in response to oxidative and nitrosative stress. *Eur. J. Biochem*. 2000; 267:4928-4944.

Kojo S. Vitamin C: Basic metabolism and its function as an index of oxidative stress. *Curr. Med. Chem.* 2004; 11:1041–1064.

Kovacic P, Jacintho JD. Mechanisms of carcinogenesis: Focus on oxidative stress and electron transfer. *Curr Med Chem.* 2001; 8:773-796.

Landis GN, Tower J. Superoxide dismutase evolution and life span regulation. *Mech. Ageing Dev.* 2005; 26:365–379.

Landmesser U, Dikalov S, Price SR et al. Oxidation of tetrahydrobiopterin leads to uncoupling of endothelial cell nitric oxide synthase in hypertension. *J. Clin Invest* 2003; 111, 1201–1209.

Li JM, Shah AM. Endothelial cell superoxide generation: regulation and relevance for cardiovascular pathophysiology. Am. J. Physiol. Regulatory Integrative Comp. *Physiol.* 2004; 287: R1014–R1030.

Liochev SI and Fridovich I. The role of O2 in the production of HO: In vitro and in vivo. *Free Radic Biol Med*. 1994; 16: 29-33.

Liochev SI and Fridovich I. The Haber-Weiss cycle – 70 years later: An alternative view. *Redox Rep*. 2002; 7:55-57.

Makropoulos V, Bruning T, Schulze-Osthoff K. Selenium-mediated inhibition of transcription factor NF-kappa B and HIV-1 LTR promoter activity. *Arch. Toxicol*. 1996; 70:277-283.

Masella R, Di Benedetto R, Vari R, Filesi C, Giovannini C. Novel mechanisms of natural antioxidant compounds in biological systems: Involvement of glutathione and glutathione related enzymes. *J. Nutr. Biochem*. 2005; 16:577-586.

Mates JM, Perez-Gomez C, De Castro IN. Antioxidant enzymes and human diseases. Clin. Biochem. 1999; 32:595-603.

Meneshian A, Bulkley GB. The physiology of endothelial xanthine oxidase: from urate catabolism to reperfusion injury to inflammatory signal transduction. *Microcirculation.* 2002; 9, 161-175.

McCord JM, Fridovich I. Superoxide dismutase an enzymatic function for erythrocuprein Hernocuprein. *J. Biol. Chem.* 1969; 244:6049-6055.

Miller DM, Buettner GR, Aust SD. Transition metals as catalysts of "autoxidation" reactions. *Free Radic. Biol. Med.* 1990; 8:95-108.

Miller ER, Pastor-Barriuso R, Dalal D, Riemersma RA, Appel LJ, Guallar E. Meta-analysis: High-dosage Vitamin E supplementation may increase all-cause mortality. *Ann. Intern. Med.* 2005; 142:37-46.

Mittal CK, Murad F. Activation of guanylate cyclase by superoxide-dismutase and hydroxyl radical – Physiological regulator of guanosine 3',5'- monophosphate formation. *Proc. Natl. Acad Sci. USA.* 1977; 74:4360-4364.

Muller FI, Liu Y, Van Remmen H. Complex III releases superoxide to both sides of the inner mitochondrial membrane. *J. Biol. Chem.* 2004; 279:49064-49073.

Nakamura H, Nakamura K, Yodoi J. Redox regulation of cellular activation. *Ann. Rev. Immunol.* 1997; 15:351-369.

Nogueira CW, Zeni G, Rocha JBT. Organoselenium and organotellurium compounds: Toxicology and pharmacology. *Chem. Rev.* 2004; 104:6255-6285.

Packer L, Suzuki YJ. Vitamin-E and alpha-lipoate - Role in antioxidant recycling and activation of the NF-kappa-b transcription factor. Mol. *Aspects Med.* 1993; 14:229-239.

Pastore A, Federici G, Bertini E, Piemonte F. Analysis of glutathione: Implication in redox and detoxification. *Clin. Chim. Acta.* 2003; 333:19-39.

Pieper AA, Verma A, Zhang J, Snyder SH. Poly (ADP-ribose) polymerase, nitric oxide and cell death. *Trends Pharmacol Sci.* 1999; 20:171-181.

Pryor WA. Vitamin E and heart disease: Basic science to clinical intervention trials. *Free Radic. Biol. Med.* 2000; 28:141–164.

Rae TD, Schmidt PJ, Pufahl RA, O'Halloran TV. Undetectable intracellular free copper: The requirement of a copper chaperone for superoxide dismutase. *Science.* 1999; 284:805-808.

Ridnour LA, Isenberg JS, Espey MG, Thomas DD, Roberts DD, Wink DA. Nitric oxide regulates angiogenesis through a functional switch involving thrombospondin-1. *Proc. Natl Acad Sci. USA.* 2005; 102:13147-13152.

Sayre LM, Smith MA, Perry G. Chemistry and biochemistry of oxidative stress in neurodegenerative disease. *Curr. Med. Chem.* 2001; 8:721-738.

Sharoni Y, Danilenko M, Dubi N, Ben-Dor A, Levy J. Carotenoids and transcription. *Arch. Biochem. Biophys.* 2004; 430:89-96.

Schrauzer GN. Interactive effects of selenium and chromium on mammary tumor development and growth in MMTV-infected female mice and their relevance to human cancer. Biol. Trace Element Res. 2006; 109:281-292.

Shen D, Dalton TP, Nebert DW, Shertzer HG. Glutathione redox state regulates mitochondrial reactive oxygen production. *J. Biol. Chem.* 2005; 280:25305-25312.

Smith AR, Shenvi SV, Widlansky M, Suh JH, Hagen TM. Lipoic acid as a potential therapy for chronic diseases associated with oxidative stress. *Curr. Med. Chem.* 2004; 11:1135-1146.

Valko M, Morris H, Mazur M, Rapta P, Bilton RF. Oxygen free radical generating mechanisms in the colon: Do the semiquinones of Vitamin K play a role in the aetiology of colon cancer? *Biochem Biophys Acta*. 2001; 1527:161-166.

Valko M, Izakovic M, Mazur M, Rhodes CJ, Telser J. Role of oxygen radicals in DNA damage and cancer incidence. *Mol. Cell Biochem*. 2004; 266:37-56.

Valko M, Morris H, Cronin MTD. Metals, toxicity and oxidative stress. *Curr. Med. Chem*. 2005; 12:1161-1208.

Valko M, Rhodes CJ, Moncol J, Izakovic M, Mazur M. Free radicals, metals and antioxidants in oxidative stress-induced cancer. *Chem. Biol. Interact*. 2006; 160:1-40.

Valko M, Leibfritz D, Moncol J, Cronin MM, Telser J. Free radicals and antioxidants in normal physiological functions and human disease. *Int. J. Biochem. Cell Biol*. 2007; 39:44-84.

Vorbach C, Harrison R and Capecchi MR. Xanthine oxidoreductase is central to the evolution and function of the innate immune system. *Trends Immunol*. 2003; 24:512-517.

In: Oxidative Stress: Clinical and Biomedical Implications ISBN: 978-1-60021-888-0
Editors: B. M. Matata, M. M. Elahi, pp. 39-57

Chapter III

The Emerging Role of Reactive Oxidant Species on Regulating Pathways of Biomaterials Biocompatibility

***Bashir M. Matata*[*1] *and Maqsood M. Elahi*[2]**

[1]The Cardiothoracic Centre NHS Trust- Liverpool,
L14 3PE United Kingdom
[2]Centre for Developmental Origins of Health and Disease,
University of Southampton, Princess Anne Hospital,
Southampton SO16 5YA, UK

Abstract

The clinical utilisation of materials and devices involving contact with blood, has promoted a growing interest in blood-biomaterial interactions and monitoring of the blood response. Most applications have been in the form of implantable or extracorporeal devices. The extracorporeal situations that are commonly considered include hemodialysis and related procedures, membrane plasma separation, cardiopulmonary bypass, artificial heart valves, artificial heart and assist devices and arterial grafts. However, the development of medical devices for cardiovascular applications has suffered due to lack of understanding of why vascular wall and blood cells act non-physiologically when exposed to biomaterials. One plausible explanation might be the chemical environment associated with disease. Given that, particular interest has been on the response of blood to biomaterials used in the construction of these devices and an improved understanding of cellular and sub-cellular mechanisms in view of the disease environment.

An important feature in using biomaterial and blood response is the contrast in behavior between reactive oxygen and nitrogen species (ROS/RNS). ROS/RNS are mainly produced through normal cellular metabolism and rendered harmless by

* Correspondence: Dr Bashir M. Matata, PhD, AIBMS. Head of Clinical Trials Unit. The Cardiothoracic Centre NHS Trust- Liverpool, Liverpool, L14 3PE. United Kingdom.Email: matata_bashir@hotmail.com; Tel: + 44 (0) 151 293 2292. Fax: + 44 (0)151 293 2283

enzymatic systems. But in the absence of adapted endothelial surface or during a disease process per se, these systems may act aberrantly and either fails to convert ROS and RNS to harmless substances or by producing more oxidants.

Recent indirect evidence suggests that even the implantation of biomedical materials may also be responsible for triggering such aberrant pathways that could lead to the eventual failure of biocompatibility of specially modified artificial surfaces. The understanding of how the vascular/ blood environment may be changed at the subcellular level by the presence of a biomaterial is critical. In this chapter, we aim to review and discuss the current thinking within vascular biology regarding ROS/ RNS and activation pathways, and how they may impact the host response and the biocompatible vascular cellular environment.

Keywords: Oxygen species, recirculation, proinflammatory cytokines, antioxidant capacity.

Introduction

The application of procedures in medicine and surgery that require exposure of blood to foreign surfaces has increased markedly over recent times. The demand for engineered devices to treat cardiovascular disease is increasing due to an aging population and sedentary lifestyle [1]. As our understanding of the interface between synthetic scaffolding and biological materials has increased, so has our ability to use them to create novel environments. The application of blood-contacting devices, however, has not been without inherent difficulties, primarily due to the risk from local thrombosis on the material or on the other hand, bleeding risk due to antithrombotic agents and consumption of hemostatic factors [2].

Biomaterial scientists and pathologists have conducted much of the research in cardiovascular device engineering. However, contributions from medically-oriented basic science, notably vascular biology, may increase our understanding of the mechanisms of cardiovascular device failure. Cardiovascular devices are implanted in the context of disease (atherosclerosis, hypertension, and diabetes) or trauma. Thus, the specific disease etiology and how it affects the implant at a molecular and signaling level is critically important. Supportive evidence suggests that at low concentrations, reactive oxygen and nitrogen species (ROS/RNS) act as signaling molecules in multiple endothelial and smooth muscle pathways, including the signaling for cellular proliferation, migration, and apoptosis [3-6]. However, at elevated concentrations, they may alter cellular behavior resulting in inflammatory signaling and there is a differential response to stress by blood, which may directly reflect upon the induction of a proinflammatory state. This mechanism may involve the early activation of neutrophils, the most likely source of free radicals during this biomaterial interact and elevated inflammatory mediators and cytokines result in leukocyte and platelet activation along with leukocyte recruitment [7-8].

In an overall relationship between blood, biomaterial and inflammatory factors (Figure-1), a more compatible response can result from alteration to the biomaterial, the cellular response or both. Here, in this chapter we specifically focus and discuss the major oxidative players in cardiovascular disease and how this may impact cardiovascular device engineering *in vivo* and *in vitro*.

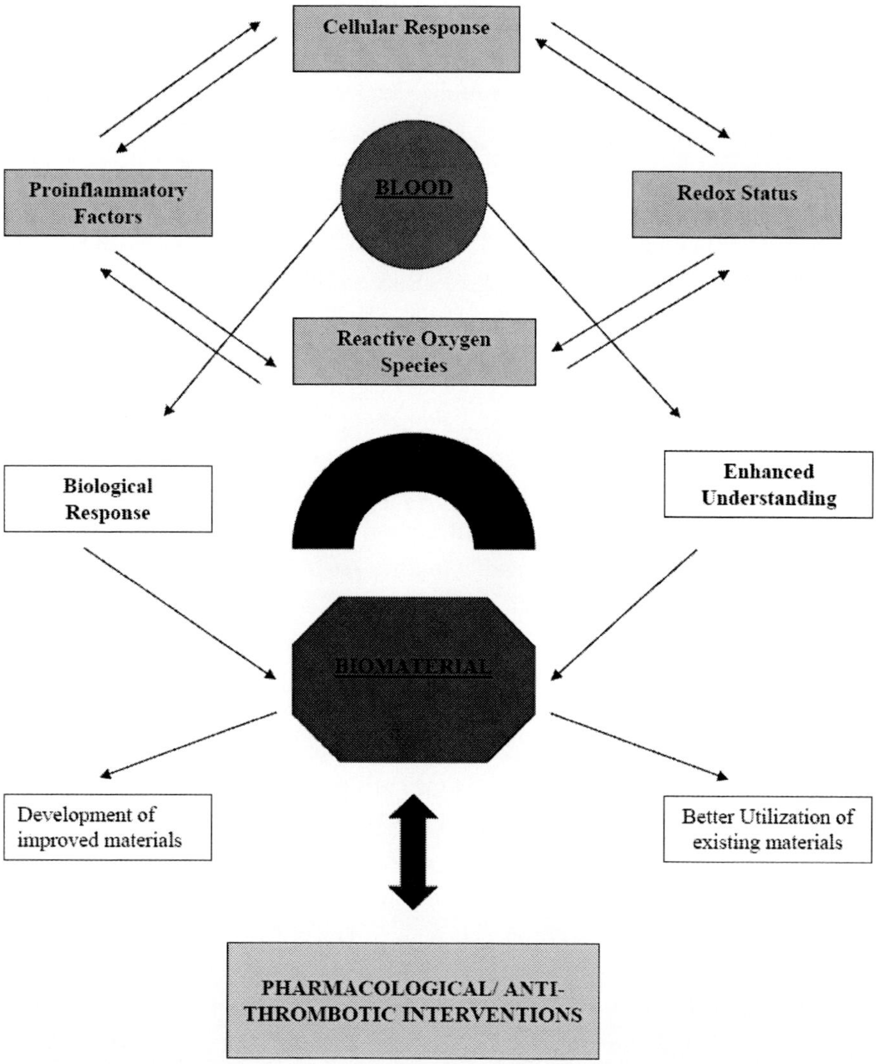

Figure 1. Blood-biomaterial interaction, a multilateral relationship. This comprised of a crosstalk between the cascade and cellular systems of the blood in different established and emerging treatment modalities. And a general objective for the future biomaterial research in utilisation and development of improved materials.

Cardiovascular Devices

Design of cardiovascular devices has increased in sophistication by understanding the effects of intimate contact between host cells and implanted synthetic materials. Whether this contact is short-lived as in the use of degradable material or permanent, the changes in vascular cell behavior can have extreme effects on the function of the device. One such change is the relationship between pro-and antioxidants which are dependant on a complex array of signaling maintained between the intra- and extracellular spaces. One of the challenges of cardiovascular device engineering is to develop systems in which host cells that are seeded or have migrated onto the device may maintain an oxidatively neutral environment.

With regard to biomaterials, evidence in the literature has suggested that oxidative stress induced by extracorporeal circulation (ECC)/ biomaterials is involved in the pathogenesis of an underlying systemic inflammatory response experienced perioperatively by most patients [9]. Emerging data increasingly supports that oxidative stress may be a possible cause of ECC-induced inflammation due to the early activation of polymorphonuclear cells that leads to enhanced activity of the respiratory burst of oxidants [10-11]. This relationship may suggest a direct effect of free radicals on the inflammatory reaction induced by ECC however, the underlying mechanism and kinetics of this link remains still elusive. It is now known that the main source of oxidants is the oxidative burst produced by polymorphonuclear neutrophils (PMN) and monocytes inflammatory response to artificial device implantation [12]. Chronic inflammation can expose surrounding endothelial and smooth muscle cells to abnormally high levels of superoxide,[13] hydrogen peroxide,[14] and nitric oxide[15]. This, in turn, can result in altered endothelial cell behavior, resulting in expression of inflammatory mediators and cytokines [16]. Attraction of monocytes and other inflammatory cells then results.

With regards to cardiological revascularisation interventions, we [17] have previously investigated that whether off-pump coronary bypass graft operations on the beating heart under normothermic conditions reduces the systemic oxidative stress and inflammatory reaction seen in patients operated under cardiopulmonary bypass (CPB). We reported significant increases from prior anesthesia values of lipid hydroperoxides (190% at 4 hours), protein carbonyls (250% at 0.5 hours) and nitrotyrosine (510% at 0.5 hours) in the CPB group, but they were abolished or significantly reduced in the off-pump group. Complement C3a and elastase levels were rapidly increased upon the institution of CPB, and this was followed by increases in IL-8, TNF-alpha, and sE-selectin. In contrast, the rise of these factors was blunted in patients operated without CPB thus deducing that the use of CPB significantly increases oxidative stress and the inflammatory reaction in patients undergoing coronary bypass graft operation. Certainly, this also suggests that there may be a role that a chronically host blood-biomaterial interaction play in modulating the oxidative environment.

This novel observation was lent support by our recent work [18] where 50 ml of blood from 25 ischemic heart disease (IHD) patients with stable angina and 25 sex and age matched healthy controls was recirculated at 37^0C for 4 h in an extracorporeal circuit in the presence and absence of NG-Nitro-L-arginine methyl ester HCl (L-NAME, 10mM), uric acid (1mM) and SOD mimetic manganese tetrakin porphyrin (Mn-SOD, 40μM) and hydroxyl radical scavenger N-2 mercaptopropionyl glycine (N2-MPG, 500μM). The findings suggested that scavenging of $ONOO^-$ by uric acid and NOS inhibition by L-NAME significantly reduced IL-6, IL-8 and TNF-α production, however Mn-SOD and N2-MPG were only effective in reducing IL-8 to the same level as the nitrosative species. Interestingly, inhibitors of ROS production on the other hand, stimulated as much or more cytokine production in healthy blood compared with the untreated samples. Results demonstrated that there is a ROS-dependent mechanism of cytokine regulation in the human and that each species imparts a differential pattern of cytokine release in blood from healthy subjects or IHD patients. This may be mediated by the variation in transcriptional responses initiated via auto-oxidation, nitration or phosphorylation of redox-sensitive regulator molecules when come in contact with extracorporeal circulation circuit. Similarly, Coyle *et al* [19] reported that seeding of endothelial cells on various biomaterials leads to increased intracellular levels of superoxide. In most severe cases, endothelial cells became nonviable on the materials.

In view of these reports it has recently been suggested that vascular cell division may be tied to inflammatory-based oxidative stress [20-21]; specifically, the vascular endothelial growth factor which acts on NADPH-oxidase in endothelial cells to activate nuclear factor-κB, thus triggering cell division. Also, inflammatory levels of hydrogen peroxide stimulate nonphagocytic NADPH-oxidase to produce superoxide in smooth muscle cells [22] and endothelial cells [23]. Hitherto, it is yet unclear whether these two mechanisms are connected in vascular cells in contact with biomaterials.

Major Oxidative Stress Factors

Many oxidants play major role in cellular signaling and are produced as a consequence of mitochondrial function. Multiple sources of ROS and RNS have been identified in vascular cells. Of theses three key vascular oxidants, superoxide ($O_2^{\cdot -}$), hydrogen peroxide (H_2O_2), and peroxynitrite ($ONOO^-$), as well as the hydroxyl radical ($OH^{\cdot}$), hypochlorite (HOCl), and lipid radicals act as major source of oxidant damage in vascular cells [24]. With onset of cardiovascular disease the imbalance which is normally associated with increased $O_2^{\cdot -}$ [25-26], and H_2O_2 [27-28], $OH^{\cdot}$ [29-30], $ONOO^-$ [31-33], and HOCl [34] damage microvascular endothelial cells an important predictor of endovascular damage. It is also linked with transplant vasculopathy, often associated with chronic allograft nephropathy. Large bursts of nitric oxide in infiltrating monocytes/ macrophages modulated by inducible nitric oxide synthase are considered pivotal in driving these mechanisms [35].

We have suggested the effect of inducible nitric oxide synthase and tumor necrosis factor-alpha (TNF-α) interaction on progressive damage to glomerular and vascular structures during renal grafts rejection [35]. Selective inhibition of inducible nitrous oxide synthase and $ONOO^-$ at micromolar concentrations can induce a dose-dependent release of proinflammatory cytokines and the nuclear translocation of the nuclear factor (NF)-kappa B transcription factor [36]. This finding indicates that phosphorylation and nitration of tyrosine residues in IkappaB-alpha molecules are correlated with NF-kappaB translocation and cytokine release. In addition, the DNA binding activity of the NF-kappa B from the nuclear extracts may also correlate with its nuclear translocation [36]. These studies suggest that $ONOO^-$ plays an essential role in the mechanism of proinflammatory cytokine release by monocytes and that Rel/NF-kappaB activation is the obligatory pathway in the blood mediated aberrant response. This may explain a potential mechanism in therapy for ameliorating endothelial dysfunction and leads to a good venue to explore in transplant graft vasculopathy.

Imbalance of Antioxidants in Blood-Biomaterial Interaction

Antioxidant systems comprised of mainly, superoxide dismutase (SOD), CAT, and glutathione peroxidase that provide the regulation of oxidants in cellular signaling and also balance the level of oxidants during acute oxidative stress [37]. Elevated levels of ROS and RNS reduce the activity and expression of antioxidants, notably manganese superoxide dismutase (MnSOD) [38]. This is of importance during vascular disease where oxidant regulation is lost and cells are exposed to severe oxidative stress. The increased protein tyrosine nitration and lipid peroxidation under cardiovascular condition is now firmly established and represents a shift from the signal-transducing physiological actions of free radicals to oxidative and potentially pathogenic actions [39].

Little is known about the exact role of ROS in blood and how this is related to the pathophysiology of biomaterials. We [40] have previously studied this effects in relation with free radicals production and antioxidant capacity (AOC) in an *ex vivo* model of extracorporeal circulation for 4h at normothermia (37^0C). Interesting but intriguing observations suggested that the overall production of O_2^- and $ONOO^-$ were significantly greater in blood obtained from cardiovascular disease patients This increase in O_2^- and $ONOO^-$ production was associated with a substantial depletion in AOC in IHD patients. This may point out that the levels of ROS produced, completely overwhelms antioxidant mechanisms and probably alter the function of enzymes such as SOD, glutathione peroxidase and catalase in an extracorporeal circulation environment. This may also be a reflection of the existence of a state of upregulation of pro-oxidant enzymes genes leading to an increased production of ROS.

Indeed this hypothesis is supported by evidence from the literature that has shown that there is a differential response of the cardiac isoforms Nox-2 and Nox-4 in the upregulation of gp91phox present in the NADH/NADPH pathway of superoxide production within the cardiac myocytes and the mitochondrial membrane. The reduced ROS production may be as a result of release of antioxidant molecules from erythrocytes or an upregulation of activities of antioxidant enzymes [reviewed in ref 41].

Results obtained from these studies recommend that there is a differential response to stress by blood from healthy subjects compared with cardiovascular disease patients, which may directly reflect upon the induction of a proinflammatory state. This mechanism may involve chemical modification of redox-sensitive transcription factors an effect which is strongly related to the concentration of $ONOO^-$. Specifically, the complexity of $ONOO^-$ reactivity within the cells is clearly of scientific interest and warrants further research to elucidate its role on membrane associated enzymes that are involved in the development of oxidative stress in disease and healthy states.

Contact Phase Activation

The complement is a major constituent of the humoral branch of the immune system and plays a pivotal role in contact phase activation pathways [42]. The soluble and cell bound proteins that constitute the complement system interact in highly regulated cascades to serve four important functions; cell lysis, opsonisation, removal of harmful immune complexes from the circulation and mediating the immune response (inflammation and release of immunoregulatory molecules) by interacting with other immune cells. The complement system is a complex cascade of proteins that are activated in 3 distinct ways, designated the classical, alternative and lectin pathways. Activation of classical pathway is initiated by C1q binding to antigen-antibody complexes, whereas activation of the alternative or lectin pathway is antibody independent [42]. The alternative pathway is activated by spontaneously cleaved C3 or by C3b generated by one of the other pathways [43]. The lectin pathway is activated by mannose binding lectin (MBL) interacting with its ligand leading to activation of 2 MBL-associated serine proteases (MASPs) [44-45].

As well as its purpose in complement activation, MBL also functions to promote opsono-phagocytosis, control of inflammatory response (by modulating cytokine release from monocytes), and promotion of apoptosis (demonstrated by its binding to apoptotic T-cells) [46]. In activation pathways (AP) MBL acts together with MASPs in a similar fashion to the C1q, C1r and C1s components of the classical pathway [47]. To date three MASPs have been identified; MASP-1, MASP-2 and MASP-3 as well as a non-enzymatic protein known as the MBL-associated protein of 19 kDa (MAp19). Of these four proteins it is MASP-2 that has been identified as combining with MBL to play a role in the activation of complement components C4 and C2 [48]. The activity of the MBL-MASP-2 complex is triggered by contact with carbohydrate moieties found on microbial surfaces. This leads to the activation of the MASP-2 portion of the complex, which goes on to cleave C4 and C2 in a similar fashion to the $C1qr_2s_2$ complex of the classical pathway, forming C4bC2a (C3 convertase). The subsequent steps in the MBL pathway up to the cleavage of C5 to C5b and beyond are identical to the classical pathway [46, 49].

Many researchers have reported that exposure of biomaterial surfaces to plasma or blood either *in vitro* or *in vivo* can trigger the AP as demonstrated by generation of AP activation products [50-51]. Some studies have also demonstrated complement and/ or lectin pathway (CP and/or LP) activation. Lhotta and colleagues [52] have previously shown that hemodialysis in patients with C4 deficiency leads to activation of complement, as a result of contact between the biomaterial and the patients' plasma, but the activation was much slower than in plasma from normal individuals. This finding reported that the biomaterial can trigger a rapid activation by the CP and/or LP (C4-dependent), followed later by AP activation.

It is well established that different biomaterial surfaces have different complement-activating properties. Physical properties such as hydrophobicity and hydrophilicity are parameters that affect the activating ability. Hydrophobic surfaces are more potent activators than hydrophilic ones, and incorporation of chemical groups such as NH_3, OH or COOH influences the activation of complement [53]. Biomaterial surfaces with free OH and NH_3 groups are generally regarded as more prone to activate complement than others, since these groups are essential for the covalent binding of C3b [51]. However, these studies have not taken into account the fact that plasma proteins immediately bind to and cover the surface of a biomaterial surface once it has made contact with blood plasma or other body fluids.

We [54] studied the effect of ROS on complement activation pathways in normal human serum and mannan binding lectin (MBL) deficient serums. We have shown that normal human blood taken from healthy volunteers when treated with ROS generators (1mM authentic peroxynitrite ($ONOO^-$), 10 μM/10 μM xanthine/xanthine oxidase mixtures to generate O_2^-, and 100 μM H_2O_2) during 30 minutes of extracorporeal circulation at normothermia demonstrates a concentration-dependent modification of protein residues by ROS and that ROS treatment of normal and MBL deficient serums inhibited C4 deposition originating from both MBL and classical pathways [54]. This study advocates a very important issue applicable to the biomaterial pathophysiology and that is ROS plays an important task in regulating AP independently of the classical and MBL pathways and that increased activities may be attributable to a direct effect on the alternate pathway.

Another type of surface that alters the configuration of ROS/ AP is the gas surface that is necessary for oxygenation of the blood in the oxygenator of the CPB circuit. It is being reported that contact between the surface of the gas (oxygen, nitrogen or air) and whole blood leads to activation of complement based ROS *in vitro* and *in vivo* [55-56]. This activation is a result of the generation of iC3 (hydrolyzed C3 with a C3b-like conformation) at the interface between fluid and gas. The mechanism is probably similar to that occurring on the material's surface. However, the iC3 generated on the gas surface is immediately displaced into the fluid phase and is able to generate a soluble C3 convertase. This convertase generates C3a but not soluble C5b-9 complexes (sC5b-9). These studies have demonstrated that absorbed proteins may expose sites that are specific for the activated protein, and they point to a mechanism by which other proteins such as IgG may be activated when they come in contact with material surfaces.

Methods for Studying Solid Phase/Blood Proteins Biocompatibility Interactions

The interactions with human serum and plasma can be studied by using techniques that detect changes in the molecular mass associated with a surface. For-example the successful application of hemodialysis membranes depends on the use of drugs such as heparin, low molecular-weight heparin, antiplatelet agents and recombinant hirudin [57]. New approaches have been to extend the traditional role of antithrombotic agents by the introduction of multienzyme inhibitors and kallikrein inhibitors, aprotinin [58-59] to inhibit the contact-phase activation. This leads to an important overall relationship between blood, the biomaterial and the pharmacological agent, with the possibility of altered blood response resulting from changes to the biomaterial, the pharmacological agent or both. Another point of consideration is the adsorption by biomaterials of the pharmacological agents and the possible influence on biological activity.

We have focused on the determination of factor XII- like activity (FXIIA) as an indicator of contact phase activation in the supernatant phase and at the membrane surface after plasma-membrane contact using incubation test cell [60-61] The finding from these studies demonstrated the strong influence of the interaction of other plasma constituents on the membrane surface and as such the binding and the subsequent activation of factor XII may be altered possibly due to competitive binding and steric hindrance. On the addition of anticoagulants such as heparin, low molecular-weight heparin, citrate and hirudin, no significant differences were observed in plasma supernatant phase FXIIA. However, each anticoagulant appears to have a distinct influence on the magnitude of plasma membrane-bound FXIIA. On the addition of even aprotinin, no significant difference was observed in plasma membrane FXIIA either. These findings offer many suggestions regarding contact phase activation with respect to the *in vitro* evaluation of membranes and blood response to them.

Models for Blood–Biomaterial Interactions

Taken together, these findings have allowed us to construct a model to explain how ROS/AP play a role on a biomaterial surface. Several experimental models investigating the pathophysiology of systemic inflammatory response during therapeutic blood filtration have been described in the literature such as cell culture preparations and *in vitro* blood recirculation circuits. However, these have failed to establish the association between the variety of biological mechanisms involved in a disease process, the blood and its components [62-63]. Since the various cascade systems and cell types in the blood are known to interact, the true effect of a biomaterial surface on the blood can only be known when whole blood is used for interaction studies. We currently make use of three in vitro systems, (1) the extracorporeal blood circulation model, (2) the six well incubation test cell and (3) controlled flow syringe pump test system to study the sequence of events that occur when biomaterial surfaces come into contact with whole blood. In order to minimize the inhibition of complement and coagulation, we use either low amounts of heparin (less than 0.5 IU/mL), or specific thrombin inhibitors.

The miniature extracorporeal recirculation circuit [64] consists of a pump, a test cell (139 x 110 x 25 mm), a blood reservoir and an oxygenator. These components are connected via polyvinylchloride (PVC) and porous platinum silicon tubing of internal diameter of 1.6 mm. Blood is pumped around the circuit through these tubes using a Watson-Marlow MH101 UR peristaltic pump under flow conditions of 5 mL/minute. 12 mL heparinised (10 U/mL) solution composed of 17.5 ml plasmalyte, 12.5 ml gelofusine, 5 ml 10% mannitol and 1 ml 8.4% sodium bicarbonate, at pH 7.4 is used to prime the circuit and the test cell was incubated in a water bath to maintain the temperature to 37°C. The miniature extracorporeal circuit maintained clinically acceptable results regarding blood gases, hemolysis and hemodilution. This *ex vivo* human model with relative small circuit area may serve as a useful bridge between cell culture studies and animal models investigating clinically pertinent research questions.

The modified 6 well incubation test cell is composed of a base plate made of polymethymethacrylate (Perspex), a piece of soft silicone rubber, a 6 well polystyrene plate opened on the top and bottom and three bars of clamping and some covers [60]. To set up the 6 well plate incubation test cell, the material to be tested is soaked overnight in 0.9% saline solution. The material is then placed in between the polystyrene 6 well plate and the silicone rubber sheet fixed on the Perspex base plate. The test cell was then clamped together with the three bars. Since the silicone rubber is soft, blood does not leak from the bottom after clamping. During blood test experiments, the test cell is usually set up before blood collection.

The controlled flow syringe pump test system consists of 80 hollow fibers potted into a Perspex (polymethylmethacrylate) outer jacket 144 mm in length using medical grade polyurethane resin. Each hollow fiber tube sheet extends to a length of 3mm from the jacket end giving a total fiber length of 150 mm and an internal surface of 75.4 cm^2 (for a fiber internal diameter of 200 um). The fiber length exposed to dialysate or rinse solution is 134 mm. Blood headers are adapted from 5ml polypropylene syringe barrels and bonded to the outer jacket using a cyanoacrylate adhesive and primer system. Male luer fittings are similarly bonded into 4mm holes in the outer jacket to form dialysate inlet and outlet ports [65].

As investigations of blood material interactions require a focus on *in vitro* test procedures, these models of work allow us the discrimination between the host blood and biomaterial response and a greater control over experimental conditions.

Contact Phase Activation by Bio-Artificial Surfaces

Extracorporeal circulation, the diversion of blood from the circulation of the patient to a device outside the body, is a feature of many treatment modalities, including hemodialysis, hemofiltration, CPB, extracorporeal membrane oxygenation (ECMO), plasmapheresis, leukapheresis and thrombapheresis. Principally, whole blood is led through the device as unseparated blood or plasma. The plasma is produced either by separating whole blood using a membrane/filter, which also triggers incompatibility reactions, or by centrifugation. Here we will concentrate on hemodialysis and CPB, since the ROS mediated complement-related side effects induced by these treatment modalities include most of those seen with other extracorporeal treatments.

The activation-mediated tissue injury associated with extracorporeal treatment has to a great extent been linked to bio-incompatibility reactions. AP were thought to be related to the so-called “first-use syndrome,” in which the patients experienced an anaphylactoid reaction during the first use of a dialyzer, but this reaction was attenuated with re-use [51]. Today these problems are less frequent, but adverse reactions still occur. Of equal concern in uremic patients undergoing maintained hemodialysis is amyloidosis and accelerating arteriosclerosis [66]. The risk of myocardial infarction in these patients is 5–10 times higher than in healthy individuals [67]. Many factors contribute to this increased risk: The uremic condition itself leads to disturbed lipoprotein levels, acidosis, toxins, free radicals and infections, all of which can add to the accelerated arteriosclerosis. Chronic inflammation, such as that found in patients with rheumatoid arthritis, systemic lupus erythematosus (SLE) or chronic rejection has been shown to be associated with accelerating arteriosclerosis and cardiovascular disease. It is likely that the chronic whole-body inflammation triggered by hemodialysis is the other major contributor to arteriosclerosis in uremic patients. This inflammation is to a great extent driven by activation pathways. The dialysis membranes used today activate complement to varying degrees. Cellulose membranes are strong complement activators, while polymers and blends thereof (such as polyarylethersulfone, polyamide and polyvinylpyrrolidone) are less activating [68]. Activation of complement generates fragments (C3b, iC3b and C3dg) and soluble peptides (C3a and C5a) that bind to the hemodialysis membrane and can trigger inflammation and release proinflammatory cytokines such as IL-1β, TNF-α, IL-6 and the highly potent chemokine IL-8.

Another effect of AP is the up regulation of receptors on leukocytes (e.g., CD11b/CD 18 and CD35), which together with down-regulation of L-selectin makes the cells very adhesive and prone to interact with platelets and endothelial cells in the vascular walls [69]. This cellular activation also leads to increased levels of acute-phase proteins such as CRP, allowing these molecules to bind to activated endothelial cells and activate complement. CRP and complement components such as C4, C3 and C9 are present in arteriosclerotic plaques together with monocytes/macrophages [70].

CPB procedures have increased dramatically over the past decade as a result of vascular bypass surgery. The CPB procedure is, however, associated with a number of side effects related to the attendant trauma, the contact made by the blood with the material surface and the gas surface, the neutralization of the heparin by protamine, and the ischemia/reperfusion triggered by the blood in the bypassed tissues (lungs and heart). The resulting cellular and humoral defence's reactions are termed the systemic inflammatory response syndrome [reviewed in ref 71]. Included in this syndrome are hemostatic disorders that take the form of bleeding, platelet activation and the generation of macro- and microthrombi. The consequence of these disorders is a high incidence of thrombotic events, including postoperative myocardial infarction and stroke that can produce serious neurological symptoms. Many of these problems result from complement activation, an effect that is supported by *in vivo* studies using anti-C5 antibodies in which complement inhibition was found to attenuate postoperative myocardial injury, cognitive deficits and blood loss [17]. Complement activation occurs on both the biomaterial surfaces exposed in the extracorporeal circuit and on the gas surface [56] that is exposed to the blood in the oxygenator of the CPB circuit. Complement is also activated by the platelets in the clotting blood, and a final spike is generated when soluble heparin is neutralized with protamine [17]. Together, these processes create a period of complement activation that lasts for at least the duration of the CPB procedure and provoke a ROS mediated inflammation cascade. Complement affects both leukocytes and platelets. An *in vitro* study [72] showed that platelets expose P-selectin and that platelet/leukocyte (PMNs) complexes are generated during CPB. Both these phenomena are totally abrogated by anti-C5 antibodies and are likely to contribute to the disturbed hemostatic balance during CPB.

Modification of the surface by the addition of a heparin coating has improved the biocompatibility of extracorporeal devices [73- 74]. Although initially intended to reduce the requirements for systemic heparinization, heparin coating has also been shown to markedly attenuate inflammatory reactions, including the response of AP.

Blood-Biomaterial Response in Associated Risk Factors

Most biomaterials, including those that are implanted into soft and hard tissues (dental implants, prostheses), come in contact with whole blood when blood is present during the implantation procedure. It is therefore reasonable to assume that the inflammatory reaction against the material is greatly influenced by the initial contact with the blood. Having said that of particular interest is the subset of diabetic patients undergoing cardiovascular interventions. These patients intrinsically are different to those without diabetes. We know now that their cardiovascular risks leading to their very specific atherosclerotic pathophysiology are different to non-diabetic patients [75]. Diabetics and those with metabolic syndrome including insulin resistance are prone to endothelial dysfunction [76]. These patients have exaggerated systemic and local inflammatory responses and are more thrombogenic. Their inflammatory markers such as sCRP, sCD40L and cytokines such as MCP-1, TNF-α, IL1β are elevated and endothelial dysfunction produces altered vascular homeostasis, mainly reflected as enhanced vasoconstriction [77]. The molecular basis of the latter can be explained by the impaired insulin receptor function involving the phosphatidylinositol kinase (PI3) and enhanced mitogen-activated protein (MAP) kinase pathway. PPARγ-dependent pathways for nitric oxide production are also impaired in diabetics [78]. Previously we [36] demonstrated that NO-mediated pathway induce a dose-dependent release of proinflammatory cytokines in human monocytes through the modulation of nuclear factor-kappa-B-DNA binding activity. However, recently, we reported that the mechanism of release of proinflammatory cytokines by blood granulocytes especially in diabetes and that participation in diabetes related complications is modulated by NO bioavailability [79]. This explains the prevailing dilemma in the use of extracorporeal circulation in such subsets of patients undergoing cardiac surgery because the metabolism of NO is altered with significant rise in nitrate/nitrite (NOx) [80]. Interestingly, infusion of GTN, an NO donor, can significantly reduced plasma NOx in such patients undergoing CPB supported cardiac surgery without reducing plasma cGMP levels thus suggesting that this elevation can be blunted by the infusion of GTN in a rapid and reversible manner [80].

Pharmacological Interventions

Several pharmacological agents that regulate or modulate AP activity have been identified in *in vitro* assays, but to date there are no specific inhibitors of in routine clinical use. However, several recombinant protein candidates are under development: One is a recombinant form of complement receptor 1 known as soluble complement receptor 1 (sCR1; TP-10). This C3 convertase blocking agent was first shown to markedly reduce experimental myocardial infarction and is now in clinical trial [81]. Another is a humanized monoclonal anti-C5 antibody (pexelizumab, eculizumab) that blocks the activation of C5 [82]. Thus, both inhibitors prevent the formation of C5a and C5b-9, but the anti-C5 antibody does not affect the activation of C3, suggesting that the unwanted effects of C3 activation are not affected but the main antimicrobial defences related to C3 is preserved. Both derivatives have been tested in *in vivo* animal models of disease and in patient clinical trials, and an efficient inhibition of AP has been confirmed [81, 83]. A conjugate of sCR1 and the selectin ligand sLe(X), combining inhibition of complement and cell adhesion, was found to be even more efficient than sCRl alone [84]. Use of a shorter, membrane-targeted form of sCRl (APT070) has also been used in a surface target-bound approach to inhibit complement activation [85]. Attention has also been paid to recombinant forms of DAF and MCP, a conjugate of these two, the complement activation blocker-2 (CAB-2; [86] and membrane-targeted CD59 [83].

However, none of these reagents has yet reached clinical trials. Because of their large molecular size (240 and 26 kDa [scFv fragment], respectively), soluble CR1 and the anti-C5 antibody have the disadvantage that they must be administered parenterally. Emphasis is now being placed on developing smaller active versions of these molecules with longer half lives and on decreasing their cost. These considerations are most important for clinical applications that require relatively long-term administration.

We have investigated the effect of nitric oxide supplementation on bypass-induced oxidative stress and inflammatory reaction in patients with and without diabetes undergoing elective coronary bypass graft surgery. And reported that CPB significantly increased lipid hydroperoxides, protein carbonyls, protein nitrotyrosine, complement C3a, elastase, soluble E-selectin, interleukin 8, and tumor necrosis factor alpha in both groups. However, infusion of nitroglycerin significantly reduces oxidative stress in patients with diabetes and differentially affects the inflammatory response to bypass both in patients with and in those without diabetes. The results have important implications with respect to the use of nitric oxide donors during cardiopulmonary bypass operation [11]. More recently, we have been able to demonstrate the treatment of human blood with different concentrations of the NOS inhibitor revealed marked depletion of the inflammatory response when recirculated in an extracorporeal circuit [86]. Hence the inhibition of enzymes in blood that generate ROS may represent a novel approach for the effective reduction of inflammatory cascade in patients undergoing cardiovascular mediated devices host- biomaterial response catch.

Conclusion

The clinical consequences of ROS/RNS and AP-associated biocompatibility reactions during extracorporeal circulations call for an increased knowledge of how activation pathways are triggered by the biomaterial surface, as a means for developing materials with lower or no complement-activating properties. Given the complexity of the situation, with multiple interactions occurring among the various cascade systems and cellular systems of the blood, full protection will most likely require the additional use of soluble inhibitors. The inhibitors under development today will be very expensive to use on a daily basis for applications such as the treatment of uremic and diabetic patients. Moreover, with an older population and increased prevalence of cardiovascular disease, further developments in cardiovascular device engineering will be needed. A more detailed understanding of the molecular mechanisms of disease is necessary in order to increase the success of cardiovascular devices.

Oxidative stress is an important component and studies are on-going to further understand the effects of cardiovascular disease on vascular cells at the cellular and molecular level. Oxidative stress, endothelial cell dysfunction, and cardiovascular disease are inherently linked and ROS and RNS play an important role. ROS and RNS contribute to the development and progression of hypertension and atherosclerosis.

There are many different sources of ROS and RNS in vascular cells and all are potential targets for disease management. At this time, NADPH oxidase may be one of the most promising due to its significant role in both inflammatory response (neutrophils and monocytes) and vascular cell dysfunction (atherosclerosis and hypertension). To successfully modify devices, the challenge is discovering the role of different oxidative signaling pathways and using them to reduce the likelihood of severe side effects with treatment. In addition, more research is needed under shear with endothelial cells to observe the function of ROS and RNS and their signaling pathways in a more biologically relevant milieu as many previous studies in vascular biology were under static culture. Several studies discussed here indicate variable ROS and RNS response in endothelial and smooth muscle cells under different shear and static conditions.

References

[1] American Heart Association (AHA). *Heart Disease and Stroke Statistics - 2005 Update*. Dallas, Texas: American Heart Association; 2004.

[2] Courtney JM, Irvine L, Jones C, Mosa SM, Robertson LM, Srivastava S. Biomaterials in medicine--a bioengineering perspective. *Int. J. Artif. Organs* 1993; 16: 164-71.

[3] Rhee SG. Redox signaling: Hydrogen peroxide as intracellular messenger. *Exp. Mol. Med.* 1999; 31: 53-59.

[4] Griendling KK, Sorescu D, Lassegue B, Ushio-Fukai M. Modulation of protein kinase activity and gene expression by reactive oxygen species and their role in vascular physiology and pathophysiology. *Arterioscler. Thromb. Vasc. Biol.* 2000; 20: 2175-2183.

[5] Brown MR, Miller FJ, Jr, Li WG, Ellingson AN, Mozena JD, Chatterjee P, Engelhardt JF, Zwacka RM, Oberley LW, Fang X, Spector AA, Weintraub NL. Overexpression of human catalase inhibits proliferation and promotes apoptosis in vascular smooth muscle cells. *Circ. Res.* 1999; 85: 524-533.

[6] Patel RP, Moellering D, Murphy-Ullrich J, Jo H, Beckman JS, Darley-Usmar VM. Cell signaling by reactive nitrogen and oxygen species in atherosclerosis. *Free Radic. Biol. Med.* 2000; 28: 1780-1794.

[7] Elahi MM, Naseem KM, Matata BM. Nitric Oxide in blood. The nitrosative-oxidative disequilibrium hypothesis on the pathogenesis of cardiovascular disease. *FEBS J.* 2007; 274: 906-23.

[8] Elahi MM, Matata BM. Is there a role of free radicals in the systemic inflammatory reaction? *J. Cardiothoracic-Renal Res*. 2006; 1: 131-133.

[9] Asimakopoulos G, Smith PLC, Ratnatunga CP, Taylor KM. Lung injury and acute respiratory distress syndrome after cardiopulmonary bypass. *Ann. Thorac. Surg.* 1999; 68:1107-1115.

[10] Elahi MM, Matata BM. Free Radicals in Blood: Evolving Concepts in the Mechanism of Ischaemic Heart Disease. *Arch. Biochem. Biophys.*, 2006; 450: 78-8811.

[11] Matata BM, Galinanes M. Cardiopulmonary bypass exacerbates oxidative stress but does not increase proinflammatory cytokine release in patients with diabetes compared with patients without diabetes: regulatory effects of exogenous nitric oxide. *J. Thorac. Cardiovasc. Surg.* 2000;120:1-11.

[12] Anderson JM. Inflammatory response to implants. *ASAIO Trans* 1988; 34: 101-107.

[13] Johnston RB, Jr, Lehmeyer JE, Guthrie LA. Generation of superoxide anion and chemiluminescence by human monocytes during phagocytosis and on contact with surface-bound immunoglobulin G. *J. Exp. Med.* 1976; 143: 1551-1556.

[14] Halliwell B. Production of superoxide, hydrogen peroxide and hydroxyl radicals by phagocytic cells: A cause of chronic inflammatory disease? *Cell Biol. Int. Rep.* 1982; 6: 529-542.

[15] Vane JR, Mitchell JA, Appleton I, Tomlinson A, Bishop-Bailey D, Croxtall J, Willoughby DA. Inducible isoforms of cyclooxygenase and nitric-oxide synthase in inflammation. *Proc. Natl. Acad. Sci. USA* 1994; 91: 2046-2050.

[16] Granger DN, Vowinkel T, Petnehazy T. Modulation of the inflammatory response in cardiovascular disease. *Hypertension* 2004; 43: 924-931.

[17] Matata BM, Sosnowski AW, Galinanes M. Off-pump bypass graft operation significantly reduces oxidative stress and inflammation. *Ann. Thorac. Surg.* 2000; 69: 785-791.3.

[18] Elahi MM, Courtney JM, Matata BM. The Interaction between Reactive Oxidant Species and Proinflammatory Cytokines in Human Blood during Extracorporeal circulation. *Filtration* 2005; 1: 89-94.

[19] Coyle CH, Mendralla S, LaNasa S, Kader KN. Endothelial cell seeding onto various biomaterials causes superoxide-induced cell death. *J. Biomater. Appl.* (in press).

[20] Ushio-Fukai M. Redox signaling in angiogenesis: Role of NADPH oxidase. *Cardiovasc. Res*. 2006; 71: 226-235.

[21] Ushio-Fukai M, Tang Y, Fukai T, Dikalov SI, Ma Y, Fujimoto M, Quinn MT, Pagano PJ, Johnson C, Alexander RW. Novel role of gp91(phox)-containing NAD(P)H oxidase in vascular endothelial growth factor-induced signaling and angiogenesis. *Circ. Res.* 2002; 91: 1160-1167.

[22] Li WG, Miller FJ, Jr, Zhang HJ, Spitz DR, Oberley LW, Weintraub NL. H(2)O(2)-induced O(2) production by a non-phagocytic NAD(P)H oxidase causes oxidant injury. *J. Biol. Chem*. 2001; 276: 29251-29256.

[23] Coyle CH, Martinez LJ, Coleman MC, Spitz DR, Weintraub NL, Kader KN. Mechanisms of H_2O_2-induced oxidative stress in endothelial cells. *Free Radic. Biol. Med.* 2006; 40: 2206-2213.

[24] Iuchi T, Akaike M, Mitsui T, Ohshima Y, Shintani Y, Azuma H, Matsumoto T. Glucocorticoid excess induces superoxide production in vascular endothelial cells and elicits vascular endothelial dysfunction. *Circ. Res*. 2003; 92: 81-87.

[25] Zhang DX, Zou AP, Li PL. Ceramide-induced activation of NADPH oxidase and endothelial dysfunction in small coronary arteries. *Am. J. Physiol. Heart Circ. Physiol.* 2003; 284: H605-H612.

[26] Fridovich I. Superoxide dismutases: Regularities and irregularities. *Harvey Lect* 1983; 79: 51-75.

[27] Hassan HM, Fridovich I. Chemistry and biochemistry of superoxide dismutases. *Eur. J. Rheumatol. Inflamm.* 1981; 4: 160-172.

[28] Wolin MS. Interactions of oxidants with vascular signaling systems. *Arterioscler Thromb. Vasc. Biol.* 2000; 20: 1430-1442.

[29] Yang MX, Cederbaum AI. Role of cytochrome b5 in NADH-dependent microsomal reduction of ferric complexes, lipid peroxidation, and hydrogen peroxide generation. *Arch. Biochem. Biophys*. 1995; 324: 282-292.

[30] Wolin MS, Davidson CA, Kaminski PM, Fayngersh RP, Mohazzab HK. Oxidant-nitric oxide signaling mechanisms in vascular tissue. *Biochemistry* (Mosc) 1998; 63: 810-816.

[31] Davidson CA, Kaminski PM, Wolin MS. NO elicits prolonged relaxation of bovine pulmonary arteries via endogenous peroxynitrite generation. *Am. J. Physiol.* 1997; 273(2, Part 1): L437-L444.

[32] Wink DA, Mitchell JB. Chemical biology of nitric oxide: Insights into regulatory, cytotoxic, and cytoprotective mechanisms of nitric oxide. *Free Radic. Biol. Med.* 1998; 25: 434-456.

[33] Taniyama Y, Griendling KK. Reactive oxygen species in the vasculature: Molecular and cellular mechanisms. *Hypertension* 2003; 42: 1075-1081.

[34] Wassmann S, Wassmann K, Nickenig G. Modulation of oxidant and antioxidant enzyme expression and function in vascular cells. *Hypertension* 2004; 44: 381-386.

[35] Elahi MM, Matata BM, Hakim NS. Quiescent Interplay between Inducible Nitric Oxide Synthase and Tumor Necrosis Factor-α: Influence on Transplant Graft Vasculopathy in Renal Allograft Dysfunction. *Exp. Clinical Transpl,* 2006; 1: 445-450.

[36] Matata BM, Galinanes M. Peroxynitrite is an essential component of cytokines production mechanism in human monocytes through modulation of nuclear factor-kappa B DNA binding activity. *J. Biol. Chem*. 2002; 277: 2330-5.

[37] Wassmann S, Wassmann K, Nickenig G. Modulation of oxidant and antioxidant enzyme expression and function in vascular cells. *Hypertension* 2004; 44: 381-386.

[38] Macmillan-Crow LA, Thompson JA. Tyrosine modifications and inactivation of active site manganese superoxide dismutase mutant (Y34F) by peroxynitrite. *Arch. Biochem. Biophys*. 1999; 366: 82-88.

[39] Radi R. Nitric oxide, oxidants, and protein tyrosine nitration. *Proc. Natl. Acad. Sci. USA.* 2004; 101: 4003-8.

[40] Elahi MM, Matata BM. Blood-Dependent Redox Activity During Extracorporeal Circulation in Health and Disease. *The Cardiology*, 2005;1:156-157.

[41] Matata BM, Elahi MM: Peroxynitrite: an intriguing double-agent in biology and cardiovascular disease. In: *Recent Research Developments in Biophysics.* SG Pandalai, ed., (Vol.3-2004; Part I: Pages 93-112), Transworld Research Communications, Gayathri Publishers.

[42] Walport MJ. Complement: first two parts. *N. Engl. J. Med.,* 2001; 344: 1058-1066.
[43] Xu Y, Narayana SVL, Volanakis JE. Structural Biology of the Alternative Pathway Convertase. *Immunological Reviews* 2001; 180: 123-135.
[44] Theil S, Vorup-Jensen T, Stover CM et al. A second serine protease associated with mannan-binding lectin that activates complement. *Nature,* 1997; 386: 506-510.
[45] Petersen SV, Thiel S, Jensenius JC. The Mannan-Binding Lectin Pathway of Complement Activation: *Biology and Disease Association. Molecular Immunology* 2001; 38: 133-149.
[46] Turner MW. The Role of Mannose-Binding Lectin in Health and Disease. *Molecular Immunology* 2003; 40: 423-429.
[47] Gadjeva M, Thiel S, Jensenius JC. The Mannan-Binding-Lectin Pathway of the Innate Immune Response. *Current Opinion in Immunology* 2001; 13: 74-78.
[48] Schwaeble W, Dahl MR, Thiel S, Stover C, Jensenius JC. The Mannan-Binding Lectin-Associated Serine Proteases (MASPs) and Map19: Four Components of the Lectin Pathway Activation Complex Encoded by Two Genes. *Immunology* 2002; 205: 455-466.
[49] Pangburn MK, Rawal N. Structure and Function of Complement C5 Convertase Enzymes. *Biochemical Society Transactions* 2002; 30: 1006-1010.
[50] Hed J, Johansson M, Lindroth M. Complement activation according to the alternate pathway by glass and plastic surfaces and its role in neutrophils adhesion, *Immunol. Lett.*1984; 8: 295–299.
[51] Chenoweth DE, Complement activation in extracorporeal circuits, *Ann. N.Y. Acad. Sci.* 1987; 516: 306–313.
[52] Lhotta K, Wurzner R, Kronenberg F, Oppermann M, Konig P. Rapid activation of the complement system by cuprophane depends on complement component C4, *Kidney Int.* 1998; 53: 1044–1051.
[53] Ekdahl KN, Nilsson B, Golander CG, Elwing H, Lassen B, Nilsson UR, Complement activation on radiofrequency plasma modified polystyrene surfaces, *J. Colloid Interface Sci.* 1993; 158: 121–128.
[54] Elahi MM, Elahi M, Matata BM. Reactive Oxidant Species Augment Complement Function in Human Blood independently of Classical and MBL-Pathways. *Int. J. Mol. Med. Adv. Science,* 2005; 1: 382-391.
[55] Ekdahl KN, Nilsson B, Pekna M, Nilsson UR, Generation of iC3 at the interface between blood and gas, *Scand. J. Immunol.* 1992; 35: 85–91.
[56] Pekna M, Nilsson L, Nilsson-Ekdahl K, Nilsson UR, Nilsson B. Evidence for iC3 generation during cardiopulmonary bypass as the result of blood–gas interaction, *Clin. Exp. Immunol.* 1993; 91: 404–409.
[57] Talbot MD, Ambler J, Butler KD et al. Recombinant desulphatohirudin (CGP 39393) anticoagulant and antithrombotic properties in vivo. *Thromb Haemost* 1989; 61: 71-80.
[58] Miyamoto H, Hirose H, Matsuda H. Analysis of complement activation profile during cardiopulmonary bypass and its inhibition by FUT-175. *Trans ASAIO* 1985; 31: 508-510.
[59] Heller W, Fuhrer G, Gallimore MJ, Michael M, Hoffmeister HE. Changes in the kallekrin-kinin system after different dose regimen of aprotinin during

cardiopulmonary bypass. In Abe K, Moriya H, Fujii S eds. Kinins V part B. *Advances in Experimental Medicine and Biology*. New York: Plenum, 1989: 43-48.

[60] Matata BM, Wark S, Sundaram S, Courtney JM, Gaylor JD, Bowry SK, Vienken J, Lowe GD. In vitro contact phase activation with hemodialysis membranes: role of pharmaceutical agents. *Biomaterials.* 1995;16:1305-12.

[61] Matata BM, Courtney JM, Sundaram S, Wark S, Bowry SK, Vienken J, Lowe GD. Determination of contact phase activation by the measurement of the activity of supernatant and membrane surface-adsorbed factor XII (FXII): its relevance as a useful parameter for the in vitro assessment of hemodialysis membranes. *J. Biomed. Mater. Res*. 1996; 31: 63-70.

[62] Courtney JM, Matata BM, Yin HQ, Esposito A, Mahiout A, Taggart DP, Lowe GD. The influence of biomaterials on inflammatory responses to cardiopulmonary bypass. *Perfusion.* 1996;11: 220-8.

[63] Courtney JM, Sundaram S, Matata BM, Gaylor JD. Biomaterials in cardiopulmonary bypass. *Perfusion.* 1994; 9: 3-10.

[64] Matata BM, Elahi MM. A novel *ex vivo* miniature model of extracorporeal circulation for investigating the pathophysiological changes and therapeutic Strategies. *J. Cardiothoracic-Renal Res.,* (in print).

[65] Matata BM, Yin HQ, Courtney JM, Gaylor JD, Lamba NM, Lowe GD, Suzuki K, Kimura H, Izumi K, Klinkmann H. In vitro blood compatibility evaluation of hollow fiber membrane using a controlled flow system: a comparative study. *Int. J. Artif. Organs*. 1996; 19: 582-9.

[66] Amore A, Coppo R. Immunological basis of inflammation in dialysis, *Nephrol. Dial. Transplant.* 2002; 17: 16–24.

[67] Santoro A, Mancini E. Cardiac effects of chronic inflammation in dialysis patients, *Nephrol. Dial. Transplant*. 2002; 17: 10–15.

[68] Deppisch R, Gohl H, Smeby L. Microdomain structure of polymeric surfaces-potential for improving blood treatment procedures, *Nephrol. Dial. Transplant*. 1998; 13: 354–1359.

[69] Nilsson B, Larsson R, Hong J, Elgue G, Ekdahl KN, Sahu A, Lambris JD. Compstatin inhibits complement and cellular activation in whole blood in two models of extracorporeal circulation, *Blood* 1998; 92: 1661–1667.

[70] Torzewski M, Rist C, Mortensen RF, Zwaka TP, Bienek M, Waltenberger J, Koenig W, Schmitz G, Hombach V, Torzewski J. C-reactive protein in the arterial intima: role of C-reactive protein receptor-dependent monocytes recruitment in atherogenesis, Arterioscler. *Thromb. Vasc. Biol*. 2000; 20: 2094–2099.

[71] Elahi MM, Khan JS, Matata BM. Deleterious Effects of Cardiopulmonary Bypass in Coronary Artery Surgery and Scientific Interpretation of Off-pump's Logic. *Acute Card Care*, 2006; 8: 196-209.

[72] Rinder CS, Rinder HM, Smith BR, Fitch JC, Smith MJ, Tracey JB, Matis LA, Squinto SP, Rollins SA. Blockade of C5a and C5b-9 generation inhibits leukocyte and platelet activation during extracorporeal circulation, *J. Clin. Invest*. 1995; 96: 1564–1572.

[73] Hsu LC, Heparin-coated cardiopulmonary bypass circuits: current status, *Perfusion* 2001; 16: 417–428.

[74] Elahi MM, Khan JS. Revascularisation with Off-pump Coronary Artery Surgery: What appears New is Actually the Old Rediscovered. *Cardiovasc Revascularisation Med.* 2007; 8: 52-59.

[75] Grundy SM, D'Agostino Sr RB, Mosca L. Cardiovascular risk assessment based on US cohort studies: findings from a National Heart, Lung, and Blood Institute workshop, *Circulation* 2001; 104: 491–496.

[76] Luscher TF, Creager MA, Beckman JA, Cosentino F. Diabetes and vascular disease: pathophysiology, clinical consequences, and medical therapy: part II, *Circulation* 2003; 108: 1655–1661.

[77] Huvers FC, De Leeuw PW, Houben AJ, De Haan CH, Hamulyak K, Schouten H, Wolffenbuttel BH, Schaper NC. Endothelium-dependent vasodilatation, plasma markers of endothelial function, and adrenergic vasoconstrictor responses in type 1 diabetes under near-normoglycemic conditions, *Diabetes*1999; 48: 1300–1307.

[78] Huszka M, Kaplar M, Rejto L, Tornai I, Palatka K, Laszlo P, Udvardy M. The association of reduced endothelium derived relaxing factor-NO production with endothelial damage and increased in vivo platelet activation in patients with diabetes mellitus, *Thromb. Res.* 1997; 86: 173–180.

[79] Elahi M, Matata B. Nitric oxide-dependent pathway regulates granulocytes cytokines release in cultured human leucocytes from type-II diabetes mellitus. *Proceedings of the 13th Biennial Congress of the International Society for Free Radical Research*, 2006: 115-118.

[80] Matata BM, Galinanes M. Effect of diabetes on nitric oxide metabolism during cardiac surgery. *Diabetes*. 2001; 50: 2603-10.

[81] Keshavjee S, Davis RD, Zamora MR, de Perrot M, Patterson GA. A randomized, placebo-controlled trial of complement inhibition in ischemia-reperfusion injury after lung transplantation in human beings, *J. Thorac. Cardiovasc. Surg.* 2005; 129: 423–428.

[82] Kaplan M, Eculizumab (Alexion), Curr. Opin. Investig. *Drugs* 2002; 3: 1017–1023.

[83] Hill A, Ridley SH, Esser D, Oldroyd RG, Cullen MJ, Kareclas P, Gallagher S, Smith GP, Richards SJ, White J, Smith RA, Hillmen P. Protection of erythrocytes from human complement-mediated lysis by membrane-targeted recombinant soluble CD59: a new approach to PNH therapy, *Blood* 2006; 107: 2131–2137.

[84] Huang J, Kim LJ, Mealey R, Marsh Jr. HC, Zhang Y, Tenner AJ, Connolly Jr ES, Pinsky DJ. Neuronal protection in stroke by an sLex-glycosylated complement inhibitory protein, *Science*1999; 285: 595–599.

[85] Smith RA. Targeting anticomplement agents, *Biochem. Soc. Trans* 2001; 30: 1037–1041.

[86] Elahi MM, Matata BM. Myocardial Protection against Ischemia-Reperfusion Injury: Novel Approaches in Maintaining Homeostatic Stability in Blood. *Recent Patents on Cardiovascular Drug Discovery* 2006; 1: 291-305.

In: Oxidative Stress: Clinical and Biomedical Implications ISBN: 978-1-60021-888-0
Editors: B. M. Matata, M. M. Elahi, pp. 59-67

Chapter IV

Steady-State Titration with H2O2: Regulation of Plasma Membrane Permeability and of NF-κB Activation by H2O2

Virgínia de Oliveira-Marques[1], H. Susana Marinho[2], Luísa Cyrne[2] and Fernando Antunes[3]

[1]Grupo de Bioquímica dos Oxidantes e Antioxidantes,
Centro de Química e Bioquímica, Faculdade de Ciências da Universidade de Lisboa,
P-1749-016 Lisboa, Portugal

[2]Grupo de Bioquímica dos Oxidantes e Antioxidantes,
Centro de Química e Bioquímica and Departamento de Química e Bioquímica da Faculdade de Ciências da Universidade de Lisboa,
P-1749-016 Lisboa, Portugal

[3]Grupo de Bioquímica dos Oxidantes e Antioxidantes,
Centro de Química e Bioquímica and Departamento de Química e Bioquímica da Faculdade de Ciências da Universidade de Lisboa, P-1749-016 Lisboa, Portugal; Instituto de Investigação Científica Bento da Rocha Cabral, Cç. Bento da Rocha Cabral, 14, P-1250-04 Lisboa, Portugal

Abstract

Techniques used to deliver H_2O_2 to cells are analyzed. The commonly used bolus addition, the use of glucose oxidase and the steady-state titration are compared. Bolus addition often disrupts cellular homeostasis due to the high doses applied; glucose oxidase overcomes this draw-back by allowing a continuous delivery of H_2O_2, but like the bolus addition it is usually applied as a non-calibrated technique because the dose delivered per cell is not measured. The steady-state titration overcomes these disadvantages being an adequate technique to study the signaling actions of low H_2O_2 concentrations. The dose delivery per cell is measured and calibrated on a daily basis allowing comparison of results between laboratories. The regulation of plasma membrane permeability by H_2O_2 and the induction of NF-κB by H_2O_2 illustrate these concepts. H_2O_2 does not cross biomembranes freely, and in addition it regulates its permeability across these membranes during adaptation. Contradictory published data on whether H_2O_2 stimulates or inhibits NF-κB activation result from the poor techniques employed to deliver H_2O_2; under controlled conditions of H_2O_2 delivery NF-κB activation is stimulated and both pro- and anti-inflammatory genes are activated. We propose that H_2O_2 is a regulator of inflammation.

Introduction

The biology of hydrogen peroxide (H_2O_2), the most abundant oxidant produced endogenously in cells, constitutes a very active topic of research. After identification of the production of H_2O_2 *in vivo*, H_2O_2 was viewed as a toxic species that could trigger oxidative degradation of biomolecules, namely through Fenton chemistry, producing the highly reactive hydroxyl radical. Subsequently, as soon as H_2O_2 intracellular production was characterized by Chance and co-workers [1, 2] it was found that H_2O_2 could mimic insulin actions [3] and activate guanylate cyclase [4]. With time the paradigm of H_2O_2 as a dangerous species that must be eliminated for cells to survive changed and at the end of the eighties a large body of data suggested that H_2O_2 could also act as a signaling species. Today it is accepted that at low concentrations H_2O_2 has important cellular roles, such as, signal transduction, control of proliferation and apoptosis, and redox regulation.

Methodology Used to Expose Cells to H_2O_2

Today most researchers still use the tools for H_2O_2 delivery that were developed when H_2O_2 was considered a deleterious species: H_2O_2 is given to cells as a single high initial dose (bolus addition), often in the range 0.1 to 1 mM. Such doses are several orders of magnitude higher than both the physiological intracellular concentrations (10 to 100 nM) [2] or the extracellular concentrations found during inflammation (up to 15 μM) [5, 6]. Therefore, they cause an abrupt and acute non-physiological shock, disrupt cellular homeostasis and, most importantly, elicit cellular responses that may not be related to those induced by the low concentrations of H_2O_2 found *in vivo* [7]. In addition, the bolus addition is inadequate to study signaling processes in the time scale of hours. As exemplified in Figure 1a, a bolus addition is rapidly removed by the cell system, and at cell densities commonly used, about 80-90 % of H_2O_2 is exhausted within 1h. To overcome these drawbacks, glucose oxidase, an enzyme that catalyzes the oxidation of glucose producing H_2O_2, is used as a continuous source of H_2O_2 (Figure 1b). However, the use of glucose oxidase and of the bolus addition are both non-calibrated procedures to expose cells to H_2O_2 and therefore, reproducibility of experiments at a quantitative level is difficult. This is due to the fact that for a given dose of H_2O_2 or glucose oxidase, the actual dose of H_2O_2 delivered per cell is unknown because it depends on several factors, such as the cellular capacity to remove H_2O_2, the cell density and the consumption of H_2O_2 by the growth medium (Figure 1a and b).

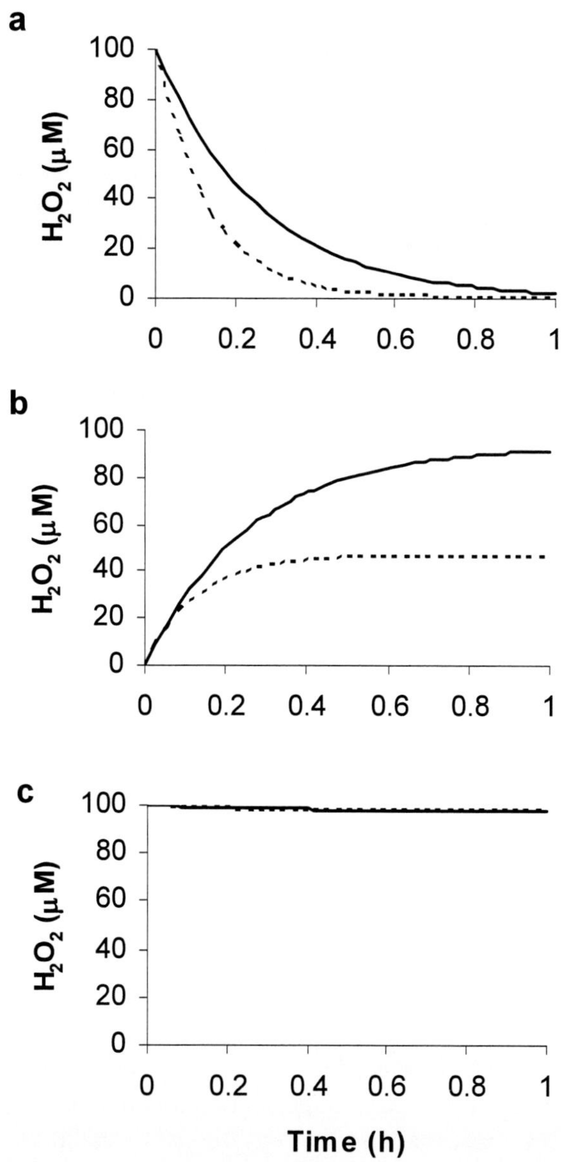

Figure 1. Simulation of H_2O_2 delivery to Jurkat t-cells by three alternative methods: (a) bolus addition; (b) glucose oxidase; (c) steady-state titration. To exemplify the dependence of H_2O_2 profiles on experimental conditions, two cell densities were studied: 1 million cells/ml (——) and 2 million cells/ml (------). Other factors such as type of cells or incubation medium can produce similar variations in H_2O_2 profiles. Minimal model and experimental kinetic parameters used were taken from ref. [13].

To investigate the cellular actions of physiological levels of H_2O_2 and its involvement in normal cellular processes and in disease, it is fundamental to study the actions of low H_2O_2 levels in a setting that mimics the cellular production of H_2O_2 *in vivo*. The *steady-state titration* method is uniquely suited for this task. In this method, H_2O_2 is initially added at the desired level together with a dose of glucose oxidase that compensates the consumption of H_2O_2 by the system [7], thus keeping the H_2O_2 concentration in a steady-state (Fig 2). The amount of glucose oxidase is calibrated on a daily basis in order to maintain the desired concentration of H_2O_2. Thus, the concentration of H_2O_2 reached in the assay is independent of the cell type and other assay conditions (Figure 1c). Similar profiles to those observed in Figure1 were observed experimentally in MCF-7 cells [8].

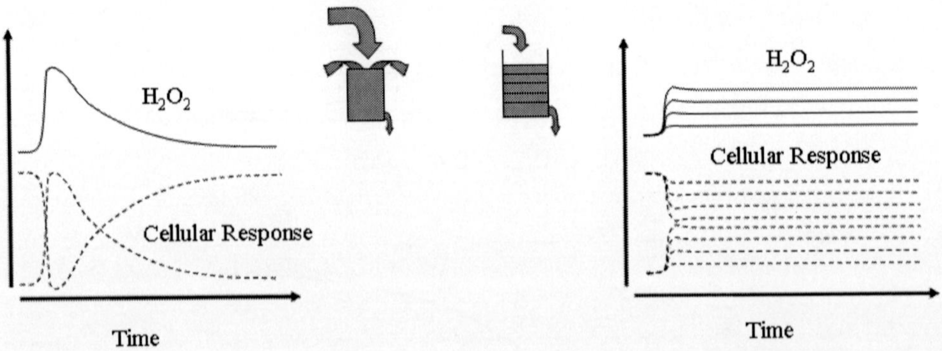

Bolus addition		*Steady-state incubation*
No	Is cellular homeostasis maintained?	Yes
No	Is the state *in vivo* mimicked?	Yes
No	Is the cellular response relevant physiologically?	Yes
Low	Reproducibility?	High

Figure 2. Hydrodynamic analogue of the delivery of H_2O_2 to cells. Bolus addition "drowns" the cell in H_2O_2 while the steady-state titration represents a mild stimulus that mimicks the endogenous production of H_2O_2.

H_2O_2 Regulates Membrane Permeability

The application of the steady-state titration requires the previous measurement of the consumption of H_2O_2 in the experimental set up used. During studies on the adaptation of *S. cerevisiae* to H_2O_2, we came across a puzzling observation. Adaptation is a general phenomenon by which a non-lethal dose of H_2O_2 triggers a cellular response that increases the resistance to subsequent lethal doses. The mechanism of adaptation includes a several-fold induction of the activity of antioxidant enzymes that remove H_2O_2 [9]. Surprisingly, we observed that in spite of this induction, the consumption of H_2O_2 during adaptation in *S. cerevisiae* did not change [10]. The general belief widely reported in the literature is that H_2O_2 diffuses freely across biomembranes [2] and so an increase of the activity of intracellular H_2O_2-removing enzymes should increase the cellular consumption of H_2O_2. On the other hand, the concept of catalase latency, which has been known since the seminal studies on sub-cellular fractioning made during the fifties [11, 12], goes against this common belief. Catalase latency describes the phenomenon in which an increase in catalase activity is observed when the compartment that entraps the enzyme is disrupted. In the disrupted compartment H_2O_2 crosses freely the membrane barrier justifying the higher activity. Thus, H_2O_2 diffusion across biomembranes can be rate-limiting and H_2O_2 gradients are formed across membranes that separate the site of production from the site of degradation of H_2O_2. Gradients across the plasma membrane when cells are exposed to extracellular H_2O_2 were recently found in mammalian cell lines [13], *E. coli* [14], and *S. cerevisiae* [10]. In addition, the unaltered H_2O_2 cell consumption during adaptation could be explained by a decrease in the permeability of the plasma membrane towards H_2O_2 that would compensate the increase in the intracellular H_2O_2-removing enzymes.

In fact, in *S. cerevisiae* the plasma membrane does not play a passive role, but its permeability towards H_2O_2 is under active regulation being part of the cellular response to oxidative stress, and playing an active role in H_2O_2 adaptation mechanisms: 1) In *S. cerevisiae*, H_2O_2 or cycloheximide decrease plasma membrane permeability to H_2O_2 by approximately 50 % [10]. 2) During the life cycle of *S. cerevisiae* cell permeability to H_2O_2 decreases by a factor of five [15].

The mechanism underlining these changes may involve alterations in the membrane sterol composition, because deletion of a single gene of the sterol pathway increases significantly the susceptibility of cells to H_2O_2 [10] and H_2O_2 represses the ergosterol biosynthesis pathway through Pof14, a new member of the F-box protein family [16]. Also aquaporins may facilitate the transport of H_2O_2 in some circumstances thus contributing for membrane permeability to H_2O_2 [17, 18]. Other mechanisms and membrane components are expected to be involved in a complex regulatory network that controls membrane permeability to H_2O_2.

Regulation of NF-κB by H_2O_2

The activation of NF-κB by H_2O_2 further illustrates the importance of delivering H_2O_2 to cells in a controlled and calibrated manner. The nuclear factor-κB (NF-κB) is an eukaryotic transcription factor with a central role in the inflammatory and immune responses [19]. It can be activated by a wide range of stimuli such as cytokines, bacterial lipopolissacharide, growth factors and viral infections. Initially it was proposed that H_2O_2 was the common mediator molecule responsible for the activation of NF-κB: the variety of agents able to activate NF-κB would do so by increasing the intracellular levels of H_2O_2 [20]. Initially, this role for H_2O_2 was confirmed by the inhibition of NF-κB activation by TNF-α or PMA in the presence of antioxidants such as *N*-acetyl-*L*-cysteine and pyrrolidine dithiocarbamate [21, 22]. Recently, the inhibition of TNF-α-induced NF-κB activation by these antioxidants was shown to be independent of their antioxidant properties [23]. Moreover, the activation of NF-κB by H_2O_2 is not a universal phenomenon because several agents are able to induce NF-κB without observable changes in the intracellular redox state, and H_2O_2 itself fails to induce NF-κB in several cell lines. Also, contradictory results were observed concerning the role of the cellular redox state on NF-κB activation: oxidative conditions can either increase or decrease NF-κB activation [21, 24-27]. These discrepancies were solved by postulating that oxidative conditions in the cytoplasm favors NF-κB activation and transport to the nucleus, and that reducing conditions in the nucleus are required for NF-κB binding to DNA, which was confirmed by several studies [28, 29]. So, the balance between oxidants and antioxidants in the cytoplasm and in the nucleus, which is cell type specific, may explain the contradictory observations concerning the activation of NF-κB by H_2O_2.

Recently, we analysed the inconsistencies for the role of H_2O_2 on NF-κB activation by applying both the steady-state titration and the bolus addition approaches for H_2O_2 delivery into MCF-7 and HeLa cells. H_2O_2 steady-state titration stimulated NF-κB activation by TNF-α, while an H_2O_2 bolus addition had the opposite effect (Figure 3). The differences can be explained by the high oxidative environment imposed by the bolus addition method. Moreover, a pre-incubation with H_2O_2 in steady-state followed by the addition of TNF-α lead as well to an inhibitory effect on NF-κB activation, confirming that a high oxidative load caused by H_2O_2 inhibits NF-κB activation [8]. Thus, contradictory findings may be explained by the lack of control of H_2O_2 delivery which makes comparison of published results by different laboratories difficult. The same bolus addition in two different experimental settings may actually constitute two very different H_2O_2 stimuli.

The mechanism of H_2O_2 action on NF-κB activation pathway is not known, but some differences with typical inducers have been observed. Typically, after exposure to a stimulus IκB-α, which is the most studied inhibitory subunit, is targeted for phosphorylation at two serine residues (Ser32 and Ser36) by the IκB kinase complex (IKK complex), followed by ubiquitination and degradation by the proteasome.

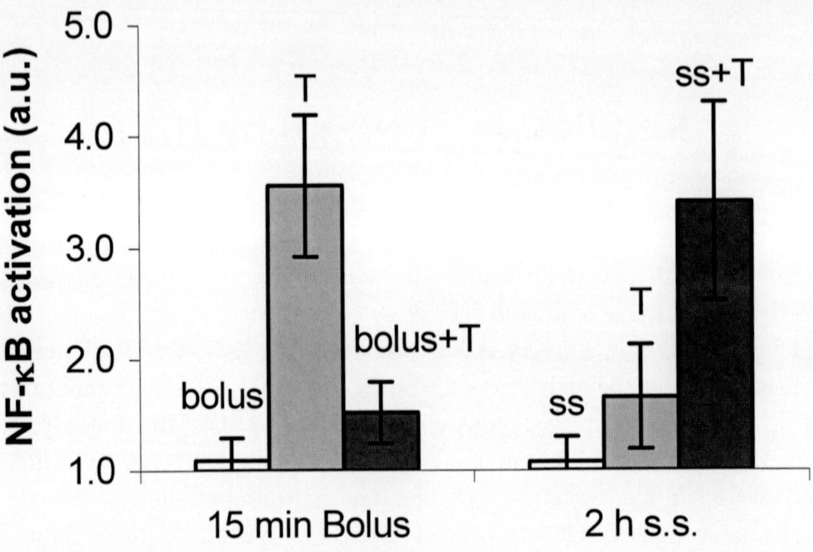

Figure 3. NF-κB activation by TNF-α is modulated by H_2O_2 and dependent on the method used for H_2O_2 addition. HeLa cells were prepared 48 h before the experiment to achieve a density of 1.5×10^6 cells and then exposed to H_2O_2, either as a bolus addition of 1 mM or a steady state of 25 μM, and to TNF−α (0.37 ng/mL) individually or simultaneously. Nuclear proteins were extracted following p65 subunit analysis by western blot.

However in the presence of H_2O_2 or pervanadate a phosphorylation at the tyrosine 42 residue has been described [30, 31]. It is controversial if IκB-α phosphorylated at Tyr42 is targeted for degradation or not [30, 31], but under the controlled conditions of a low steady-state H_2O_2 concentration, IκB-α degradation is not observed [8].

Concerning the role of H_2O_2 as a stimulator of NF-κB activation, incubation both with a low H_2O_2 steady-state concentration and with TNF-α up-regulated the expression of a set of pro-inflammatory genes, such as IL-8, MCP-1, TLR2 and TNF-α [8]. Also heme-oxygenase 1 and IL-6, which are anti-inflammatory proteins, were also stimulated suggesting a dual regulatory role for H_2O_2 on NF-κB activation [8]. This dual role proposed for H_2O_2 follows a typical design for control systems in cells in which the activation of a pathway (in this case the stimulation of pro-inflammatory genes via NF-κB) produces an inhibitor (in this case, the anti-inflammatory genes) which blocks the activation process. The pairs NF-κB/(IkB,A20), p53/Mdm2, HSF1/hsp70 exemplify such a design [32].

Conclusion

Delivery of H_2O_2 into cells by using either bolus or glucose oxidase addition is not adequate to study the regulatory roles of low endogenous H_2O_2 concentrations. Here, we exemplified how the use of these methods can lead to contradictory results for NF-κB activation by H_2O_2. The same phenomenon is expected to occur for other pathways. Exposure to low steady-state concentrations of H_2O_2 emerges as an appropriate method to investigate the biochemistry and cell biology of H_2O_2 under the paradigm of H_2O_2 being a signaling molecule.

Acknowledgements

Work supported by grant POCTI/BCI/42245/2001 from FCT – Portugal. VOM and FA acknowledge fellowships BD/16681/2004 and BPD/11487/2002 from FCT – Portugal.

References

[1] Cadenas, E., Ragan, C. I. and Stoppani, A. O. M. (1977). Production of superoxide radicals and hydrogen peroxide by NADH-ubiquinone reductase and ubiquinol-cytochrome c reductase from beef-heart mitochondria. *Arch. Biochem. Biophys.*, 180, 248-257.

[2] Chance, B., Sies, H. and Boveris, A. (1979). Hydroperoxide metabolism in mammalian organs. *Physiol. Rev.*, 59, 527-605.

[3] Czech, M. P., Lawrence, J. C., Jr. and Lynn, W. S. (1974). Evidence for the involvement of sulfhydryl oxidation in the regulation of fat cell hexose transport by insulin. *Proc. Natl. Acad. Sci. USA.*, 71, 4173-4177.

[4] White, A. A., Crawford, K. M., Patt, C. S. and Lad, P. J. (1976). Activation of soluble guanylate cyclase from rat lung by incubation or by hydrogen peroxide. *J. Biol. Chem.*, 251, 7304-7312.

[5] Test, S. T. and Weiss, S. J. (1984). Quantitative and temporal characterization of the extracellular H_2O_2 pool generated by human neutrophils. *J. Biol. Chem.*, 259, 399-405.

[6] Liu, X. and Zweier, J. L. (2001). A real-time electrochemical technique for measurement of cellular hydrogen peroxide generation and consumption: evaluation in human polymorphonuclear leukocytes. *Free Radic. Biol. Med.*, 31, 894-901.

[7] Antunes, F. and Cadenas, E. (2001). Cellular titration of apoptosis with steady-state concentrations of H_2O_2. Sub-micromolar levels of H_2O_2 induce apoptosis through Fenton chemistry independent of cellular thiol state. *Free Radic. Biol. Med.*, 30, 1008-1018.

[8] Oliveira-Marques, V., Cyrne, L., Marinho, H. S. and Antunes, F. (2007). A quantitative study of NF-kappaB activation by H_2O_2: relevance in inflammation and synergy with TNF-a. *J. Immunol., in press*.

[9] Izawa, S., Inoue, Y. and Kimura, A. (1996). Importance of catalase in the adaptive response to hydrogen peroxide: analysis of acatalasaemic *Saccharomyces cerevisiae. Biochem. J.,* 320, 61-67.

[10] Branco, M. R., Marinho, H. S., Cyrne, L. and Antunes, F. (2004). Decrease of H_2O_2 plasma membrane permeability during adaptation to H_2O_2 in *Saccharomyces cerevisiae*. *J. Biol. Chem*., 279, 6501-6506.

[11] Clayton, R. K. (1959). Permeability Barriers and the Assay of Catalase in Intact Cells. Biochim. *Biophys. Acta*, 36, 35-39.

[12] De Duve, C. (1965). The separation and characterization of subcellular particles. *Harvey Lect. Ser.*, 59, 48-87.

[13] Antunes, F. and Cadenas, E. (2000). Estimation of H_2O_2 gradients across biomembranes. *FEBS Lett.,* 475, 121-126.

[14] Costa Seaver, L. and Imlay, J. A. (2001). Hydrogen peroxide fluxes and compartmentalization inside growing *Escherichia coli*. *J. Bacteriol.*, 183, 7182-7189.

[15] Sousa-Lopes, A., Antunes, F., Cyrne, L. and Marinho, H. S. (2004). Decreased cellular permeability to H_2O_2 protects *Saccharomyces cerevisiae* cells in stationary phase against oxidative stress. *FEBS Lett*., 578, 152-156.

[16] Tafforeau, L., Le Blastier, S., Bamps, S., Dewez, M., Vandenhaute, J. and Hermand, D. (2006). Repression of ergosterol level during oxidative stress by fission yeast F-box protein Pof14 independently of SCF. *EMBO J.*, 25, 4547-4556.

[17] Bienert, G. P., Schjoerring, J. K. and Jahn, T. P. (2006). Membrane transport of hydrogen peroxide. Biochim. Biophys. *Acta-Biomembranes*, 1758, 994-1003.

[18] Bienert, G. P., Moller, A. L., Kristiansen, K. A., Schulz, A., Moller, I. M., Schjoerring, J. K. and Jahn, T. P. (2007). Specific aquaporins facilitate the diffusion of hydrogen peroxide across membranes. *J. Biol. Chem*., 282, 1183-1192

[19] Ghosh, S., May, M. J. and Kopp, E. B. (1998). NF-kappaB and Rel proteins: evolutionarily conserved mediators of immune responses. *Annu. Rev. Immunol.,* 16, 225-60.

[20] Schreck, R., Rieber, P. and Baeuerle, P. A. (1991). Reactive oxygen intermediates as apparently widely used messengers in the activation of the NF-kappaB transcription factor and HIV-1. *EMBO J.*, 10, 2247-2258.

[21] Staal, F. J. T., Roederer, M., Herzenberg, L. A. and Herzenberg, L. A. (1990). Intracellular Thiols Regulate Activation of Nuclear Factor-kappaB and Transcription of Human Immunodeficiency Virus. *Proc. Natl. Acad. Sci. USA,* 87, 9943-9947.

[22] Schreck, R., Meier, B., Mannel, D. N., Droge, W. and Baeuerle, P. A. (1992). Dithiocarbamates as potent inhibitors of nuclear factor kappa B activation in intact cells. *J. Exp. Med.*, 175, 1181-1194.

[23] Hayakawa, M., Miyashita, H., Sakamoto, I., Kitagawa, M., Tanaka, H., Yasuda, H., Karin, M. and Kikugawa, K. (2003). Evidence that reactive oxygen species do not mediate NF-kappaB activation. *EMBO J.*, 22, 3356-3366.

[24] Flohe, L., BrigeliusFlohe, R., Saliou, C., Traber, M. G. and Packer, L. (1997). Redox regulation of NF-kappa B activation. *Free Radic. Biol. Med.*, 22, 1115-1126.

[25] Toledano, M. B. and Leonard, W. J. (1991). Modulation of Transcription Factor NF-kappaB Binding Activity by Oxidation-Reduction *In vitro*. *Proc. Natl. Acad. Sci. USA*, 88, 4328-4332.

[26] Brennan, P. and Oneill, L. A. J. (1995). Effects of oxidants and antioxidants on Nuclear factor-kappaB activation in 3 different cell-lines - Evidence against a universal hypothesis involving oxygen radicals. *Biochim. Biophys. Acta-Gene Structure and Expression*, 1260, 167-175.

[27] Matthews, J. R., Wakasugi, N., Virelizier, J. L., Yodoi, J. and Hay, R. T. (1992). Thioredoxin regulates the DNA binding activity of NF-kappa B by reduction of a disulphide bond involving cysteine 62. *Nucleic Acids Res.*, 20, 3821-3830.

[28] Brennan, P. and Oneill, L. A. J. (1996). 2-mercaptoethanol restores the ability of nuclear factor kappa B (NF kappa B) to bind DNA in nuclear extracts from interleukin 1-treated cells incubated with pyrollidine dithiocarbamate (PDTC) - Evidence for oxidation of glutathione in the mechanism of inhibition of NF kappa B by PDTC. *Biochem. J.*, 320, 975-981.

[29] Jornot, L., Petersen, H. and Junod, A. F. (1997). Modulation of the DNA binding activity of transcription factors CREP, NF kappa B and HSF by H_2O_2 and TNF alpha. Differences between *in vivo* and *in vitro* effects. *FEBS Lett.*, 416, 381-386.

[30] Schoonbroodt, S., Ferreira, V., Best-Belpomme, M., Boelaert, J. R., Legrand-Poels, S., Korner, M. and Piette, J. (2000). Crucial role of the amino-terminal tyrosine residue 42 and the carboxyl-terminal PEST domain of I kappa B alpha in NF-kappa B activation by an oxidative stress. *J. Immunol.*, 164, 4292-4300.

[31] Takada, Y., Mukhopadhyay, A., Kundu, G. C., Mahabeleshwar, G. H., Singh, S. and Aggarwal, B. B. (2003). Hydrogen peroxide activates NF-kappa B through tyrosine phosphorylation of I kappa B alpha and serine phosphorylation of p65: evidence for the involvement of I kappa B alpha kinase and Syk protein-tyrosine kinase. *J. Biol. Chem.*, 278, 24233-24241.

[32] Lahav, G., Rosenfeld, N., Sigal, A., Geva-Zatorsky, N., Levine, A. J., Elowitz, M. B. and Alon, U. (2004). Dynamics of the p53-Mdm2 feedback loop in individual cells. *Nat. Genet.*, 36, 147-150.

In: Oxidative Stress: Clinical and Biomedical Implications ISBN: 978-1-60021-888-0
Editors: B. M. Matata, M. M. Elahi, pp. 69-83

Chapter V

Reactive Oxygen Species as a Mediator of Cardioprotection

Dipak K. Das and Manika Das
Cardiovascular Research Center,
University of Connecticut School of Medicine,
Farmington, CT 06030-110, USA

Abstract

This review focuses on the possible role of reactive oxygen species as signaling molecules that regulate cadiomyocyte life and death during ischemia and reperfusion. Ischemic preconditioning (IPC) is a most powerful endogenous mechanism for myocardial protection against ischemia/reperfusion injury. It is now apparent that reactive oxygen species (ROS) generated in the mitochondrial respiratory chain act as a trigger of IPC. ROS mediate signal transduction in the early phase of IPC through the posttranslational modification of redox-sensitive proteins. ROS-mediated activation of Src tyrosine kinases serves a scaffold for interaction of proteins recruited by G protein-coupled receptors and growth factor receptors that is necessary for amplification of cardioprotective signal transduction. Protein kinase C (PKC) plays a central role in this signaling cascade. A crucial target of PKC is the mitochondrial ATP-sensitive potassium channel, which acts as a trigger and a mediator of IPC. Mitogen-activated protein (MAP) kinases (extracellular signal-regulated kinase, p38 MAP kinase, and c-Jun NH (2)-terminal kinase) are thought to exist downstream of the Src-PKC signaling module, although the role of MAP kinases in IPC remains undetermined. The late phase of IPC is mediated by cardioprotective gene expression. This mechanism involves redox-sensitive activation of transcription factors through PKC and tyrosine kinase signal transduction pathways that are in common with the early phase of IPC. The effector proteins then act against myocardial necrosis presumably through alleviation of oxidative stress. Elucidation of complex cardioprotective signaling processes will help in the development of more effective pharmacological approaches for prevention of myocardial ischemia/reperfusion injury.

Introduction

The first quarter-century of ROS research focused on damage inflicted on cellular constituents. Then, in the mid-1970s, ROS was shown to activate guanulate cyclase, causing formation of the well characterized second messenger cyclic guanosine monophospahte [1]. Gradually, as other second messenger systems were shown to be modulated by ROS, a biologic role for ROS in cell function became accepted. Involvement of oxygen free radicals in mitogenic stimulation of cell as well as growth factor- and cytokine-induced signal transduction strongly suggest that the reactive oxygen species function as second messengers [2]. Moreover, the mitogenic signals mediated through the generation of reactive oxygen species, activate many transcription factors including nuclear transcription factor NFkB and genes including antioxidant enzymes and bcl-2 [3]. Indeed, it is now known that ROS act as important mediators in signal transduction processes involved in multiple aspect of cardiac cellular function.

A large number of degenerative diseases, including coronary heart disease, have been linked with the overproduction of oxygen-derived free radicals. The results of many studies, including our own, have demonstrated that excessive production of reactive oxygen species in concert with drastic reduction of antioxidant reserve play a crucial role in the pathophysiology of ischemic heart disease [4]. Myocardial ischemia and reperfusion cause the cardiomyocytes to face conditions that shift their redox status to undergo a drastic change subjecting them to oxidative stress [5]. Interventions with oxygen free radical scavengers or antioxidant therapy have been found to be cardioprotective against ischemic reperfusion injury [6].

Recent studies have documented that coronary heart diseases cause cardiomyocyte death not only by necrosis but also by apoptosis [7]. Reperfusion of ischemic myocardium results in apoptotic cell death and DNA fragmentation [8]. In concert, ischemia/reperfusion is associated with the induction of a number of both pro- and anti- apoptotic genes and transcription factors [9]. Ebselen, a glutathione peroxidase (GSHPx) mimic, was found to reduce cardiomyocyte apoptosis [10]. The hearts from the transgenic mice over expressing GSHPx-1 gene were resistant to ischemia/reperfusion injury while those from the GSHPx-1 knockout animals, devoid of any copy of GSHPx-1, was extremely vulnerable to the cellular injury as compared to wild-type controls [11].

The results mentioned above clearly indicate that cardiomyocyte death induced by ischemia/reperfusion is redox-regulated and that reactive oxygen species not only trigger the cell death but also function as intracellular signaling molecules. This review will discuss the intracellular signaling pathways potentiated by the reactive oxygen species that lead to the induction of gene expression and apoptotic cell death.

Evidence In Support of a Role of Oxidants in Preconditioning

Ischemic preconditioning is the manifestation of the earlier stress response that occurs during repeated episodes of brief ischemia and reperfusion and can render the myocardium to a subsequent, potentially lethal ischemic injury. A possible role for oxygen radicals in preconditioning was first suggested by Murry *et al.* [12], who showed that administration of superoxide dismutase (SOD), an enzyme that removes superoxide anions, during the reperfusion period following the initial "preconditioning" ischemia, could prevent the phenomenon. The authors hypothesized that myocardial reperfusion after the first, short ischemic episode results in the generation of low amounts of oxygen free radicals, not sufficient to causes cell necrosis, but which could modify cellular activities and thus induce preconditioning. Other investigators have subsequently obtained findings indicating that oxygen radicals may be involved in preconditioning. Administration of oxygen radical scavengers, such as mercapto propionyl-glycineor SOD, was able to blunt the protective effect of ischemic preconditioning on infarct size [13, 14]. In another study, treatment with *N*-acetylcysteine negated the beneficial effect of preconditioning on post-ischemic recovery of contractile function [15]. Similarly, the beneficial effect of preconditioning on reperfusion induced arrhythmias was prevented by SOD administration during the preconditioning ischemia [16].

Mechanism of Action of Oxidants in Preconditioning

Several mechanisms might explain the effects of oxygen radicals. Recently, it has become appreciated that exposure of cells to mild oxidative stress can reversibly modify several cellular activities, in the absence of cell damage, but secondary to changes in the activity of various enzymes and other cellular components [17, 18, 19]. Among others, oxidant stress can modify some of the cellular activities that have been implicated *in vivo* as mediators of the preconditioning phenomenon. It could thus be hypothesized that reperfusion after the initial, "preconditioning" ischemic episode results in the generation of relatively low amounts of oxygen radicals, insufficient to cause cell necrosis, but enough to modify cellular activities and thus induce preconditioning.

Protein Kinase C and Tyrosine Kinase

A possibility that should be taken into account is the effect of free radicals on protein kinase C. As many of the stimuli involved in preconditioning share the ability to activate this enzyme [20, 21], and as preconditioning can be prevented by inhibition of protein kinase C in various experimental models [22, 23], activation of protein kinase C is currently held as a central mediator of ischemic preconditioning. With respect to the effects of oxygen radicals, studies in various cell types demonstrate that mild oxidative conditions can activate protein kinase C [24, 25, 26].This effect is presumably linked to an increase of the Ca/phospholipids independent form of protein kinase C [25] and is accompanied by translocation of the inactive form of the enzyme from the cytoplasm to the cell membrane, where protein kinase C exerts its activity [26]. The hypothesis that protein kinase C activation and/or translocation might be a mechanism of action of oxygen radicals in preconditioning received support by our observation that the preconditioning effect induced by oxygen radicals in the absence of ischemia was significantly attenuated by protein kinase C inhibition [27]. However, as inhibition of protein kinase C did not completely prevent the beneficial effects of oxygen radicals, other mechanisms were thought to b be involved in this phenomenon. It has been suggested that important steps in the signaling pathway of preconditioning are represented by activation of phospholipase D [28, 29], tyrosine kinase [30, 31], and mitogen-activated protein kinase activation [32, 31], with complex interactions among these pathways that have not been entirely elucidated [33, 34]. Again, oxidative stress has been proposed as a possible activator of these enzymes in various cell types [35, 36, 37], and it is thus possible that an influence on these enzymes might con-tribute to oxygen radical-induced preconditioning, independent of protein kinase C activation.

The Src family of tyrosine kinases appears to play a crucial role in the ROS mediated transactivation of receptor tyrosine kinase [38]. The Src family of tyrosine kinase is known to be activated by oxidative stress [39, 40]. Several fold increase in activation of Src family kinases has been reported during hypoxia/ reoxygenation and ischemia reperfusion [40] or in response to activation of a variety of G protein-coupled receptors with their respective ligands [41]. Src kinases are known to interact with many signaling proteins including PKC, and phosphatidylinositol-3- kinase [42]. Such protein-protein interaction forms a signaling module that may be important in integrating cardioprotective signal transduction in IPC [43].

Role of MAP Kinases

The mitogen-activated protein (MAP) kinases, a serine/threonine protein kinase family, play an essential role in mediating intracellular signal transduction events. In response to extracellular stimulation, MAP kinases are rapidly activated and in turn regulate cellular functions by inducing the phosphorylation of proteins, such as an oncogene product c-jun, S6 ribosomal protein kinase, and MAP kinase activated protein kinase 2 [44, 45]. MAPKAP kinase 2 has been implicated in a novel mammalian stress activated signal transduction pathway initiated by a variety of mitogens, pro-inflammatory cytokines, or environmental stresses, where it regulates its substrate molecules by serine/threonine phosphorylation [46]. In the case of rat heart, a mitogen-activated protein kinase cascade has already been identified [47]. These authors have demonstrated that MAPK isoforms p42 MAPK and P44 MAPK and two peaks of MEK were activated by more than 10-fold in perfused hearts or ventricular myocytes exposed to PMA for 5 min. In our own study, we identified the participation of MAP kinase cascades in the ischemia adapted rat hearts [48]. The results of our study demonstrated that a kinase cascade involving tyrosine kinase-phospholipase D-MAP kinases-MAPKAP kinase 2 is triggering after ischemic stress.

The intracellular signaling mechanisms that lead to adaptation require one or more members of MAP kinase cascades. Among the three distinct MAP kinase families, stress-activated protein kinase (SAPK), also known as c-Jun NH_2-terminal kinases (JNK), and p38 MAP kinase are known to be regulated by extracellular stresses including environmental stress, oxidative stress, heat shock and UV radiation [49]. JNKs and p38 MAP kinase appear to be involved in distinct cellular functions, because they possess different cellular targets and are located on different signaling pathways. Thus, JNK kinases activate c-Jun while p38 MAP kinase stimulates MAPKAP kinase. A recent study demonstrated that ischemic adaptation triggered a tyrosine kinase-regulated signaling pathway leading to the translocation and activation of p38 MAP kinase and MAPKAP kinase 2 [50].

The results of a recent study documented that 30 min of ischemia followed by 2 hr of reperfusion increased the induction of JNK1, c-JUN and p38 MAP kinase proteins [51]. Ischemic adaptation also enhanced these kinases compared to control. However, subsequent ischemia/reperfusion-mediated increase in JNK1, p38MAPK and c-Jun was blocked by this adaptation. Fifteen minutes of perfusion with anisomycin increased the amount of both JNK1 and p38 MAP kinases as well as c-Jun in the heart, all of which were decreased in amount after subsequent ischemia/reperfusion. Curcumin, a JNK and c-Jun inhibitor blocked the ischemia/ reperfusion- and adaptation -mediated increase in JNK1 and c-Jun while it had no effect on p38 MAP kinase. SB 203580, an inhibitor for P38 MAPK phosphorylation, on the other hand, was equally effective in reducing the amount of p38 MAP kinase, but exerted no effects on JNK1 and c-Jun. Ischemia/reperfusion-mediated increased myocardial infarction was reduced by treating the hearts with anisomycin, a dual activator of P38 MAPK and JNK. The cardioprotective effects of preconditioning were abolished by either curcumin or SB 203580. Thus, it appears that activation of SAPKs are obligatory for myocardial adaptation to ischemia, but the activation is only transient, and the activities rapidly come down to near baseline levels after subsequent ischemia and reperfusion [51].

Evidence suggests that MAPKAP kinase 2 is a crucial step leading to gene expression and myocyte adaptation resulting in adaptive cardioprotection. This unique protein kinase is highly expressed in heart muscle suggesting that it also may be expressed and functioning in the myocardium in response to stress. MAP KAP kinase 2 also has been shown to have increased activity when subjected to oxidative stress as well as heat shock. [50,47]. This increased activity of MAPKAP kinase 2 in association with heat shock protein gives rise to the hypothesis that this kinase may be one of the critical factors involved with ultimate transcription of proteins leading to adaptive protection of the heart. In cultured myocytes, the activity of MAPKAP kinase 2 was found to increase when the myocytes were subjected to oxidative stress as well as heat shock [47]. Heat shock proteins (HSP) are early targets of phosphorylation by a variety of stress conditions. HSP-27, HSP 32 and HSP-70 can be induced by oxidative stress and ischemic preconditioning [4]. Heat shock proteins have been found to be not only cardioprotective through reduction of infarct size but evidence also suggests that they are important in delayed protection against infarction, the so called "second window" of protection.

ATP-Dependent Potassium Channels

Oxygen radicals might also activate other cardio-protective mechanisms. In this regard, recent studies have shown that oxidants can increase the open state of ATP-dependent K channels in patch-clamped myocytes [52] and other cell types [53], and in isolated hearts [54]. Opening of the ATP-dependent K channels exerts a strong cardio-protective effect, and some studies have suggested that it might play a major role in the pathogenesis of preconditioning [55,56].This possibility is supported by the preliminary observation that the preconditioning effect of hydrogen peroxide can be prevented by glybenclamide, an inhibitor of ATP-dependent K channels [57].

De Novo Protein Synthesis and Gene Expression

Expression of new proteins has been found to play a role only in the late phase of preconditioning, as at least 12 hour are necessary for translation of RNA into proteins [58]. This has been confirmed by the observation that ischemic preconditioning causes a rapid increase in myocardial protein synthesis, which is necessary for development of late protection against myocardial stunning [59]. A brief ischemic episode may increase the synthesis of several proteins, including heat-shock proteins, antioxidant enzymes, nitric oxide synthase, proto-oncogenes, a fatty acid transport protein, and several mitochondrial genes [60, 61]. Although the importance of nitric oxide synthase in delayed myocardial protection has been demonstrated [62], the precise role of most of these proteins in this phenomenon has not been elucidated [63, 58], and there is still considerable controversy about the relative importance of antioxidant defenses as discussed in the subsequent paragraphs. It has been suggested that preconditioning may increase gene expression by influencing transcription factors. A brief ischemic episode is accompanied by activation of nuclear transcription factor-*k*B (NF-*k*B), and preconditioning may be prevented by NF-*k*B inhibition [64, 65].

Oxygen radicals play a major role inactivation of NF-*k*B [66], and induction of the synthesis of heat-shock proteins [67, 68]. In this regard, NF-*k*B activation during preconditioning is prevented by removal of oxygen radicals [69], and Korea *et al.* [70] have shown that exposure of isolated rat hearts to low concentrations of oxygen radicals results in increased expression of heat-shock protein mRNA.

Redox Sensing Transcription Factors

NFκB and AP-1 are two well-known redox-sensitive transcription factors. NFκB is a critical regulator for gene expression induced by diverse stress signals including mutagenic, oxidative and ischemic stresses. Activation of NFκB is likely to be involved in the induction of gene expression associated with the ischemic adaptation. This nuclear transcription factor is a member of Rel transcription factor family which is involved in the regulation of stress defense mechanisms. Survey of literature suggests that NFkB is an oxidative stress responsive transcription factor, and that reactive oxygen intermediates play a crucial role in the activation of the factor. Recent studies from our laboratory demonstrated ischemic adaptation translocated and increased the binding of NFκB in the heart [71].NFκB binding activity is very low in non-ischemic control hearts. Reperfusion of ischemic myocardium significantly increases the translocation of NFκB from cytosol to nucleus. Perfusion of the heart with DMTU inhibits NFκB translocation from cytosol to nucleus [72]. NFκB binding activity is increased in the ischemia adapted hearts.

AP-1 is another redox-sensitive signaling molecule which also plays an important regulatory role in cellular responses to stress induced by external factors including UV radiation, phorbol esters, and TNFα [73]. The binding site of AP-1 is recognized by Jun family member homodimers and Jun/Fos family member heterodimers. The balance between Jun and FOS is very critical for gene expression. Stress induced by ischemia/reperfusion was previously shown to induce the activation of c-Jun [74]. AP-1 is a redox-sensitive signaling molecule which plays an important regulatory role in cellular responses to stress induced by external factors. The AP-1 transcription factor complex is composed of a group of proteins encoded by the Jun and FOS families which bind to the AP-1 consensus sequences. Regulation of AP-1 in response to external stimuli is mediated by members of the MAPK family. AP-1 regulates the activation of transcription of a variety of genes. Electrophoretic mobility shift assay indicated increased AP-1 binding activity in the ischemic/reperfused rat heart compared to control perfused group [75]. DMTU inhibited this binding activity significantly. Similar effects were also observed for ischemia adapted group. Myocardial adaptation induced by cyclic episodes of short term ischemia and reperfusion results in the activation of NFκB but not AP-1.

Redox Sensing Genes and Proteins

In various cell lines, wild-type-P53 (transcription factor, pro-apoptotic gene) induced by DNA damage has been shown to induce programmed cell death or apoptosis. It is reported that P53 protein functions as an active transcription factor in lesioned brain [76]. How activation of p53 promotes apoptosis is unclear, but it might involve Bax, a series of p53-inducible genes or signaling through Fas-related pathways [77]. There are other p53 effectors which include caspases that execute apoptotic cell death [78]. Several studies on the master controller gene of apoptosis, p53, reveal conflicting results [79]. Our recent study showed an increase in p53 activity in the ischemic/reperfused myocardium [75]. Such an increase in p53 activity was prevented by adapting the heart to ischemia.

Another mechanism that controls the expression of stress-responsive genes is the activation of STAT. These proteins are tyrosine phosphorylated by JAKs in a redox sensitive manner following the binding of cytokine to its receptor. And by a variety of receptor and nonreceptor protein tyrosine kinase [80, 81]. Once they are phosphorylated, STAT protein transactivate STAT-responsive genes. The JAK-STAT pathway has been implicated in cardiac hypertrophy, apoptosis and inflammation [82, 83]. Activation of STAT3 has been reported to limit apoptosis in rat models of myocardial infraction [84]. STAT1 and STAT3 have also been found to exert proapoptotic and antiapoptotic effects respectively, in cultured neonatal cardiomyocyte [85]. More recently it has also demonstrated that IPC induces isoform selective activation of JAK1, JAK2, STAT1, STAT3 and activation of JAK-stat pathway is responsible for up regulation of iNOS, which plays an essential role in the development of late IPC [86,87].

Bcl-2 is a well-known anti-death gene which functions as an intracellular antioxidant. Recent analysis of the bcl-2 gene family reveals complex network regulating apoptosis. Within this bcl-2 gene family, some of the candidates can suppress apoptosis, while the others can induce apoptosis [88]. Among the proteins coded by the genes of this family, Bcl-2 and Bcl-x_L act as cell death repressors [89], whereas Bax and an alternatively sliced Bcl-x product, Bcl-xs, promote cell death [90]. When in excess over Bcl-2, Bax counteracts the repressive action of Bcl-2 on apoptosis. Similarly, excess Bcl-xs antagonize the function of Bcl-xL. Thus, a critical balance between the Bcl-2, Bax and Bcl-$x_{L/S}$ molecules may determine the fate of cells in response to cytotoxic agents or environmental stress. It has been reported that p53 activate the transcription of the Bax gene via p53–response elements while down regulating Bcl-2 expression at the same time [77].Apoptosis initiated by various different stimuli can be blocked by over expressing Bcl-2. For example, the activation of Bcl-2 was associated with the inhibition of apoptosis in the adapted myocardium [75]. A downregulation of this anti-death gene occurred in concert with the significant amount of apoptosis in the ischemic/reperfused myocardium.

Redox Signaling

Unlike ROS being the causative factor of cellular injury, redox signaling utilizing the same ROS for cell survival. The biochemical basis of the redox signaling is that oxidizing conditions are maintained by the stabilizing disulfides in the extracellular surface while the intracellular environment is maintained in the reduced state with the help of free sulfhydryl groups. The principle disulfide reductase responsible for maintaining the inside of the cell in the reduced state is a low molecular weight redox-active protein with two cysteine residues in its active sites, thioredoxin [91]. Thioredoxin is ubiquitously present in mammalian cells including heart. Several reports exist indicating that thioredoxin is induced by oxidative stress [92]. Thioredoxin appears to play a crucial role in redox regulation of the ROS signaling during ischemia/reperfusion. Thioredoxin may be an important component of the cellular defense against cardiac injury. Oxidized thioredoxin was found to be released into plasma of the patients undergoing cardiopulmonary bypass surgery [93]. Endurance training by swimming accompanied by a reduction of ischemia/reperfusion-induced oxidative stress with a concomitant increase in thioredoxin reductase resulted in protection against myocardial ischemia/reperfusion injury [94]. Thioredoxin attenuated hypoxia-reoxygenation injury of murine endothelial cells in a thiol-free condition, suggesting thioredoxin protection of myocardial injury through a novel redox signaling pathway [95].

Thioredoxin are critical for redox regulation of protein function and signaling via thiol redox control. Thioredoxins are reduced by electrons from NADPH via thioredoxin reductase. A recent study has demonstrated that thioredoxin (Trx-1) plays an important role in redox signaling and in transmitting survival signal in ischemic myocardium [96]. Trx-1 is downregulated after ischemia/reperfusion but upregulated in the preconditioned myocardium. Inhibition of Trx-1 abrogated cardioprotective effects of preconditioning as evidenced from impaired post-ischemic ventricular recovery increased myocardial infarct size, and cardiomyocyte apoptosis. The role of Trx-1 in transmitting survival signal was supported further from the results that transgenic mouse hearts with extra copies of Trx-1 were resistant to apoptotic cardiomyocyte death.

It appears from the above discussion; survival and death of the cardiomyocytes depend critically on their redox state. (Figure 1) While sufficient antioxidant reserve warrants a reducing environment for the survival of the myocytes, the reserve is exhausted during ischemic heart disease leading to a change in redox state. Such changes in the redox state determine whether the cardiomyocytes should die, and if they die, whether they die of necrosis or apoptosis. Ultimate survival or death depends on the redox signaling and transcription regulation of several redox-sensing genes and DNA repair processes.

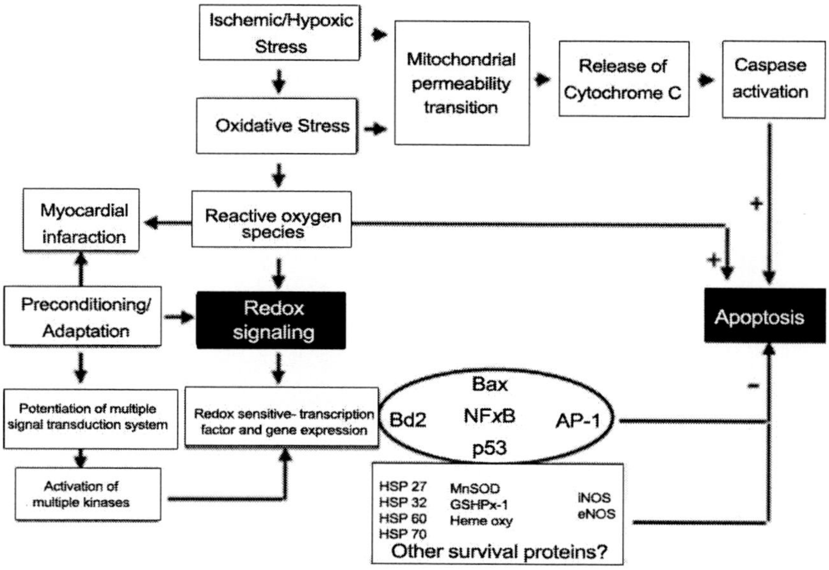

Figure 1. Generation of anti-death signal through redox regulation. (Reproduced from: Das DK and Maulik N. Archives of Biochemistry and Biophysics, 420: 305-311, 2003).

References

[1] Mittal CK, Murad F. Activation of guanylate cyclase by super oxide dismutase and hydroxyl radical : a physiological regulat of guanosine 3', 5'-monophospahte formation. *Proc. Nats. Acad. Sci. USA.* 74, 4360-4, 1977

[2] Rosette C, and Karin M. Ultraviolet light and osmotic stress: Activation of the JNK cascade through multiple growth factor and cytokine receptors. *Science* 274: 1194-1197, 1996.

[3] Dalton TP, Shertzer HG, Puga A. Regulation of gene expression by reactive oxygen. *Annu. Rev. Pharmacol. Toxicol.* 39: 67-101, 1999.

[4] Das DK, and Maulik N. Evaluation of antioxidant effectiveness in ischemia reperfusion tissue injury methods. *Methods Enzymol.* 233: 601-610, 1994.

[5] Meerson FZ, Kagan VE, Kozlov P, Beekina LM, Khipenko YV. The role of .lipid peroxidation in pathogenesis of ischemic damage and the antioxidant protection of the heart. *Basic Res Cardiol* 77: 465-472, 1982.

[6] Tosaki A, Droy-Lefaix MT, Pali T, Das DK. Effects of SOD, catalase and a novel anti-arrhythmic drug, EGB 671, on reperfusion-induced arrhythmias in isolated rat hearts. *Free Rad. Biol. Med.* 14: 361-370, 1993.

[7] Olivetti G, Abbi R, Quaini F, Kajstura J, Cheng W, Nitahara JA, Quaini E, Loreto C Di, Beltrami CA, Krajewski S, Reed J C, Anversa P. Apoptosis in the failing heart, *New Eng. J. Med.* 336:1131-1141, 1997.

[8] Maulik N, Yoshida T, and Das DK. Oxidative stress developed during the reperfusion of ischemic myocardium induces apoptosis. *Free Rad. Biol. Med.* 24: 869-875, 1998.

[9] Nishio Y, Kashiwagi A, Taki H, Shinozaki K, Maeno Y, Kojima H, Maegawa H, Haneda M, Hidaka H, Yasuda H, Horiike K, Kikkawa R. Altered activities of transcription factors and their related gene expression in cardiac tissues of diabetic rats. *Diabetes* 47: 1318-1325, 1998.

[10] Das DK, Maulik N. Apoptosis in ischemia reperfusion injury. In: *Biological Oxidants and Antioxidants* edited by L. Packer, ASH Ong AOCS Press, IL. Pp 165-177, 1998.

[11] Maulik N, Yoshida T, and Das DK. Regulation of cardiomyocyte apoptosis in ischemic reperfused mouse heart by glutathione peroxidase. *Mol Cell Biochem* 196: 13-21, 1999.

[12] Murry CE, Richard VJ, Jennings RB, and Reimer KA. Preconditioning with ischemia: is the protective effect mediated by free radical induced myocardial stunning?(Abstract). *Circulation*78(Suppl II): 77,1988.

[13] TanakaM,FujiwaraH,Yamasaki K, and SasayamaS. Superoxide dismutase and *N*-2-mercaptopropionyl glycine attenuate infarct size limitation effect of is-chaemic preconditioning in the rabbit. *Cardiovasc. Res.* 28: 980–986,1994.

[14] Tang X-L, Qiu Y, Turrens JF, Sun JZ, and Bolli R.Late preconditioning against stunning is not mediated by increased antioxidant defenses in conscious pigs. *Am. J. Physiol.* 273:H1651–H1657,1997.

[15] Chen W, Gabel S, Steenbergen C, and Murphy E. A redox-based mechanism for cardioprotection induced by ischemic preconditioning in perfused rat heart. *Circ. Res.* 77: 424–429,1995.

[16] Osada M, TakedaS, Sato T, Komori S, and TamuraK. The protective effect of preconditioning on reperfu-sion-induced arrhythmias is lost by treatment with superoxide dismutase. *Jpn Circ. J.* 58: 259–263, 1994.

[17] Burke TM, and Wolin MS. Hydrogen peroxide elicits pulmonary arterial relaxation and guanylate cyclase activation. *Am. J. Physiol.* 252:H721–H732, 1987.

[18] Golino P, Ragni M, Cirillo P, Avvedimento VE, Feli-ciello A, Esposito N, Scognamiglio A, Trimarco B, Iaccarino B, CondorelliM, ChiarielloM, and Ambro-sioG. Effects oftissue factorinduced by oxygenfree radicals on coronary flow during reperfusion. *Nat. Med.* 2: 35–40, 1996.

[19] Reeves JP, Bailey CA, and Hale CC. Redox modifica-tion of sodium–calcium exchange activity in cardiac sarcolemmal vesicles. *J. Biol. Chem.* 561: 4948–4955, 1986.

[20] Kloner RA, Bolli R, Marban E, Reinlib L, and Brauwald E. Medical and cellular implications of stunning, hibernation and preconditioning. *Circulation* 97:1848–1867, 1998.

[21] Yellon DM, Baxter GF, Garcia-Dorado D, Heusch G, and Sumaray MK. Ischemic preconditioning: present position and future directions. *J. Cardiovasc. Res.* 37: 21–33,1998.

[22] Speechly-Dick ME, Mocanu MM, and Yellon DM. Proteinkinase C. Its role in ischemicpreconditioning inthe rat. *Circ. Res.* 75: 586–590,1994.

[23] Speechly-Dick ME, Grover GJ, and Yellon DM. Does ischemic preconditioning in the human involve pro-tein kinase C and the ATP-dependent K^1 channel? Studies of contractile function after simulated is-chemia in an atrial in vitro model. *Circ. Res.* 77: 1030–1035,1995.

[24] Brawn MK, Chiou WJ, and Leach KL. Oxidant-in-duced activation of proteinkinase Cin UC11MGcells. *Free Radic. Res.* 22:23–37, 1995.

[25] Gopalakrishna R,and Anderson WB. Ca^{21}-and phos-pholipid-independent activation of protein kinase C by selectiveoxidative modification ofthe regulatory domain. *Proc. Natl. Acad. Sci. USA* 86: 6758–6762, 1989.

[26] Von Ruecker AA, Han-Jeon BG, Wild M, and BidlingmaierF.Protein kinase C involvement in lipid peroxidation and cell damage induced by oxygen de-rived radicals in hepatocytes. *Biochem. Biophys. Res. Commun.*163: 836–842, 1989.

[27] Tritto I, D'Andrea D, Eramo N, Scognamiglio A, De Simone C, Violante A, Esposito A,Chiariello M,and Ambrosio G. Oxygen radicals can induce precondi-tioning in rabbit hearts. *Circ. Res.* 80: 743–748, 1997.

[28] Cohen MV, Liu Y, Liu GS, Wang P, Weinbrenner C, Cordis GA,DasDK, and Downey JM. Phospholipase D plays a role in ischemic preconditioning in rabbit heart. *Circulation* 94: 1713–1718, 1996.

[29] Tosaki A, Maulik N, Cordis G, Trifan OC, Popescu LM, and Das DK. Ischemic preconditioning triggers phospholipase D signaling in rat heart. *AmJ Physiol* 273: H1860–H1866, 1997

[30] Imagawa JI, Baxter GF, and Yellon DM. Genistein, a tyrosine kinase inhibitor, blocks the "second window ofprotection"48hafter ischemic preconditioningin the rabbit. *J. Mol. Cell Cardiol.* 29: 1885–1893, 1997.

[31] Ping P, Zhang J, Zheng YT, Li RC, Dawn B, Tang XL, Takano H, Balafanova Z, and Bolli R. Demonstration ofselective protein kinase C-dependent activation of Src and Lck tyrosine kinases during ischemic pre-conditioning in conscious rat. *Circ. Res.* 85: 542–550, 1999.

[32] Maulik N, Sato M, Price BD, and Das DK. An essen-tial role of NF-*k*B in tyrosine kinase signaling of p38 MAP kinase regulation of myocardial adaptation to ischemia. *FEBS Lett.* 429:365–369, 1998.

[33] Brooks G, and Hearse DJ. Role of protein kinase C in ischemic preconditioning: player or spectator? *Circ. Res.* 79: 627–630,1996.

[34] Vahlhaus C, Schulz R, Post H, Rose J, and Heusch G. Preventionof ischemic preconditioningonly by com-bined inhibition of protein kinase C and protein ty-rosine kinase in pigs. *J. Mol. Cell Cardiol.* 30: 197–209, 1998.

[35] BauskinAR, Alkalay I, and Ben-Neriah Y.Redoxreg-ulation of a protein tyrosine kinase in the endoplas-mic reticulum. *Cell* 66:684–696, 1991.

[36] Brummel JH, Burkhardt AL, Bolen JB, and Grinstein S.Endogenous reactive oxygen intermediatesactivate tyrosine kinases in human neutrophils. *J. Biol. Chem.* 271: 1455–1461, 1996.

[37] Natarajan V, Vespa S, Shamlal R, Al-Hassani M, Ramasarma T, Ravishiankar HN, and Scribner WN. Tyrosine kinase andcalcium dependentactivation of endothelial cell phospholipase D by diperoxyvandate. *Mol. Cell Biochem.* 183: 113–124, 1998.

[38] Krieg T, Qin Q, Maintosh EC, Cohen MV and Downey JM. Ach and adenosine activate PI-3-kinase in rabbit hearts through transactivation of receptor tyrosine kinase. Am J *Physiol Heart Circ. Physiol.* 283: H2322-H2330, 2002.

[39] Nakashima I, Kato M, Akhand AA, Suzuki H, Takeda K, Hossani K and Kawamoto Y. Redox linked signal transduction pathways for protein tyrosine kinase activation. *Antioxi Redox Signal.* 4: 517-531, 2002.

[40] Seko Y, Tobe K, Takahashi N, Kaburagi Y, Kadoyaki T and Yazaki Y. Hypoxia and Hypoxia reoxygenation activate Src family Tyrosine kinase and p21ras in cultered cardiac myocytes. *Biochem. Biophys. Res. Commun.* 226: 530-535, 1996.

[41] Cheng YH, Pouyssegur J, Courtneidge SA and Van Obberghen –Schilling E. Activation of Src family kinase activity by G protein coupled thrombin receptor in growth –responsive fibroblast. *J. Boil. Chem.* 269: 27372-27377, 1994.

[42] Song C, Vondriska TM, Wang GW, Klein JB, Cao X, Zhang J and Ping P. Molecular conformation dictates signaling module formation : example odf PKC epsilon and Src tyrosine kinase. *Am. J. Physiol. Heart Circ. Physiol.* 292: H1166-H1171, 2002.

[43] Vondriska TM, Zhang j, Song C, Tang XL, Cao X, Baines CP, Pass JM, wang S, Bolli R and ping P. Protein Kinase C epsilon –Src modules direct signal transduction in nitric oxide –induced cardioprotection: Complex formation as a means for cardioprotective signaling. *Circ. Res.* 88: 1306-1313, 2001.

[44] Anderson NG, Maller JI., Tonks NK, and Sturgill TW. Requirement for integration of signals from two distinct phosphorylation pathways for activation of Mapkinase. *Nature (Lond)* 343: 651-653, 1990.

[45] Seger R, and Krebs, EG. The MAPK signaling cascade. *FASEB J.* 9, 726-735, 1995.

[46] Cobb MH, and Goldsmith EJ. How MAP kinases are regulated? *J. Biol. Chem.* 270: 14843-14846, 1995.

[47] Zu YL, Ai Y, Gilchrist A, Maulik N, Watras J, Sha'afi RI, Das DK, Huang CK. High expression and activation of MAP kinase-activated protein kinase 2 in myocardium. *J. Mol. Cell Cardiol.* 29: 2150-2168, 1997.

[48] Maulik N, Watanabe M., Zu YL, Huang CK, Cordis GA, Schley JA, and Das DK.. Ischemic preconditioning triggers the activation of MAP kinases and MAPKAP kinase 2 in rat hearts. *FEBS Lett.* 396, 233-237, 1996.

[49] Kyriakis J.M, Banerjee P, Nikolakaki E, Dai T, Rubie EA, Ahmad MF, Avruch J, and Woodgett JD. The stress-activated protein kinase subfamily of c-Jun kinases. *Nature* 369: 156-160, 1994.

[50] Maulik N.,Yoshida T, Zu YL.,Sato M., Banerjee A, and Das DK. Ischemic preconditioning triggers tyrosine kinase signaling: potential role for MAPKAP kinase 2. *Am. J. Physiol.* 275: H1857-H1864, 1998.

[51] Sato M, Cordis GA, Maulik N, and Das DK. SAPKs regulation of ischemic preconditioning. *Am. J. Physiol* (in press).

[52] Tokube K,Kiyosue T, and AritaM. Openings of ATP-sensitive potassium channels by oxygen free radicals producedbyxanthine oxidase reaction. *Am. J. Physiol.* 271: H478–489,1996

[53] Kuo SS, Saad AH, Koong AC, Hahn GM, nd Giaccia AJ. Potassium channel activation in response to low doses of gammairradiation involves reactive oxygen intermediates in nonexcitatory cells. *Proc. Natl. Acad. Sci. USA* 90: 908–912,1993.

[54] Ambrosio G, Tritto I, Shang UL, Eramo N, D'Andrea D, EliaPP, Scognamiglio AS, ChiarielloM, and Becker LC. Oxygen radicals may exertnegative ionotropic ef-fects via K^{1}-channel stimulation (Abstract). *Circulation* 90:I-428, 1994.

[55] Gross GJ,and Auchampach JA,Blockade ofATP-sen-sitive potassium channels prevents myocardial pre-conditioningin dogs. *Circ. Res.* 70: 223–233,1992.

[56] Toombs CF, Moore TL, Shebushi RJ. Limitation of in-farct size by ischemic preconditioning is reversible with glibenclamide. *Cardiovasc. Res.* 27: 617–622,1993.

[57] Pathak SK, Qian Y-Z, Hess ML, and Kukreja RC. Hy-drogenperoxide preconditions rabbit hearts viaacti-vation of ATP-sensitive potassium channels. (Ab-stract). *Circulation* 92(Suppl I): I–717, 1995.

[58] Yellon DM, Baxter GF, Garcia-Dorado D, Heusch G, and Sumaray MK. Ischemic preconditioning: present position and future directions. *J. Cardiovasc. Res.* 37: 21–33,1998.

[59] Rizvi A, Tang XL, Qiu Y, Xuan YT, Jadoon AK, and Bolli R. Increased protein synthesis is necessary for the late preconditioning against myocardial stunning. *Am. J. Physiol.* 277:H874–H884, 1999.

[60] Marber MS, Latchman DS, Walker DM, and Yellow DM. Cardiac stress protein elevation 24 hours after brief ischemia or heat stress is associated with re-sistance tomyocardial infarction. *Circulation* 88: 1264– 1272, 1993.

[61] Maulik N, and Das DK. Molecular cloning, sequenc-ing and expression analysis by ischemic precondi-tioning and oxidative stress. *Mol. Cell Biochem.* 160/161: 241–247, 1996.

[62] Takano H, Manchikalapudi S, Tang XL, Qiu Y, Rizvi A, JadoonAK, Zhang Q, and Bolli R. Nitricoxide syn-thase is the mediator of late preconditioning against myocardial infarctionin conscious rabbits.*Circulation* 98: 441–449,1998.

[63] TanakaM,FujiwaraH,Yamasaki K, and SasayamaS. Superoxide dismutase and *N*-2-mercaptopropionyl glycine attenuate infarct size limitation effect of is-chaemic preconditioning in the rabbit. *CardiovascRes* Yokota R,HasegawaK, FujiwaraT,and SasayamaS. Ischemic preconditioning elevates cardiac stresspro-tein but does not limit infarct size 24 or 48 h later in rabbits. *Am. J. Physiol.* 267:H1476–H1482, 1994.

[64] Maulik N, Sato M, Price BD, and Das DK. An essen-tial role of NF-*k*B in tyrosine kinase signaling of p38 MAP kinase regulation of myocardial adaptation to ischemia. *FEBS Lett.* 429:365–369, 1998.

[65] YamashitaN, HoshidaS, Taniguchi N,Kuzuya T,and Masatsugu H. A "second window of protection" occurs 24 h after ischemic preconditioning in the rat heart. *J. Mol. Cell Cardiol.* 30:1181–1189,1998.

[66] Schreck R, Rieber P, and Baeuerle PA. Reactive oxy-gen intermediates as apparently widely used mes-sengers in the activation of the NF-*k*B transcription factor and HIV-1. *EMBOJ* 10: 2247–2258, 1991.

[67] Becker J, Mezger W, Courgeon AM, and Best-Belpomme M. On the mechanism of action of H2O2 in the cellular stress. *Free Radic. Res. Commun* 12–13: 455–460, 1991.

[68] Omar R, and Pappolla M. Oxygen free radicals as in-ducersof heatshock proteinsynthesisin cultured hu-man neuroblastoma cells: relevance to neurodegen-erative disease. *Eur. Arch. Psychiatry Clin. Neurosci.* 242: 262–267, 1993.

[69] Bolli R, Jeroudi MO, Patel BS, Dubose CM, Lai EK, Roberts R, and McCay PB. Direct evidence that oxy-gen-derived free radicals contribute to postischemic myocardial disfunction in the intact dog. *Proc. Natl. Acad. Sci. USA* 86: 4695–4699,1989.

[70] Kukreja RC, Kontos MC, Loesser KE, Batra SK, Qian YZ, Gbur CJ Jr, Naseem SA, Jesse RL, and Hess ML. Oxidant stress increases HSP70 mRNA in isolated perfused rat heart. *Am. J. Physiol.* 267: H2213–H2219, 1994.

[71] Maulik N, Goswami S, Galang N, and Das DK. Differential regulation of bcl-2, AP-1 and NFkB on cardiomyocyte apoptosis during myocardial ischemic stress adaptation. *Febs Lett.* 443: 331-336, 1999.

[72] Das DK, Maulik N, Sato M, and Ray P. Reactive oxygen species function as second messenger during ischemic preconditioning of heart. *Mol. Cell Biol.* 1999.

[73] McMahon SB, Monroe JG. Role of primary response genes in generating cellular responses to growth factors. *FASEB J.* 62: 2707-2715, 1992.

[74] Artuc M., Karman D, Jurgovsky, K., and Schadendorf D. Humanmelonoma cell lines selected in vitro displaying various levels of drug resistance against cisplatin, fotemustine, vindestine or etoposide: modulation of protooncogene expression. *Anticancer* 17, 4359-4370, 1997.

[75] Maulik N, Sasaki H, Addya S, and Das DK. Regulation of cardiomyocyte apoptosis by redox sensitive transcription factors. *J. Mol. Cell Cardiol.* (in press).

[76] Hughes PE, Alexi T , Yoshida T, Schreiber SS, and Knusel B. Excitotoxic lesion of rat brain with quinolinic acid induces expression of p53 messenger RNA and protein and p53 inducible genes BAX and GADD-45 brain areas showing DNA fragmentation. *Neuroscience* 74: 1143-1160, 1996.

[77] Miyashita T, Krajewski S, Krajewska, Wang MH, Lin K, Liebermann D, Hoffmann B, and Reed JC. Tumor suppressor p53 is a regulator of bcl-2 and bax gene expression in vitro and in vivo. *Oncogene* 9: 1799-1805, 1994.

[78] Thornberry NA, and Lazebnik Y. Caspases: enemies within. *Science* 281:1312-1316, 1998.

[79] Clark AR, Purdie CA, Harrision DJ, Morris RG, Bird CC, Hooper ML, and Wyllie AH. Thymocyte apoptosis induced by p53-dependent and independent pathways. *Nature* 362: 849-852, 1993.

[80] Ihle JN. STATs: Signal transducers and activators of transcription. *Cell* 84: 331-334, 2002.

[81] Schieffer B, Luchtefeld M, Braun S, Hilfiker A, Hilfiker-Kleiner D and Drexler H. Role of NAD(P)H oxidase in angiotensin II induced JAk/STAT signaling and cytokine induction. *Circ. Res.* 87:1195-1201, 2000.

[82] Mc Whinney CD, Hunt RA, Conrad KM, Dostal DE. and Baker KM. The type I angiotensin II receptor couples to STAT1 and STAT 3 activation through JAK2 kinase in neonatal rat cardiac myocytes. *J. Mol. Cell cardiol.*29: 2513-2524, 1997.
[83] Sheng Z, Knowlton K, Chen J, Hoshijima M, Brown JH and Chien KR. CT-1 inhibition of cardiac myocyte apoptosis via a mitogen activated protein kinase dependent pathway. *J. Biol. Chem.* 272: 5783-5791, 1997.
[84] Negoro S, Kunisada K, Tone E, Funamoto M, Oh H, Kishimoto T, and Yamauchi-Takihara K. Activation OF JAK/ STAT pathway transduces cytoprotective signal in rat acute myocardial infraction. *Cardiovasc. Res.* 47:797-805, 2000.
[85] Stephanou A, Brar BK, Scarabelli TM, Jonassen AK, Yellon DM, Marber MS, Knight RA and Latchman DS. Ischemia-induced Stat-1 expression and activation play a crutial role in cardiomyocyte apoptosis. *J. Biol. Chem.* 273:10002-10008, 2000.
[86] Booli R. The late phase of preconditioning. *Circ. Res.* 87: 972-983, 2000.
[87] Boli R, Manchikalapudi S, Tang XL, Takano H, Qiu Y, Guo Y, Zhang Q and Jadoon AK. The protective effect of late preconditioning against myocardial stunning in conscious rabbits is mediated by nitric oxide synthase. *Circ. Res.* 81: 1094-1107, 1997.
[88] Williams GT., and Smith CA. Molecular regulation of apoptosis: genetic controls on cell death. *Cell* 74:777-779, 1993.
[89] Nunez G., London L, Hockenbery D, Alexander M, Mckearn JP and Korsmeyer SJ. Deregulated Bcl-2 gene expression selectively prolongs survival of growth factor-deprived hemopoietic cell lines. *J. Immunol.* 144:3602-3610, 1999.
[90] Olivetti G, Abbi R, Quaini F, Kajstura J, Cheng W, Nitahara JA, Quaini E, Loreto C Di, Beltrami CA, Krajewski S, Reed J C, Anversa P. Apoptosis in the failing heart, *New Eng. J. Med.* 336:1131-1141, 1997.
[91] Arner ESJ, and Holmgren A. Physiological functions of thioredoxin and thioredoxin reductase. *Eur. J. Biochem.* 267: 6102–6109, 2000.
[92] Prieto-Alamo MJ, Jurado J. and Gallardo-Madueno R. Transcriptional regulation of glutaredoxin and thioredoxin pathways and related enzymes in response to oxidative stress. *J. Biol. Chem.* 275 : 13398–13405, 2000.
[93] Nakamura H, Vaage J. and Valen G. Measurements of plasma glutaredoxin and thioredoxin in healthy volunteers and during open-heart surgery. *Free Radicals Biol. Med.* 24:1176–1186, 1998.
[94] Kihlstrom M. Protection effect of endurance training against reoxygenation-induced injuries in rat heart. *J. Appl. Physiol.* 68:1672–1678, 1990.
[95] Isowa N, Toshimura T. and Kosaka S. Human thioredoxin attenuates hypoxia-reoxygenation injury of murine endothelial cells in a thiol-free condition. *J. Cell Physiol.* 182: 33–40, 2000.
[96] Turoczi T, Chang VW, Engelman RM, Maulik N, Ho YS. and Das DK. Thioredoxin redox signaling in the ischemic heart: an insight with transgenic mice overexpressing Trx1. *J. Mol. Cell. Cardiol.* 35: 695–704, 2003.

In: Oxidative Stress: Clinical and Biomedical Implications ISBN: 978-1-60021-888-0
Editors: B. M. Matata, M. M. Elahi, pp. 85-102

Chapter VI

Pyruvate: A Physiologic Energy Substrate and Antioxidant for Cardiopulmonary Bypass

Devin C. Flaherty*[†], Robert T. Mallet[†] and Albert H. Olivencia-Yurvati *[‡]
Departments of ‡ Surgery and †Integrative Physiology
University of North Texas Health Science Center
Fort Worth, Texas, USA

Abstract

Cardiopulmonary bypass (CPB) is necessary for many complex cardiac surgeries including valve replacement, correction of congenital cardiac anomalies and difficult coronary artery revascularizations. Contact of blood with foreign surfaces within the extracorporeal circuit, hypothermia, microemboli, ischemia, and surgical trauma elicit a massive systemic inflammatory response. If unchecked, this inflammation inflicts end-organ damage culminating in significant postoperative morbidity and mortality. Reactive oxygen and nitrogen metabolites are generated upon reperfusion and released by activated neutrophils, initiating the inflammatory cascade. Cardioplegic arrest and reperfusion impose significant oxidative stress on cardiac tissue, which inactivates myocardial enzymes and produces cardiac edema. Pyruvate, a naturally occuring metabolic intermediate, energy substrate and powerful antioxidant, blunts intestinal and hepatic inflammation during endotoxic and septic shock. The antioxidant actions of pyruvate are proposed to interrupt systemic and intramyocardial inflammatory cascades, ameliorate injury to the myocardium and, thus, enhance post-surgical recovery of cardiac mechanical function as well as stabilize atrial rhythm. This proposed mechanism has been tested in swine subjected to cardioplegic arrest on CPB. Relative to control cardioplegia, pyruvate-fortified cardioplegia prevented oxidative stress and edema, bolstered endogenous antioxidant defenses, protected oxidant-sensitive enzymes and increased the myocardial energy state. Results of these investigations, as well as encouraging results from human studies, provide crucial empirical support for the use of pyruvate in cardioplegia to mitigate post-surgical inflammation and cardiac dysfunction.

Keywords: cardioplegia, cardiopulmonary bypass, inflammation, peroxynitrite, superoxide, pyruvate

Abbreviations

CPB: cardiopulmonary bypass;
dP/dt: rate of change of left ventricular intracavitary pressure;
GSH: glutathione;
GSSG: glutathione disulfide;
IL: interleukin;
LVSWI: left ventricular stroke work index;
MMP: matrix metalloproteinase;
NAC: *N*-acetylcysteine;
NOS: nitric oxide synthase;
RONS: reactive oxygen and nitrogen species;
TNF: tumor necrosis factor

Introduction

Cardiac surgical procedures often require an arrested, quiescent heart with diversion of blood from the surgical field, enabling the surgeon to perform delicate operative techniques. To arrest the heart, depolarizing cardioplegia solutions are introduced into the myocardium anterogradely via the coronary circulation, retrogradely via the coronary sinus and directly via vein grafts. Cardiopulmonary bypass (CPB) sustains peripheral organ perfusion during cardioplegic arrest. Pulsatile perfusion is provided by extracorporeal support from a heart-lung machine (Figure 1). In 2004, over 500,000 coronary artery bypass grafting procedures were performed in the United States with approximately 77% requiring CPB [59].

Unfortunately, cardiopulmonary bypass is not entirely benign. A host of factors, including blood contact with the artificial surfaces of the extracorporeal circuit, the use of heparin and protamine to control clotting, hypothermia to slow organ metabolism, anesthesia and surgical trauma, re-infusion of blood collected from the chest cavity, sepsis and endotoxemia due to intestinal mucosal hyperpermeability, and formation of microemboli in the blood combine to provoke a massive systemic inflammatory response. This systemic inflammation produces a constellation of post-surgical morbidities, collectively termed *'post-pump' syndrome* [4, 43, 55], including pulmonary insufficiency and acute respiratory failure [12, 72], renal insufficiency [4, 86], neurocognitive impairment [58], and cardiodepression [9, 61]. These complications lengthen hospitalization, increase healthcare costs and produce increased morbidity and mortality [73].

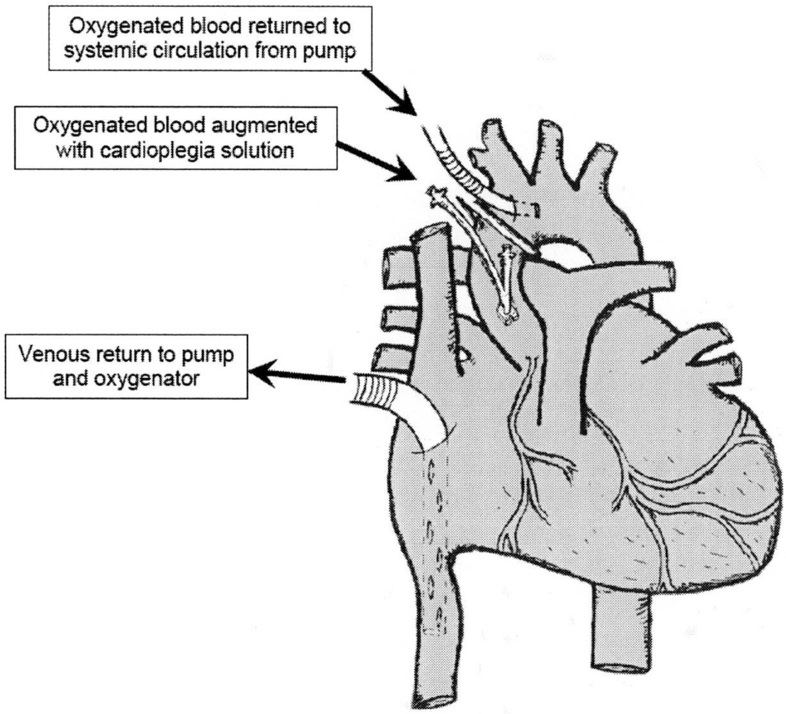

Figure 1. Cannulation of the heart and great vessels for cardiopulmonary bypass.

Pro-Inflammatory Effects of Reactive Oxygen and Nitrogen Intermediates

Cardiopulmonary bypass generates chemically aggressive oxygen and nitrogen derivatives, including superoxide ($\cdot O_2^-$), hydrogen peroxide (H_2O_2), hydroxyl radical ($\cdot OH$), nitric oxide (NO) and peroxynitrite ($ONOO^-$). These are collectively termed reactive oxygen and nitrogen species (RONS), intensify the inflammatory reaction and directly damage tissue by modifying phospholipids, enzymes and structural proteins. Massive amounts of RONS are generated upon reperfusion of the myocardium following aortic crossclamp release [19, 30]. Suboptimal perfusion of other organs during bypass generates RONS systemically, and activated neutrophils release RONS [22, 80]. Hemolysis releases hemoglobin, which catalyzes RONS formation [14]. CPB provokes systemic RONS formation in adult patients [14, 15] and children [63]. Numerous biomolecular signatures of RONS were detected in these studies, including the lipid peroxidation products malondialdehyde [14] and 8-isoprostane [54, 77], and protein modifications including carbonyls [14], and tyrosine nitration [54, 77].

Reintroduction of O_2 into myocardium following cross-clamp release causes formation of $\cdot O_2^-$, which irreversibly condenses with NO to form $ONOO^-$ [49]. NO and $ONOO^-$ play central roles in post-bypass inflammation and organ injury. CPB and inflammatory cytokines activate NO synthase (NOS), the source of NO. Myocardial NOS activity increased following CPB in dogs [53] and patients [51, 54]. Two NOS isoforms reside in myocardium: the constitutive, Ca^{2+}-dependent endothelial NOS (eNOS) and the Ca^{2+}-independent, inducible isoform (iNOS), which is activated by metabolic stress. Myocardial iNOS activity increased several fold in dogs post-bypass, [52] but changes in eNOS activity were less consistent [52, 53]. Pro-inflammatory cytokines activate iNOS and, thus, increase NO formation in myocardium; [66] moreover, $ONOO^-$ increases interleukins 6 (IL-6) and 8 (IL-8) formation, [29] which may perpetuate NOS activation. $ONOO^-$ activates neutrophils [57], promotes platelet aggregation [57], and increases circulating activities of pro-inflammatory cytokines [29]. NO and $ONOO^-$ activate matrix metalloproteinases (MMPs) [26] accordingly, the NO scavenger AMD6221 abrogated MMP activation and lowered post-bypass inotropic support requirements in dogs [5].

The central role of RONS in systemic inflammation raises the prospect that antioxidant interventions may dampen inflammation and ameliorate organ dysfunction following CPB. However, evidence of salutary effects of pharmacological antioxidants is equivocal. Intravenous and cardioplegic administration of the sulfhydryl antioxidant *N*-acetylcysteine (NAC) during bypass improved recovery of preload-recruitable stroke work in dogs [24] and prevented 8-isoprostane and 3-nitrotyrosine accumulation in left ventricular myocardium of patients [77], but NAC failed to prevent postoperative renal dysfunction in patients undergoing coronary artery revascularization on bypass [11]. Dietary supplementation with ascorbic acid and α-tocopherol for 7-10 days before surgery did not prevent myocardial injury, improve myocardial perfusion, or stabilize electrocardiographic rhythm in CPB patients [61].

Pyruvate: Metabolic Protector of the Heart

Pyruvate, a natural metabolic fuel and antioxidant in myocardium and other tissues, exerts a variety of cardioprotective actions when provided at supraphysiological concentrations [44, 48]. Pyruvate increases cardiac contractile performance and myocardial energy state [44, 46, 75], bolsters endogenous antioxidant systems [45, 47, 74], and protects myocardium from ischemia-reperfusion injury and oxidant stress [10, 45, 48]. By enhancing myocardial inotropic state, supra-physiological concentrations (*c.* 2-10 mM) of exogenous pyruvate increase left ventricular developed pressure, dP/dt, and cardiac power [10, 46, 75]. Recent studies have demonstrated pyruvate enhancement of contractile performance in human ventricular muscle preparations [28, 33] and in patients with heart failure [31, 32]. These inotropic effects have been ascribed to pyruvate's enhancement of cytosolic ATP phosphorylation potential and Gibbs free energy of ATP hydrolysis (ΔG_{ATP}) [10, 46] the thermodynamic energy source for contractile work. Unlike catecholamines, pyruvate does not elevate heart rate [10, 60] and, therefore, does not appreciably increase cardiac internal work. Consequently, pyruvate enhancement of cardiac performance neither increases myocardial O_2 demand nor depletes the heart's energy reserves [46]. Thus, pyruvate could address the objective set forth by Weisel *et al.* [82]: "*...improvements in myocardial protection should attempt to stimulate mitochondrial energy production and preserve adenine nucleotide precursors.*"

Pyruvate Acts as a Potent Metabolic Antioxidant

Aside from being a natural physiologic fuel in the heart, pyruvate also has antioxidant properties [47, 48]. Its α-keto-carboxylate structure enables pyruvate to detoxify peroxides in a direct, non-enzymatic chemical reaction in which peroxides are reduced to their conjugate alcohols and pyruvate decomposes to acetate and CO_2. [17, 18, 20] Indeed, pyruvate restored contractile function of hearts exposed to 100 μM H_2O_2, [45] a concentration which has been detected in plasma of CPB patients [13] and in abscess fluid of inflamed tissues [55]. Pyruvate detoxifies $ONOO^-$ in a similar fashion, yielding NO_2^-, CO_2 and acetate (Figure 2) [47, 81].

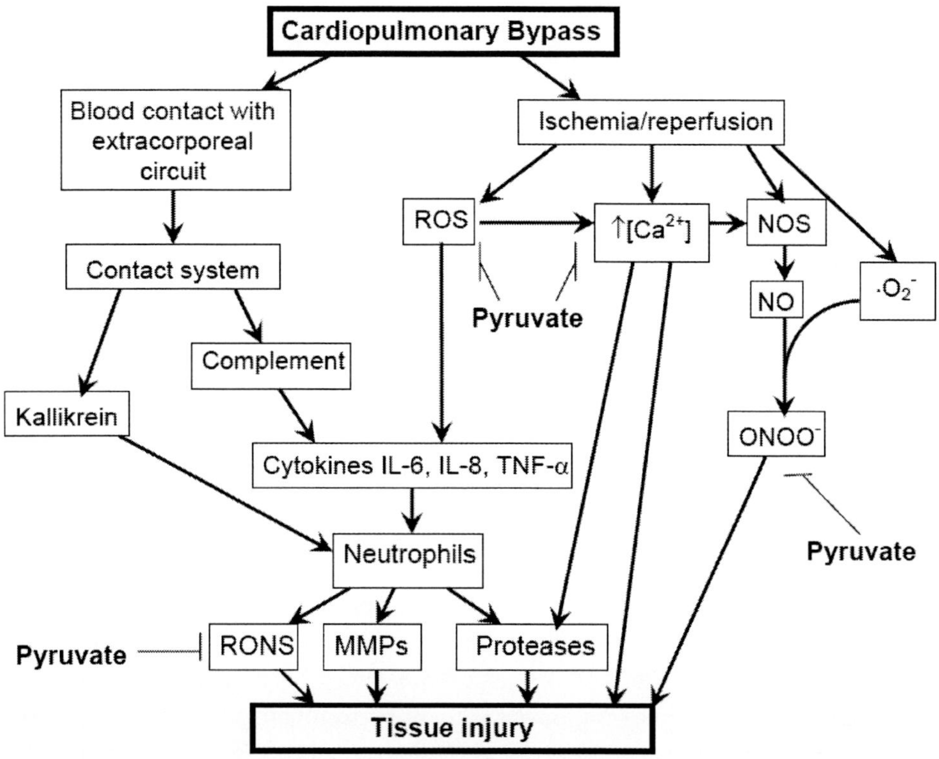

Figure 2. Inflammatory cascade: components and pyruvate intervention. Pyruvate is proposed to attenuate tissue injury by detoxifying reactive oxygen and nitrogen species (RONS) and improving cellular Ca^{2+} homeostasis, thereby suppressing intracellular Ca^{2+} overload. IL: interleukin; MMPs: matrix metalloproteinases; NO: nitric oxide; NOS: nitric oxide synthase; $ONOO^-$: peroxynitrite; TNF-α: tumor necrosis factor-α.

Mitochondrial metabolism may indirectly contribute to pyruvate's antioxidant character. Malic enzyme and pyruvate carboxylase condense pyruvate and CO_2 to generate the Krebs cycle intermediates malate and oxaloacetate, respectively [16, 37]. These anaplerotic mechanisms increase steady-state citrate content. Citrate increases NADPH formation by diverting glycolytic flux into the hexose monophosphate shunt and generating isocitrate for $NADP^+$-dependent isocitrate dehydrogenase [47]. NADPH provides reducing equivalents to maintain the redox state of GSH, the principal intracellular antioxidant [65] and source of reducing power to detoxify peroxides and $ONOO^-$. In accordance with this scenario, pyruvate increased myocardial citrate content and $NADPH/NADP^+$ ratio, and restored GSH redox state depleted by ischemia-reperfusion [74] H_2O_2 [45], and cardiac arrest [39, 67].

Recent studies have demonstrated several benefits of pyruvate's antioxidant actions. Pyruvate dampened the post-ischemic burst of $\cdot O_2^-$ formation in isolated guinea-pig hearts [6], and ameliorated H_2O_2-induced contractile impairment of isolated rat [42] and guinea-pig [45] hearts. Pyruvate-fortified crystalloid solution preserved α-tocopherol content and increased oxyradical neutralizing capacity of isolated rat hearts [21]. Pyruvate potentiated β-adrenergic enhancement of cardiac function in post-ischemic guinea-pig hearts [74, 75] and in ventricular muscle from failing human hearts [33]. This β-adrenergic potentiation lowered catecholamine dosages required to augment cardiac performance.

Depletion of cellular energy reserves during myocardial ischemia, and the massive burst of RONS formation and intracellular Ca^{2+} overload following reperfusion, inflicts lethal tissue injury culminating in myocardial infarction. By increasing energy reserves, neutralizing RONS and improving cellular Ca^{2+} homeostasis, pyruvate may mitigate ischemic injury [44, 48]. A recent investigation tested pyruvate's ability to limit infarction in *in situ* pig hearts [40] subjected to occlusion and reperfusion of the left anterior descending coronary artery. Left atrial infusion of pyruvate from 30 min before occlusion until reperfusion reduced infarct volume from 66 ± 1% to 49 ± 3% of the ischemic territory. When pyruvate treatment was extended until 60 min reperfusion, infarcts were further limited to 30 ± 2% of the ischemic region.

Inflammatory Sequelae of Cardiopulmonary Bypass

Cardiopulmonary bypass activates several cytokines, most notably IL-6, IL-8 and tumor necrosis factor alpha (TNF-α) [3, 36, 71]. These proteins, produced by vascular endothelium [36, 43, 85], white blood cells [22, 36], and fibroblasts [22] function as chemoattractants which recruit neutrophils to target tissues [3, 27, 36]. Pro-inflammatory cytokines have been implicated in the pathogenesis of post-bypass cardiodepression [76], lung injury [8] and atrial fibrillation [25].

Neutrophils extravasation into the interstitium and invasion of tissue parenchyma culminates the inflammatory cascade. Neutrophils are armed with proteases including elastase, cathepsin G, lysozymes and MMPs which, when secreted into the target tissue, degrade the extracellular matrix and damage cellular proteins [23]. To penetrate the tissue, neutrophils secrete MMP-2 and MMP-9, which degrade the extracellular matrix. These proteases have been detected in canine myocardium, lung and brain [51, 52], and porcine lung [12] after CPB. Activated neutrophils generate and release cytotoxic amounts of NO and $\cdot O_2^-$ [22, 80] which condense in a direct, diffusion-limited reaction to form the powerful reactive nitrogen intermediate $ONOO^-$ [7, 49].

Anti-Inflammatory Properties of Pyruvate and Ethyl Pyruvate

Recent studies have demonstrated the anti-inflammatory capabilities of pyruvate and its derivative ethyl pyruvate. Pyruvate stabilized cardiac function, prevented hepatic cell death and maintained GSH/GSSG during hemorrhagic shock in pigs, [56] and improved survival and increased arterial pressure in rats subjected to hemorrhagic shock and recovery [69]. Ringers ethyl pyruvate solution suppressed systemic inflammation in mice during resuscitation from hemorrhagic shock [83]. Pyruvate and ethyl pyruvate decreased intestinal mucosal injury and ameliorated mucosal hyperpermeability in mice [78] and rats [68] subjected to mesenteric ischemia/reperfusion. Moreover, ethyl pyruvate suppressed increases in plasma TNF and hepatic TNF mRNA abundance during intestinal ischemia in mice, [78] dampened the surge in plasma TNF and improved survival in a murine sepsis model [79], and suppressed lipopolysaccharide-induced IL-6 secretion, nitric oxide formation and expression of IL-6 and iNOS mRNA in cultured murine lymphocytes [70]. Unmodified pyruvate wasn't tested in the cytokine studies; moreover, the effect of pyruvate or its derivatives on neutrophil activity and MMPs is unknown. The addition of an ethyl moiety to pyruvate via an ester bond is intended to prevent pyruvate degradation before it enters cells, where esterases cleave this bond and release active pyruvate. The lower water solubility of ethyl pyruvate *vs.* its parent compound, and the potentially detrimental effects of ethanol released by esterases, may limit the concentrations of ethyl pyruvate that can be administered safely to anesthetized patients.

To assess pyruvate's effect on infiltration of inflammatory cells into pulmonary tissue, samples were taken from porcine lung 4 h post-bypass (Figure 3). Pre-bypass lung tissue (Figure 3A) contains no neutrophils or lymphocytes, and the alveolar spaces are free of inflammatory cells.

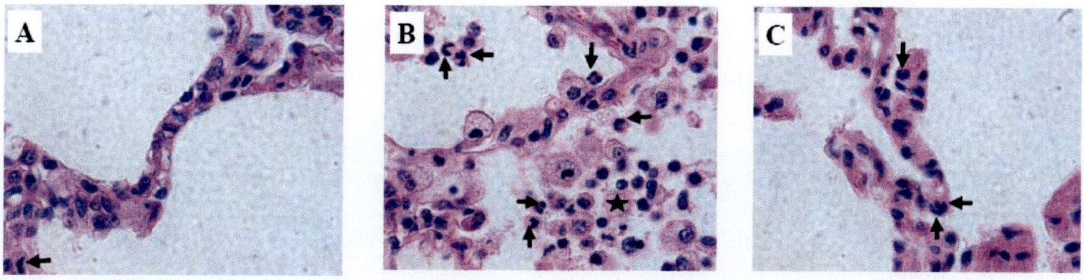

Figure 3. Pyruvate dampened pulmonary inflammatory cell infiltration in pigs on cardiopulmonary bypass. Hematoxylin-eosin staining of lung tissue taken from the left cardiac lobe (A) before instrumentation, (B) at the end of the 5 h 30 min sham protocol, and (C) after CPB with pyruvate-enhanced cardioplegia and 4 h recovery. Arrows: neutrophils; star: infiltrating lymphocyte.

Lung tissue from a sham control pig (Figure 3B) shows an increase in neutrophils (arrows) and lymphocyte infiltration (star). Many inflammatory cells are seen in the alveolar spaces, indicating parenchymal injury. Lung tissue from a pyruvate bypass pig (Figure 3C) contains fewer neutrophils and lymphocytes and less parenchymal injury than the sham control pig, suggesting that pyruvate may have suppressed inflammation elicited by thoracotomy and cannulation.

Inflammation and Atrial Fibrillation

A frequent and potentially harmful complication of cardiopulmonary bypass, atrial fibrillation occurs in approximately 30-40% of cases [34, 62]. Atrial fibrillation prolongs hospitalization [5, 50], increases post-operative morbidity and mortality [2, 34] and is the most common reason for readmission after cardiac surgery [41]. Peri- and post-operative inflammation could be the main trigger of atrial fibrillation [1, 34]. Anti-inflammatory corticosteroid therapy decreased incidence and duration of atrial fibrillation in patients [64, 84] and dogs [35] recovering from CPB; however, corticosteroids inhibit wound repair, which limits their use in patients recovering from cardiothoracic surgery. The post-bypass increases in atrial $ONOO^-$ concentrations, as well as increased pro-inflammatory cytokines within cardiac tissue are both associated with increased incidence of atrial fibrillation [7].

By detoxifying $ONOO^-$, pyruvate may limit the incidence of post-operative arrhythmias, specifically atrial fibrillation. Pyruvate may also truncate the inflammatory cascades that occur within the heart. Both of these effects ameliorate the pro-arrhythmic environment within inflamed cardiac tissue and lower the likelihood of a second cardiac event that may permanently compromise cardiac performance.

Clinical Trial: Pyruvate-Enhanced Cardioprotection during Cardiopulmonary Bypass

A randomized clinical trial tested whether a pyruvate-fortified cardioplegia solution provided cardioprotection superior to lactate-based solutions in patients undergoing elective coronary revascularization on CPB [61]. Fifteen patients received conventional lactated Ringer's-based cardioplegia, and another 15 patients received pyruvate-fortified cardioplegia. These crystalloid solutions were mixed with 4 volumes of whole blood and administered antegradely into the aortic root and retrogradely into the coronary sinus to arrest the heart.

Pre-surgical left ventricular function (stroke work) was similar in the two groups, as was function upon separation from bypass (Figure 4) [61]. Function subsequently fell in all 15 lactate cardioplegia patients, reaching a nadir at 4 h post-bypass before gradually recovering; this pattern exemplifies the temporary cardiodepression typically observed post-bypass [9]. In striking contrast, stroke work sharply increased over the first 4 h post-bypass in all 15 patients treated with pyruvate cardioplegia, and remained well above the lactate cardioplegia value throughout the first 12 h recovery. Concomitantly ten of the lactate cardioplegia patients but only 4 of the pyruvate patients required inotropic support for sluggish post-bypass cardiac performance [61].

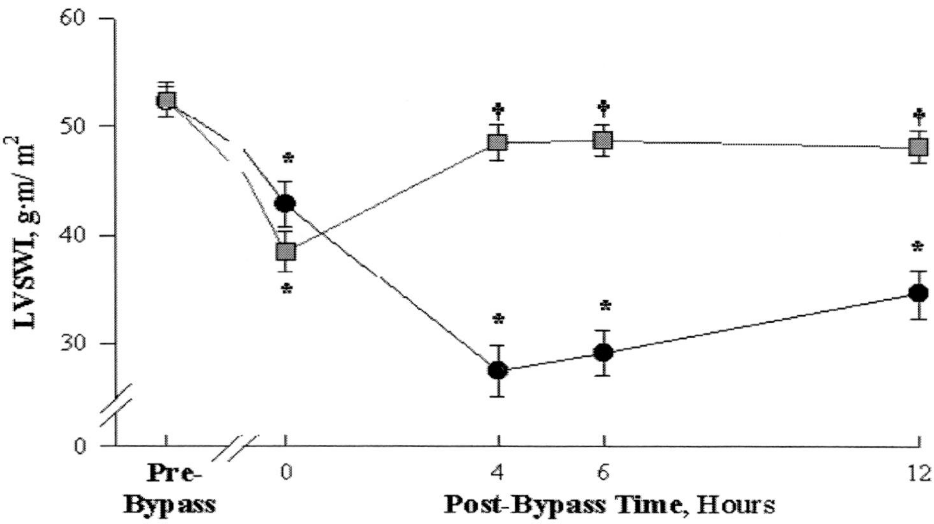

Figure 4. Left ventricular contractile performance pre- and post-CPB. Left ventricular stroke work index (LVSWI) was measured at pre-bypass baseline and at 0-12 h post-bypass in patients receiving lactate- (circles) or pyruvate-fortified (squares) cardioplegia solution during bypass. Values are means ± SEM; n = 15 per group; *P < 0.05 vs. pre-bypass; †P < 0.05 vs. lactate group. Reprinted from reference 61 with the generous permission of Elsevier Science Publishers.

Pyruvate cardioplegia prevented the increased release of cardiac proteins which occurred in the lactate group post-bypass, indicating that pyruvate mitigated myocardial ischemic injury during cardiac arrest [61]. Importantly, patients receiving pyruvate-fortified cardioplegia were discharged 5.2 ± 0.1 days post-surgery versus 6.3 ± 0.3 days in the lactate cardioplegia group. Thus, the use of pyruvate- versus lactate-fortified cardioplegia obviated β-adrenergic inotropic support and hastened postsurgical recovery, thereby shortening postsurgical hospitalization by more than a day and lowering costs by >$3,000 per patient [61]. It should be noted that these persistent benefits were produced by temporary administration of pyruvate as a cardioplegia component during 60 min of aortic cross-clamp; pyruvate was rapidly cleared from the heart upon reperfusion with normal whole blood.

Impact of Pyruvate-Fortified Cardioplegia on Oxidative Stress and Myocardial Energy State

The remarkable results of the clinical trial [61] prompted investigations of pyruvate's cardioprotective mechanisms [38, 39]. The pig was selected as the animal model for these invasive studies, based on its physiological and metabolic similarity to human, and its size, which permitted withdrawal of sufficient blood to prepare cardioplegia and to reperfuse the heart following arrest. To test the hypothesis that pyruvate mitigates oxidative stress and preserves energy state in cardioplegically arrested myocardium, *in situ* swine hearts were arrested for 60 min with a 4:1 mixture of blood and crystalloid cardioplegia solution containing 188 mM glucose alone (control) or with additional *c.* 24 mM lactate or pyruvate, then reperfused for 3 min with cardioplegia-free blood [38]. Glutathione (GSH), glutathione disulfide (GSSG), and energy metabolites phosphocreatine, creatine, and inorganic phosphate were measured in myocardium snap frozen at 45 min arrest and 3 min reperfusion to determine antioxidant glutathione redox state (GSH/GSSG) and phosphocreatine phosphorylation potential, a measure of myocardial free energy state [10, 46]. Coronary sinus 8-isoprostane, a product of lipid peroxidation, indexed oxidative stress.

Relative to control or lactate cardioplegia, pyruvate cardioplegia lowered 8-isoprostane release by approximately 40% during arrest. Lactate and pyruvate but not control cardioplegia dampened the surge of 8-isoprostane release following reperfusion. Pyruvate doubled GSH/GSSG *vs.* lactate cardioplegia during arrest, but GSH/GSSG fell in all 3 groups following reperfusion. Although phosphorylation potential was maintained in all 3 groups during arrest, pyruvate cardioplegia prevented the sharp declines in phosphorylation potential that occurred upon reperfusion in the lactate and control groups [38]. Thus, pyruvate cardioplegia mitigated oxidative stress during cardioplegic arrest and enhanced myocardial energy state upon reperfusion.

Pyruvate Dampens Oxidative Stress during Reperfusion of Cardioplegia-Arrested Myocardium

Pyruvate-fortified cardioplegia increased GSH redox state during cardiac arrest but not during the first few minutes of reperfusion, a time of intense oxyradical formation leading to enzyme inactivation and cardiac injury. Accordingly, we tested whether selective intracoronary administration of pyruvate upon reperfusion after cardioplegic arrest can protect myocardial antioxidant defenses and prevent inactivation of oxyradical-sensitive metabolic enzymes [39]. *In situ* swine hearts were arrested with glucose-fortified 4:1 blood:crystalloid cardioplegia, and then reperfused for 3 min with cardioplegia-free blood with or without *c.* 12 mM pyruvate. Left ventricular myocardium was biopsied at 45 min arrest and 3 min reperfusion for measurements of oxidative stress, energy and antioxidant metabolites, and oxyradical-sensitive enzyme activities.

8-Isoprostane content sharply increased upon reperfusion, coincident with a 70% decline in GSH redox state (Figure 5). Aconitase and glucose 6-phosphate dehydrogenase activities fell during arrest; creatine kinase and phosphofructokinase were inactivated upon reperfusion. Pyruvate suppressed 8-isoprostane formation, maintained glutathione redox state (Figure 5), and enhanced phosphocreatine phosphorylation potential, a measure of myocardial energy state, during reperfusion. Pyruvate reactivated creatine kinase, a portion of which resides in the mitochondria, and aconitase, an entirely mitochondrial enzyme, but did not protect the exclusively cytosolic enzymes glucose 6-phosphate dehydrogenase and phosphofructokinase. These results localize pyruvate's main antioxidant effect to the mitochondria, suggesting that anaplerotic flux of pyruvate into the Krebs cycle may have been the predominant antioxidant mechanism of pyruvate during myocardiaal reperfusion following cardioplegic arrest.

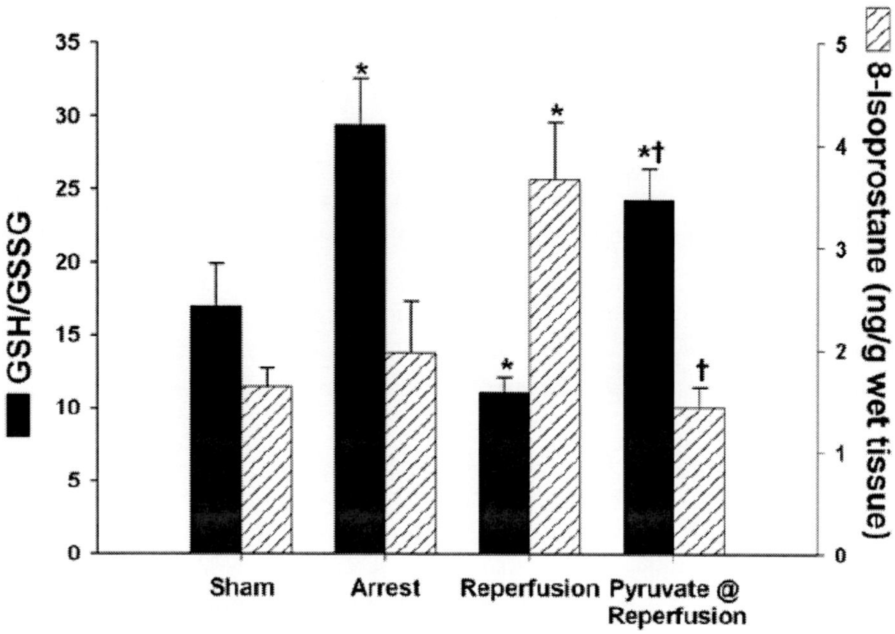

Figure 5. Myocardial antioxidant state and oxidative stress during cardioplegic arrest and reperfusion. Glutathione redox state (GSH/GSSG) and 8-isoprostane content were measured in left ventricular myocardium sampled at 45 min arrest (n = 8), at 3 min reperfusion with (n = 6) and without (n = 7) pyruvate infusion, and at 105 min post-sternotomy in non-arrested sham hearts (n = 6). Values are means ± SEM. *P < 0.05 vs. sham; †P < 0.05 vs. reperfusion. Reprinted from reference 39 with the generous permission of the Society of Thoracic Surgeons.

Conclusion

The studies summarized herein show pyruvate to be an effective, naturally occurring antioxidant capable of protecting myocardium from oxidative stress and its sequelae. Importantly, pyruvate did not produce toxic side effects at the concentrations and dosages utilized in these studies. Collectively, these results support the therapeutic application of pyruvate-fortified cardioplegia to prevent cardiac injury following cardioplegic arrest.

Acknowledgements

This work was supported by grants from the Osteopathic Heritage Foundation (G70283) and the U.S. National Heart, Lung and Blood Institute (HL-71684). DCF was supported by a fellowship from The Graduate School of Biomedical Sciences of the University of North Texas Health Science Center.

References

[1] Abdelhadi RH, Gurm HS, Van Wagoner DR, Chung MK. Relation of an exaggerated rise in white blood cells after coronary bypass or cardiac valve surgery to development of atrial fibrillation postoperatively. *Am. J. Cardiol.* 93: 1176-1178, 2004.

[2] Almassi GH, Schowalter T, Nicolosi AC, Aggarwal A, Moritz TE, Henderson WG, Tarazi R, Shroyer AL, Sethi GK, Grover FL, Hammermeister KE. Atrial fibrillation after cardiac surgery: a major morbid event? *Ann. Surg.* 226: 501-511, 1997.

[3] Anselmi A, Abbate A, Girola F, Nasso G, Biondi-Zoccai GGL, Possati G, Gaudino M. Myocardial ischemia, stunning, inflammation, and apoptosis during cardiac surgery: a review of evidence. *Eur. J. Cardiothorac. Surg.* 25: 304-311, 2004.

[4] Antunes PE, Prieto D, de Oliveira JF, Antunes MJ. Renal dysfunction after myocardial revascularization. *Eur. J. Cardiothorac Surg.* 25: 597-604, 2004.

[5] Aranki SF, Shaw DP, Adams DH, Rizzo RJ, Couper GS, VanderVliet M, Collins JJ Jr., Cohn LH, Burstin HR. Predictors of atrial fibrillation after coronary artery surgery. Current trends and impact on hospital resources. *Circulation* 94: 390-397, 1996.

[6] Bassenge E, Sommer O, Schwemmer M, Bünger R. Antioxidant pyruvate inhibits cardiac formation of reactive oxygen species through changes in redox state. *Am. J. Physiol. Heart Circ. Physiol.* 279: H2431-H2438, 2000.

[7] Beckman JS, Beckman TW, Chen J, Marshall PA, Freeman BA. Apparent hydroxyl radical production by peroxynitrite: implications for endothelial injury from nitric oxide and superoxide. *Proc. Natl. Acad. Sci. USA* 87: 1620-1624, 1990.

[8] Ben-Abraham R, Weinbroum AA, Dekel B, Paret G. Chemokines and the inflammatory response following cardiopulmonary bypass – a new target for therapeutic intervention? *Paed. Anesth.* 13: 655-661, 2003.

[9] Breisblatt WM, Stein KL, Wolfe CJ, Follansbee WP, Capozzi J, Armitage JM, Hardesty RL. Acute myocardial dysfunction and recovery: a common occurrence after coronary bypass surgery. *J. Am. Coll Cardiol.* 15: 1261-1269, 1990.

[10] Bünger R, Mallet RT, Hartman DA. Pyruvate-enhanced phosphorylation potential and inotropism in normoxic and postischemic isolated working heart: near-complete prevention of reperfusion contractile failure. *Eur. J. Biochem.* 180: 221-233, 1989.

[11] Burns KAE, Chu MWA, Novick RJ, Fox SA, Gallo K, Martin CM, Stitt LW, Heidenheim AP, Myers ML, Moist L. Perioperative *N*-acetylcysteine to prevent renal dysfunction in high-risk patients undergoing CABG surgery. *J. Am. Med. Assoc.* 294: 342-350, 2005.

[12] Carney DE, Lutz CJ, Picone AL, Gatto LA, Ramamurthy NS, Golub LM, Simon SR, Searles B, Paskanik A, Snyder K, Finck C, Schiller HJ, Nieman GF. Matrix metalloproteinase inhibitor prevents acute lung injury after cardiopulmonary bypass. *Circulation* 100: 400-406, 1999.

[13] Cavarocchi NC, England MD, Schaff HV, Russo P, Orszulak TA, Schnell WA, O'Brien JF, Pluth JR. Oxygen free radical generation during cardiopulmonary bypass: correlation with complement activation. *Circulation* 74 (suppl III): III-130-III-133, 1986.

[14] Christen S, Finckh B, Lykkesfeldt J, Gessler P, Frese-Schaper M, Nielsen P, Schmid ER, Schmitt B. Oxidative stress precedes peak systemic inflammatory response in pediatric patients undergoing cardiopulmonary bypass operation. *Free Radic Biol Med* 38: 1323-1332, 2005.

[15] Clermont G, Vergely C, Jazayeri S, Lahet JJ, Goudeau JJ, Lecour S, David M, Rochette L, Girard C. Systemic free radical activation is a major event involved in myocardial oxidative stress related to cardiopulmonary bypass. *Anesthesiology* 96: 80-87, 2002.

[16] Comte B, Vincent G, Bouchard B, Jetté M, Cordeau S, Des Rosiers C. A ^{13}C mass isotopomer study of anaplerotic pyruvate carboxylation in perfused rat hearts. *J. Biol. Chem.* 272: 26125-26131, 1997.

[17] Constantopoulos G, Barranger JA. Nonenzymatic decarboxylation of pyruvate. *Anal. Biochem.* 139: 353-358, 1984.

[18] Crestanello JA, Kamelgard J, Whitman GJR. The cumulative nature of pyruvate's dual mechanism for myocardial protection. *J. Surg. Res.* 59: 198-204, 1995.

[19] Davies SW, Underwood SM, Wickens DG, Feneck RO, Dormandy TL, Walesby RK. Systemic pattern of free radical generation during coronary bypass surgery. *Br. Heart J.* 64: 236-240, 1990.

[20] DeBoer LWV, Bekx PA, Han L, Steinke L. Pyruvate enhances recovery of rat hearts after ischemia and reperfusion by preventing free radical generation. *Am. J. Physiol. Heart Circ Physiol* 265: H1571-H1576, 1993.

[21] Dobsak P, Courderot-Masuyer C, Zeller M, Vergely C, Laubriet A, Assem M, Eicher J-C, Teyssier J-R, Wolf J-E, Rochette L. Antioxidative properties of pyruvate and protection of the ischemic rat heart during cardioplegia. *J. Cardiovasc. Pharmacol.* 34: 651-659, 1999.

[22] Downing SW, Edmunds LH Jr. Release of vasoactive substances during cardiopulmonary bypass. *Ann. Thorac. Surg.* 54: 1236-1243, 1992.

[23] Edmunds LH Jr. Inflammatory response to cardiopulmonary bypass. *Ann. Thorac. Surg.* 66: S12-S16, 1998.

[24] Fischer UM, Cox CS Jr., Allen SJ, Stewart RH, Mehlhorn U, Laine GA. The antioxidant *N*-acetylcysteine preserves myocardial function and diminishes oxidative stress after cardioplegic arrest. *J. Thorac. Cardiovasc Surg.* 126: 1483-1488, 2003.

[25] Gaudino M, Andreotti F, Zamparelli R, Di Castelnuovo A, Nasso G, Burzotta F, Iacoviello L, Donati MB, Schiavello R, Maseri A, Possati G. The -174G/C interleukin-6 polymorphism influences postoperative interleukin-6 levels and postoperative atrial fibrillation. Is atrial fibrillation an inflammatory complication? *Circulation* 108 (suppl II): II-195-II-199, 2003.

[26] Gu Z, Kaul M, Yan B, Kridel SJ, Cui J, Strongin A, Smith JW, Liddington RC, Lipton SA. S-nitrosylation of matrix metalloproteinases: signaling pathway to neuronal cell death. *Science* 297: 1186-1190, 2002.

[27] Gute DC, Ishida T, Yarimizu Koji, Korthuis RJ. Inflammatory responses to ischemia and reperfusion in skeletal muscle. *Mol. Cell Biochem.* 179: 169-187, 1998.

[28] Hasenfuss G, Maier LS, Hermann H-P, Lüers C, Hünlich M, Zeitz O, Jannsen PML, Pieske B. Influence of pyruvate on contractile performance and Ca^{2+} cycling in isolated failing human myocardium. *Circulation* 105: 194-199, 2002.

[29] Hayashi Y, Sawa Y, Nishimura M, Fukuyama N, Ichikawa H, Ohtake S, Nakazawa H, Matsuda H. Peroxynitrite, a product between nitric oxide and superoxide anion, plays a cytotoxic role in the development of post-bypass systemic inflammatory response. *Eur J. Cardiothorac. Surg.* 26: 276-280, 2004.

[30] Hayashi Y, Sawa Y, Ohtake S, Fukuyama N, Nakazawa H, Matsuda H. Peroxynitrite formation from human myocardium after ischemia-reperfusion during open heart operation. *Ann. Thorac. Surg.* 72: 571-576, 2001.

[31] Hermann H-P, Arp J, Pieske B, Kögler H, Baron S, Janssen PML, Hasenfuss G. Improved systolic and diastolic myocardial function with intracoronary pyruvate in patients with heart failure. *Eur. J. Heart Fail.* 6: 213-218, 2004.

[32] Hermann H-P, Pieske B, Schwartzmüller E, Keul J, Just H, Hasenfuss G. Haemodynamic effects of intracoronary pyruvate in patients with congestive heart failure. *Lancet* 353: 1321-1323, 1999.

[33] Hermann H-P, Zeitz O, Lehnart SE, Keweloh B, Datz N, Hasenfuss G, Janssen PML. Potentiation of beta-adrenergic inotropic response by pyruvate in failing human myocardium. *Cardiovasc Res.* 53: 116-123, 2002.

[34] Hogue CW Jr., Creswell LL, Gutterman DD, Fleisher LA. Epidemiology, mechanisms, and risks: American College of Chest Physicians guidelines for the prevention and management of postoperative atrial fibrillation after cardiac surgery. *Chest* 128: 9S-16S, 2005.

[35] Ishii Y, Schuessler RB, Gaynor SL, Yamada K, Fu AS, Boineau JP, Damiano RJ. Inflammation of atrium after cardiac surgery is associated with inhomogeneity of atrial conduction and atrial fibrillation. *Circulation* 111: 2881-2888, 2005.

[36] Kawamura T, Wakusawa R, Okada K, Inada S. Elevation of cytokines during open heart surgery with cardiopulmonary bypass: participation of interleukin 8 and 6 in reperfusion injury. *Can. J. Anaesth.* 40: 1016-1021, 1993.

[37] Khairallah M, Labarthe F, Bouchard B, Danialou G, Petrof BJ, Des Rosiers C. Profiling substrate fluxes in the isolated working mouse heart using ^{13}C-labeled substrates: focusing on the origin and fate of pyruvate and citrate carbons. *Am. J. Physiol. Heart Circ. Physiol.* 286: H1461-H1470, 2004.

[38] Knott EM, Ryou M-G, Sun J, Heymann A, Sharma AB, Lei Y, Baig M, Mallet RT, Olivencia-Yurvati AH. Pyruvate-fortified cardioplegia suppresses oxidative stress and enhances phosphorylation potential of arrested myocardium. *Am. J. Physiol. Heart Circ. Physiol.* 289: H1123-H1130, 2005.

[39] Knott EM, Sun J, Lei Y, Ryou M-G, Olivencia-Yurvati AH, Mallet RT. Pyruvate mitigates oxidative stress during reperfusion of cardioplegia-arrested myocardium. *Ann. Thorac. Surg.* 81: 928-934, 2006.

[40] Kristo G, Yoshimura Y, Niu J, Keith BJ, Mentzer RM Jr., Bünger R, Lasley RD. The intermediary metabolite pyruvate attenuates stunning and reduces infarct size in in vivo porcine myocardium. *Am. J. Physiol. Heart Circ. Physiol.* 286: H517-H524, 2004.

[41] Lahey SJ, Campos CT, Jennings B, Pawlow P, Stokes T, Levitsky S. Hospital readmission after cardiac surgery. Does "fast track" cardiac surgery result in cost saving or cost shifting? *Circulation* 98 (suppl II): II-35-II-40, 1998.

[42] Leon H, Atkinson LL, Sawicka J, Strynadka K, Lopaschuk GD, Schulz R. Pyruvate prevents cardiac dysfunction and AMP-activated protein kinase activation by hydrogen peroxide in isolated rat hearts. *Can. J. Physiol. Pharmacol.* 82: 409-416, 2004.
[43] Levy JH, Tanaka KA. Inflammatory response to cardiopulmonary bypass. *Ann. Thorac. Surg.* 75: S715-S720, 2003.
[44] Mallet RT. Pyruvate: metabolic protector of cardiac performance. *Proc. Soc. Exp. Biol. Med.* 223: 136-148, 2000.
[45] Mallet RT, Squires JE, Bhatia S, Sun J. Pyruvate restores contractile function and antioxidant defenses of hydrogen peroxide-challenged myocardium. *J. Mol. Cell Cardiol.* 34: 1173-1184, 2002.
[46] Mallet RT, Sun J. Mitochondrial metabolism of pyruvate is required for its enhancement of cardiac function and energetics. *Cardiovasc. Res.* 42: 149-161, 1999.
[47] Mallet RT, Sun J. Antioxidant properties of myocardial fuels. *Mol. Cell Biochem.* 253: 103-111, 2003.
[48] Mallet RT, Sun J, Knott EM, Sharma AB, Olivencia-Yurvati AH. Metabolic cardioprotection by pyruvate: recent progress. *Exp. Biol. Med.* 230: 435-443, 2005.
[49] Manukhina EB, Downey HF, Mallet RT. Role of nitric oxide in cardiovascular adaptation to intermittent hypoxia. *Exp. Biol. Med.* 231: 343-365, 2006.
[50] Mathew JP, Parks R, Savino JS, Friedman AS, Koch C, Mangano DT, Browner WS. Atrial fibrillation following coronary artery bypass graft surgery: predictors, outcomes, and resource utilization. *J. Am. Med. Assoc.* 276: 300-306, 1996.
[51] Mayers I, Hurst T, Puttagunta L, Radomski A, Mycyk T, Sawicki G, Johnson D, Radomski MW. Cardiac surgery increases the activity of matrix metalloproteinases and nitric oxide synthase in human hearts. *J. Thorac. Cardiovasc. Surg.* 122: 746-752, 2001.
[52] Mayers I, Hurst T, Radomski A, Johnson D, Fricker S, Bridger G, Cameron B, Darkes M, Radomski MW. Increased matrix metalloproteinase activity after canine cardiopulmonary bypass is suppressed by a nitric oxide scavenger. *J. Thorac. Cardiovasc Surg.* 125: 661-668, 2003.
[53] Mayers I, Salas E, Hurst T, Johnson D, Radomski MW. Increased nitric oxide synthase activity after canine cardiopulmonary bypass is suppressed by S-nitrosoglutathione. *J. Thorac. Cardiovasc Surg.* 117: 1009-1016, 1999.
[54] Mehlhorn U, Krahwinkel A, Geissler HJ, LaRosee K, Fischer UM, Klass O, Suedkamp M, Hekmat K, Tossios P, Bloch W. Nitrotyrosine and 8-isoprostane formation indicate free radical-mediated injury in hearts of patients subjected to cardioplegia. *J. Thorac. Cardiovasc. Surg.* 125: 178-183, 2003.
[55] Melley DD, Evans TW, Quinlan GJ. Redox regulation of neutrophil apoptosis and the systemic inflammatory response syndrome. *Clin. Sci.* 108: 413-424, 2005.
[56] Mongan PD, Capacchione J, West S, Karaian J, Dubois D, Keneally R, Sharma P. Pyruvate improves redox status and decreases indicators of hepatic apoptosis during hemorrhagic shock in swine. *Am. J. Physiol. Heart Circ. Physiol.* 283: H1634-H1644, 2002.

[57] Moro MA, Darley-Usmar V, Goodwin DA, Read NG, Zamora-Pino R, Feelisch M, Radomski MW, Moncada S. Paradoxical fate and biological action of peroxynitrite on human platelets. *Proc. Natl. Acad. Sci. USA* 91: 6702-6706, 1994.

[58] Nakamura K, Ueno T, Yamamoto H, Iguro Y, Yamada K, Sakata R. Relationship between cerebral injury and inflammatory responses in patients undergoing cardiac surgery with cardiopulmonary bypass. *Cytokine* 29: 95-104, 2005.

[59] Nathoe HM, van Dijk D, Jansen EW, Suyker WJ, Diephuis JC, van Boven WJ, de la Riviere AB, Borst C, Kalkman CJ, Grobbee DE, Buskens E, de Jaegere PP; Octopus Study Group. A comparison of on-pump and off-pump coronary bypass surgery in low-risk patients. *N. Engl. J. Med.* 348: 394-402, 2003.

[60] Ochiai K, Zhang J, Gong G, Zhang Y, Liu J, Ye Y, Wu X, Liu H, Murakami Y, Bache RJ, Ugurbil K, From AHL. Effects of augmented delivery of pyruvate on myocardial high-energy phosphate metabolism at high workstate. *Am. J. Physiol. Heart Circ. Physiol.* 286: H2237-H2242, 2004.

[61] Olivencia-Yurvati AH, Blair JL, Baig M, Mallet RT. Pyruvate-enhanced cardioprotection during cardiopulmonary bypass surgery. *J. Cardiothorac Vasc. Anesth.* 17: 715-720, 2003.

[62] Olivencia-Yurvati AH, Wallace N, Ford S, Mallet RT. Leukocyte filtration and aprotinin: synergistic anti-inflammatory protection. *Perfusion* 19 (suppl): S13-S19, 2004.

[63] Pesonen EJ, Korpela R, Peltola K, Leijala M, Sairanen H, Raivio KO, Venge P, Andersson S. Regional generation of free oxygen radicals during cardiopulmonary bypass in children. *J. Thorac. Cardiovasc Surg.* 110: 768-773, 1995.

[64] Rubens FD, Nathan H, Labow R, Williams KS, Wozny D, Karsh J, Ruel M, Mesana T. Effects of methylprednisolone and a biocompatible copolymer circuit on blood activation during cardiopulmonary bypass. *Ann. Thorac. Surg.* 79: 655-665, 2005.

[65] Schafer FQ, Buettner GR. Redox environment of the cell as viewed through the redox state of the glutathione disulfide/glutathione couple. *Free Radic. Biol. Med.* 30: 1191-1212, 2001.

[66] Schulz R, Nava E, Moncada S. Induction and potential biologic relevance of a Ca^{2+}-independent nitric oxide synthase in the myocardium. *Br. J. Pharmacol.* 105: 575-580, 1992.

[67] Sharma AB, Knott EM, Bi J, Martinez RR, Sun J, Mallet RT. Pyruvate improves cardiac electromechanical and metabolic recovery from cardiopulmonary arrest and resuscitation. *Resuscitation* 66: 71-81, 2005.

[68] Sims CA, Wattanasirichaigoon S, Menconi MJ, Ajami AM, Fink MP. Ringer's ethyl pyruvate solution ameliorates ischemia/reperfusion-induced intestinal mucosal injury in rats. *Crit. Care Med.* 29: 1513-1518, 2001.

[69] Slovin PN, Huang C-J, Cade JR, Wood CE, Nasiroglu O, Privette M, Orbach P, Skimming JW. Sodium pyruvate is better than sodium chloride as a resuscitation solution in a rodent model of profound hemorrhagic shock. *Resuscitation* 50: 109-115, 2001.

[70] Song M, Kellum JA, Kaldas H, Fink MP. Evidence that glutathione depletion is a mechanism responsible for the anti-inflammatory effects of ethyl pyruvate in cultured

lipopolysaccharide-stimulated RAW 264.7 cells. *J. Pharmacol. Exp. Ther.* 308: 307-316, 2004.

[71] Strüber M, Cremer JT, Gohrbandt B, Hagl C, Jankowski M, Völker B, Rückoldt H, Martin M, Haverich A. Human cytokine responses to coronary artery bypass grafting with and without cardiopulmonary bypass. *Ann. Thorac. Surg.* 68: 1330-1335, 1999.

[72] Taggart DP, El-Fiky M, Carter R, Bowman A, Wheatley DJ. Respiratory dysfunction after uncomplicated cardiopulmonary bypass. *Ann. Thorac. Surg.* 56: 1123-1128, 1993.

[73] Talmor M, Hydo L, Barie PS. Relationship of systemic inflammatory response syndrome to organ dysfunction, length of stay, and mortality in critical surgical illness. *Arch. Surg.* 134: 81-87, 1999.

[74] Tejero-Taldo MI, Caffrey JL, Sun J, Mallet RT. Antioxidant properties of pyruvate mediate its potentiation of β-adrenergic inotropism in stunned myocardium. *J. Mol. Cell Cardiol* 31: 1863-1872, 1999.

[75] Tejero-Taldo MI, Sun J, Caffrey JL, Mallet RT. Pyruvate potentiates β-adrenergic inotropism of stunned guinea-pig myocardium. *J. Mol. Cell Cardiol.* 30: 2327-2339, 1998.

[76] Te Velthius H, Jansen PG, Oudemans-van Straaten HM, Sturk A, Eijsman L, Wildevuur CR. Myocardial performance in elderly patients after cardiopulmonary bypass is suppressed by tumor necrosis factor. *J. Thorac. Cardiovasc. Surg.* 110: 1663-1669, 1995.

[77] Tossios P, Bloch W, Huebner A, Raji MR, Dodos F, Klass O, Suedkamp M, Kasper S-M, Hellmich M, Mehlhorn U. *N*-acetylcysteine prevents reactive oxygen species-mediated myocardial stress in patients undergoing cardiac surgery: results of a randomized, double-blind, placebo-controlled clinical trial. *J. Thorac. Cardiovasc Surg.* 126: 1513-1520, 2003.

[78] Uchiyama T, Delude RL, Fink MP. Dose-dependent effects of ethyl pyruvate in mice subjected to mesenteric ischemia and reperfusion. *Intensive Care Med.* 29: 2050-2058, 2003.

[79] Ulloa L, Ochani M, Yang H, Tanovic M, Halperin D, Yang R, Czura CJ, Fink MP, Tracey KJ. Ethyl pyruvate prevents lethality in mice with established lethal sepsis and systemic inflammation. *Proc. Nat. Acad. Sci. USA* 99: 12351-12356, 2002.

[80] Vane JR, Anggard EE, Botting RM. Regulatory functions of the vascular endothelium. *N. Engl. J. Med.* 323: 27-36, 1990.

[81] Vásquez-Vivar J, Denicola A, Radi R, Augusto O. Peroxynitrite-mediated decarboxylation of pyruvate to both carbon dioxide and carbon dioxide radical anion. *Chem Res Toxicol* 10: 786-794, 1997.

[82] Weisel RD, Mickle DA, Finkle CD, Tumiati LC, Madonik MM, Ivanov J. Delayed myocardial metabolic recovery after blood cardioplegia. *Ann. Thorac. Surg.* 48: 503-507, 1989.

[83] Yang R, Gallo DJ, Baust JJ, Uchiyama T, Watkins SK, Delude RL, Fink MP. Ethyl pyruvate modulates inflammatory gene expression in mice subjected to hemorrhagic shock. *Am. J. Physiol. Gastrointest Liver Physiol.* 283: G212-G221, 2002.

[84] Yared JP, Starr NJ, Torres FK, Bashour CA, Bourdakos G, Piedmonte M, Michener JA, Davis JA, Rosenberger TE. Effects of single dose, postinduction dexamethasone on recovery after cardiac surgery. *Ann. Thorac. Surg.* 69: 1420-1424, 2000.

[85] Zahler S, Massoudy P, Hartl H, Hähnel C, Meisner H, Becker BF. Acute cardiac inflammatory responses to postischemic reperfusion during cardiopulmonary bypass. *Cardiovasc. Res.* 41: 722-730, 1999.

[86] Zanardo G, Michielon P, Paccagnella A, Rosi P, Calo M, Salandin V, Da Ros A, Michieletto F, Simini G. Acute renal failure in the patient undergoing cardiac operation. Prevalence, mortality rate, and main risk factors. *J. Thorac. Cardiovasc Surg.* 107: 1489-1495, 1994.

In: Oxidative Stress: Clinical and Biomedical Implications ISBN: 978-1-60021-888-0
Editors: B. M. Matata, M. M. Elahi, pp. 103-132

Chapter VII

Redox Regulation of Cell Signaling

Magdalena L. Circu and Tak Yee Aw [1]
Department of Molecular and Cellular Physiology
Louisiana State University Health Sciences Center, 1501 Kings Highway,
Shreveport, Louisiana 71130-3932, USA

Abstract

In recent years, redox regulation of cell signaling has gained rapid recognition as an important modulator of cell function. Moreover, a disturbance in the cellular prooxidant-to-antioxidant balance in favor of oxidation and the resultant redox shifts has been suggested to underpin the pathogenesis of a variety of pathologies that include inflammation, cancer, and cardiovascular and degenerative diseases. Protein sulfhydryls can be oxidatively modified by S-thiolation, S-nitrosylation, or S-hydroxylation, and redox alterations of proteins during oxidative challenge have been implicated in regulating cellular defense mechanisms. Thus, redox modifications of protein thiols are emerging as pivotal mechanisms by which the cellular redox systems control cell signaling. The current chapter will focus on the discussion of compartmentation of cellular redox systems and the concepts that are salient to understanding the control of redox homeostasis, redox regulation of cell signaling, the relationship between cellular redox status and proliferative and apoptotic responses, and how changes in redox signaling alters these responses. These considerations underscore the centrality of redox in the control of cell signals and the governance of cell fate.

I. Introduction

The redox state of a cell is represented by the ratio of inter-convertible reduced and oxidized forms of the sum total of various redox couples within the cell. Schafer and Buetner [2001] have proposed that the description of redox environment is a better representation of the redox state of a cell. The redox environment is determined by a set of redox couples that are responsive to electron flow which maintain the reduced and oxidized milieu within a cell.

The cellular redox systems are comprised of the thiol redox and pyridine nucleotide redox systems. The main thiol redox couples include glutathione/glutathione disulfide (GSH/GSSG), cysteine/cystine (Cys/CySS), and reduced and oxidized thioredoxin (Trx/TrxSS) systems while the pyridine nucleotide couples include NAD^+/NADH and $NADP^+$/NADPH. Together, the oxidation-reduction status of these redox systems impact redox regulation of different protein activity that directly affect cell functionality. Within cells, the various redox systems are compartmentalized and independently controlled in the plasma membrane, cytosol, nucleus, mitochondria and endoplasmic reticulum [Hansen et. al., 2006]. The existence of these specialized redox compartments that are associated with unique cellular organelles or membranal structures is integral to the redox regulation of varied normal cell function. Depending on the concentrations of the respective redox couples, their distribution or fluxes, the compartmentation of specific redox systems may, in fact, represent a crucial and generalized mechanism for optimizing cell activity within mammalian cells. Increasing research efforts in recent years have focused on delineating the fundamental importance of redox compartmentation and the heterogeneity of redox distribution in the control of cell signaling and gene expression.

II. Compartmentation of Cellular Redox Systems and Control of Redox Homeostasis

II.1. Glutathione/Glutathione Disulfide (GSH/GSSG) Redox System and Its Subcellular Compartmentation

Among all the redox systems, the intracellular redox state is commonly represented by measure of the GSH/GSSG redox couple given its cellular abundance and its role in thiol modification of cellular proteins. Indeed, the ratio of GSH-to-GSSG has typically been used as an indicator of cellular redox state and an index of cellular oxidative stress.

II.1.1. GSH/GSSG and Cellular Redox Balance

The tripeptide, GSH (γ-glutamyl-cysteinyl glycine) is the most abundant low-molecular-weight thiol in cells that participates in antioxidant defense and regulates the intracellular thiol-disulfide redox state [Meister and Anderson, 1983; Sies, 1999]. Additionally, GSH is involved in the regulation and redox activation of transcription factors which is a part of an adaptive cellular mechanism that participates in cell signaling and stress responses [Kamata and Hirata, 1999]. As the main thiol redox system in the cell, the ratio of reduced GSH and its oxidized form, GSSG, is maintained in favor of the reduced state (>90% reduced) by a dynamic process that achieves a highly reduced environment essential for cell function. GSH homeostasis is maintained by several mechanisms including *de novo* synthesis from its precursor amino acids (glutamate, cysteine, and glycine), reduction of GSSG by glutathione reductase (GR) and uptake from exogenous GSH sources across plasma membranes. Intracellularly, the GSH redox cycle catalyzes GSSG reduction through the function of GSH peroxidase and GSSG reductase (Figure 1). The reductant for this process comes from NADPH whose availability is intricately tied to the function of the pentose phosphate pathway, mostly notably, glucose flux and the activity of the rate-limiting enzyme, glucose-6-phosphate dehydrogenase (Figure 1).

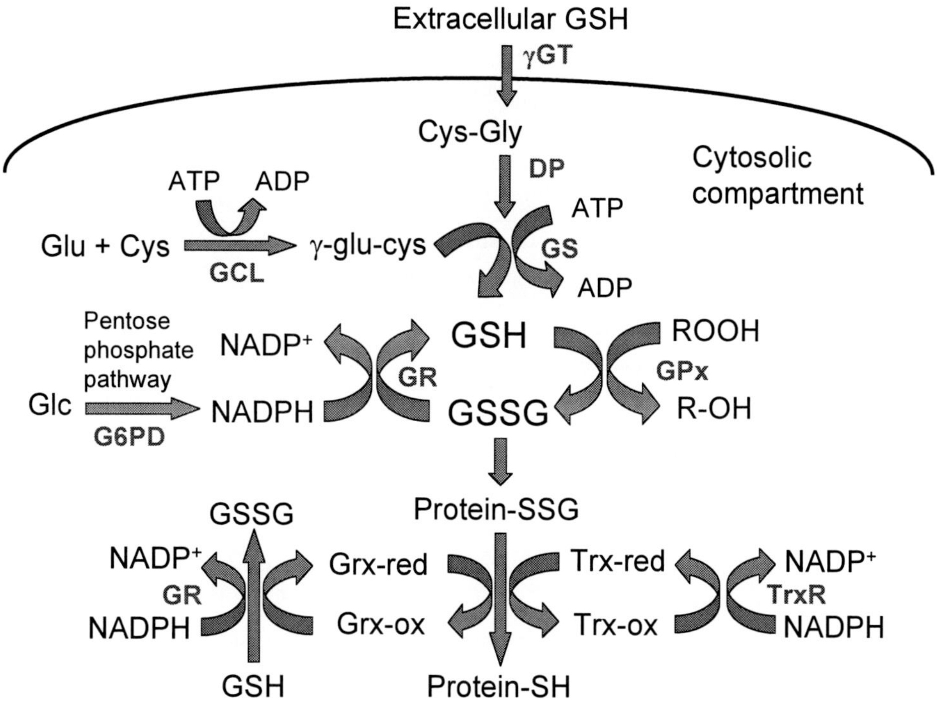

Figure 1. GSH, thioredoxin and glutaredoxin in redox homeostasis. Synthesis of GSH from its constituent amino acids (glutamate, cysteine, glycine) in the cytosol is catalyzed by glutamate-cysteine ligase and glutathione synthase at the expense of two moles of ATP. Extracellular GSH is hydrolyzed to its component amino acids by the sequential action of γ-glutamyltransferase and dipeptidase and reused in glutathione synthesis. During peroxide elimination, regeneration of GSH from GSSG is maintained by the GSH peroxidase and GSH reductase system (the GSH redox cycle). The electron donor for GSH reductase activity comes from NADPH which is produced from glucose flux through the pentose phosphate pathway. GSSG excess can form protein mixed disulfide which are reduced by thioredoxin/thioredoxin reductase or by glutaredoxin/glutathione reductase systems at the expense of NADPH. Cys: cysteine; gly: glycine; glu: glutamic acid; GCL: glutamate-cysteine ligase; GS: glutathione synthase; ATP: adenosine triphosphate; ADP: adenosine diphosphate; γ-GT: γ-glutamyltransferase; DP: dipeptidase; GPx: GSH peroxidase; GR: GSH reductase; G6PD: glucose-6-phosphate dehydrogenase; Protein-SSG: protein mixed disulfide; Trx: thioredoxin; TrxR: thioredoxin reductase; Grx: glutaredoxin.

The simple ratio of GSH-to-GSSG represents a convenient measure of cellular redox status and oxidative stress. However, a useful quantitative expression for the redox state of the GSH/GSSG pool is the redox potential (E_h), an expression that takes into consideration the stoichiometry of two GSH oxidized per GSSG formed. E_h is calculated from the Nernst equation: $E_h = E_0 + (RT/ nF)\ln([GSSG]/[GSH]^2)$ [Jones, 2002], where R is the gas constant, T is the absolute temperature, F is Faraday's constant and n is the number of electrons transferred. E_0 is the standard potential for the redox couple and is calculated at equilibrium conditions relative to a standard hydrogen electrode. Under physiological conditions, the estimates of E_h in cells and tissues for the GSH/GSSG redox system differ in different cell types and tissues which typically range between -260 to -150 mV [Jones, 2002]. E_h can be similarly calculated for other redox systems, given known cellular concentrations of their respective reduced and oxidized forms; thus expression of redox potential is a convenient way to represent the redox state of any redox couple. Furthermore, the tendency of a redox couple to donate or accept electrons from other redox-active components can be determined from the quantitative value of its E_h. For instance, a redox couple with a more negative E_h will easily donate electrons and is therefore a better reductant than a redox couple with a more positive E_h. Because the GSH/GSSG redox state has an important role in redox regulation, the loss of GSH redox balance and the accumulation of GSSG during oxidative stress will result in altered metabolic control, cell transition (e.g., proliferation or apoptosis) or cell survival [Aw, 1999; 2003].

II.1.2. Compartmentation of GSH Redox within Cells

Cellular GSH is present in millimolar concentrations and is compartmentalized within the cytosol, mitochondria, endoplasmic reticulum and nucleus as separate and distinct pools (Figure 2). The distribution of GSH among the various cellular compartments is not uniform. The *de novo* synthesis of GSH occurs in the cytosol where the majority of GSH also exists; its cytosolic concentration is estimated to be between 5-10mM [Meister and Anderson, 1983]. Generally, the cytosolic GSH pool is highly reduced; in the liver, the GSH-to-GSSG ratio under normal conditions can be in excess of 100 to 1, but during oxidative stress, this redox ratio can significantly decrease to less than 4 to 1. Although exclusively synthesized in the cytosol, GSH is distributed in the mitochondria. Due to a lack of intra-mitochondrial *de novo* GSH synthesis [Meister and Anderson, 1983], the maintenance of matrix GSH is sustained by GSH transport from the cytosol by carrier proteins that are located in the mitochondrial inner membrane [Chen and Lash, 1998; Lash, 2005]. In liver and kidney, the dicarboxylate and 2-oxoglutarate membrane carriers are two known mitochondrial GSH transporting proteins [Chen and Lash, 1998] The existence of a discrete mitochondrial GSH pool was first described in the liver comprising 10-15% of the total hepatocellular GSH pool [Jocelyn and Kamminga, 1974]. Mitochondrial GSH concentrations are similar to cytosolic GSH (5 to 10mM), but the mitochondrial GSH/GSSG redox ratio is more oxidized. Moreover, the matrix GSH compartment is metabolically separate from the cytosol with regard to synthetic rate, turnover, and sensitivity to chemical depletion [Meredith and Reed, 1982]. In this regard, the matrix GSH pool is stable with a $t_{1/2}$ of 30h compared to 2h for cytosolic GSH, and incorporates cysteine at a rate 15 times less than the cytosolic GSH pool [Meredith and Reed, 1982]. Functionally, the mitochondrial GSH pool is considered more important than the cytosolic pool for cell survival as evidenced by the sensitization of cells to oxidant-induced injury with depletion of mitochondrial GSH. Indeed, early studies have shown that loss of mitochondrial GSH is a determining factor in oxidative vulnerability [Meredith and Reed, 1982], providing the first evidence for the significance of this redox compartment in cell survival. Subsequent studies have linked the loss of mitochondrial GSH to the cytotoxic effects induced by aromatic hydrocarbons [Hallberg and Rydstrom, 1989], hypoxia [Lluis et. Al., 2005], *tert*-butylhydroperoxide [*t*BH, Olafsdottir and Reed, 1988], and ethanol intoxication [Fernadez-Checa et. al., 1991]. Studies from our laboratory in normal colonic intestinal cells showed that maintenance of mitochondrial redox homeostasis is critical to the preservation of cell survival against oxidative (menadione) challenge [Maracine and Aw, unpublished]. Studies in freshly isolated microsomes demonstrated the existence of a distinct GSH pool in the endoplasmic reticulum (ER) whose total concentration is estimated to be between 6-10 mM [Bass et al., 2004] (Figure 2) of similar pool sizes to cytosolic and mitochondrial GSH.

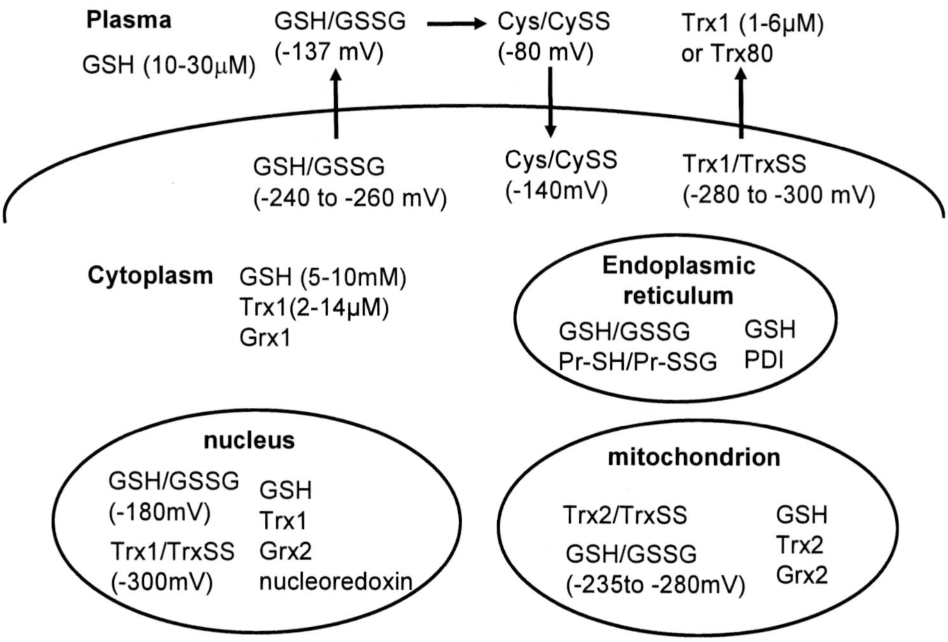

Figure 2. Cellular compartmentation of thiol/disulfide redox couples and redox proteins. Three redox couples that participate in maintaining the cellular reduced-to-oxidized environment include GSH/GSSG, CyS/CySS, and Trx/TrxSS. Each redox systems exhibit different redox potential (indicated as mV). The cysteine residues in cellular proteins are maintained in the reduced state by GSH and thiol reductases, namely, thioredoxin, glutaredoxin and sulfiredoxin. The different isoforms of the thiol reductases have different cellular localization in plasma, cytoplasm, mitochondria and nucleus. GSH: glutathione; GSSG; glutathione disulfide; Cys: cysteine; CySS: cystine; Trx: thioredoxin; TrxSS; oxidized thioredoxin; Trx80: truncated form of Trx; Grx: glutaredoxin; PDI: protein disulfide isomerase; Pr-SH: reduced protein; Pr-SSG: oxidized protein.

However, the GSH redox environment in ER is highly oxidized (GSH-to-GSSG ratio of 3:1-1:1) which favors protein oxidation [Bass et. al., 2004; Hwang et. al., 1992; Moldovan and Moldovan, 2004]. Moreover, less than 50% of the ER thiol pool (GSH + GSSG) is free; the majority of GSH is reversibly bound to proteins [Bass et. al., 2004]. This protein mixed disulfide pool may serve as a GSH reserve or as an important component of the ER redox buffering system. However, the functional significance of an ER GSH pool is not completely resolved. Frand and Kaiser [2000] found that GSH compartmentation in the ER is necessary for disulfide formation as elevated reduced GSH disrupts ER function and activates the unfolding protein response (UPR) that triggers cellular apoptosis. An independent nuclear pool of GSH has been described [Cotgreave, 2003, Figure 2]. By using a GSH-specific antibody and confocal microscopy, Soderdahl et. al. [2002] found that nuclear GSH concentrations are lower than the cytosol, suggesting that the cytosolic and nuclear redox pools are not in equilibrium. Furthermore, the thiol oxidant, diamide induces a greater preponderance of protein oxidation in the nucleus [Soderdahl et. al., 2002]. The functional significance of a distinct nuclear GSH redox pool is poorly understood, but may be speculated to play a role in the regulation of redox-sensitive transcription factors and gene expression (section III.6 below). GSH/GSSG is also present extracellularly in plasma; its redox state is, however highly oxidized (-137mV) as compared to tissues and cultured cells (E_h between -185 to -258mV). Unlike the Cys/CySS redox couple (section II.3 below) a physiological role of plasma GSH/GSSG in extracellular cell signaling is less well defined.

II.2. Thioredoxin (Trx) and Trx Redox Proteins

The regulation of cellular GSH/GSSG homeostasis is linked to the function of redox sensitive proteins. The oxidative modification of cysteine residues in specialized proteins has similar significant biological impact on cell functionality as the oxidation of cysteine in GSH. Figure 1 illustrates the relationship between GSH homeostasis and its inter-connection with the thiol redox proteins, thioredoxin and glutaredoxin and their respective participation in the maintenance of protein sulphydryl redox homeostasis.

II.2.1. Thioredoxins

Thioredoxins (Trx) are small ubiquitous redox proteins (≈12 kDa with E_h values in the range of – 280mV). Trx are involved in maintaining the thiol/disulfide redox state of proteins and, together with its oxidized form, TrxSS, constitute a key redox protein couple in mammalian cells. While plasma concentrations of Trx of 1-6μM are similar to that of plasma GSH (~5μM), its intracellular concentrations of 2-14μM are three orders of magnitude lower than that of GSH (2-10 mM) [Nakamura et. al., 1998]. Despite the lower intracellular pool, Trx modulates the activity of a variety of proteins, including antioxidant enzymes and transcription factors and catalyzes the reversible reduction of disulfide bonds in oxidized proteins at the expense of cysteine residues in its active motif site, Cys-x-x-Cys. Mammalian cells contain two forms of Trx which are differentially compartmentalized (Figure 2). Trx1 has a cytosolic location, but can translocate to the nucleus where it is involved in the redox regulation of transcription factors like AP-1, p53 and NF-kB (section III.6). The redox state of nuclear Trx1 is more reduced than cytosolic Trx1 [Watson and Jones, 2003]. Trx1 can also be secreted from cells like lymphocytes, hepatocytes, fibroblasts and endothelial cells in response to oxidative stress [Rubartelli et. al., 1992]. Part of the secreted Trx1 remains bound to the external membrane as a truncated form, Trx80, while the rest is rapidly taken up by other cells and contributes to intracellular redox control in those cells. In terms of function, extracellular Trx1 acts as a cytokine or chemokine [Pekkari and Holmgren, 2004] whereas cytosolic Trx1 functions as an inhibitor of MAPKKK and ASK-1 (apoptosis signal-regulating kinase-1) and interfere directly with cell signaling pathways. Moreover, Trx1 and Trx-like protein (TRP14), by reducing disulfide bond formed during oxidative stress, can reactivate the protein tyrosine phosphatase, PTEN, which reverses phosphoinositide 3-kinase (PI3K) signaling [Lee et. al., 2002]. Trx2 is a mitochondrial enzyme that is involved in antioxidant protection in the mitochondria and functions as a redox sensor for the mitochondrial ASK-1 [Damdimopoulos et. al., 2002]. As compared to cytosolic Trx1, mitochondrial Trx2 is relatively oxidized [Watson and Jones, 2003].

The reduction of oxidized Trx is catalyzed by thioredoxin reductase (TrxR) via electrons transferred from NADPH. Mammalian TrxR is a dimeric NADPH-dependent, FAD-containing disulfide reductase that contains a selenocysteine residue in the sequence Cys-Secys-Gly at the C terminus of each subunit. Three mammalian TrxR isoenzymes have been identified, viz., cytosolic TrxR1; mitochondrial TrxR2 and testis specific TGR (Trx GSH reductase) [Zhong et. al., 2000]. The importance of the Trx system in the control of cell function is evidenced by the observation that mice deficient in Trx1 and Trx2 (Trx1/2$^{-/-}$ null) die during embryogenesis [Nakamura, 2005]. The members of the Trx family of proteins share a common structural motif called the thioredoxin fold. This structural motif consists of a four-stranded β-sheet and three surrounding α-helices and is found in proteins from different classes like oxidoreductases (Trx, glutaredoxins and protein disulfide isomerases) or in functionally different proteins such as glutathione transferases or glutathione peroxidases. The oxidoreductase proteins have similar active sites as Trx: Cys-xxx-yyy-Cys and catalyze the reversible reduction of disulfides at cysteine residues [Lillig and Holmgren, 2007].

II.2.2. Trx Redox Proteins

Glutaredoxins. Glutaredoxins (Grx) are cellular enzymes that belong to the Trx family of proteins that are involved in disulfide-dithiol exchange and share common functions with Trx in preserving sulfhydryl groups of redox-sensitive proteins in the reduced form. Grx, similar to Trx, utilizes the two redox-active cysteine of the conserved Cys-x-x-Cys catalytic site to reduce proteins, but Grx are more active in the reduction of S-glutathionylated substrates than Trx [Johansson et. al., 2004]. As depicted in Figure 1, the reduction of oxidized Grx is catalyzed by the GSH and glutathione reductase (GR) system, a feature that is distinct from Trx wherein oxidized TrxSS is reduced by TrxR with electrons donated from NADPH. Both Trx and Grx redox systems participate in the reduction of protein disulfide bonds (Figure 1). A cytosolic Grx1 and a mitochondrial Grx2 form of Grx exit in mammalian cells (Figure 2). Functionally, cytosolic Grx1 utilizes only GSH in its reductive reaction, while mitochondrial Grx2 can reduce oxidized substrates using either GSH or by coupling to TrxR [Gladyshev et. al., 2001; Johansson et. al., 2004]. Human Grx2 was described by Lillig et. al. [2005] as the first iron-sulfur (FeS) protein; the FeS-Grx2 cluster has been proposed to function as a redox sensor for the activation of Grx2 during oxidative stress. The FeS cluster complexes with two Grx2 monomers and two molecules of GSH that are non-covalently bound to proteins and in equilibrium with GSH in solution. When the concentration of GSH decreases, the holo-Grx2 complex dissociates and yields enzymatically active Grx2. In addition, the direct reduction of Grx2 via TrxR enables Grx2 to reduce glutathionylated proteins during conditions of oxidative stress and low mitochondrial GSH concentrations [Johansson et. al., 2004]. Grx1 has been shown to be involved in redox regulation of the ser/thr kinase, Akt, which suppresses the recruitment of protein phosphatase 2A; the resultant activation of Akt inhibits stress-induced apoptosis [Murata et. al., 2003].

Nucleoredoxin (Nrx). Nrx belongs to the Trx family of proteins that is located in the nucleus (Figure 2) where it participates in redox regulation of nuclear proteins. Unlike the aforementioned Trx and Grx redox proteins, relatively little is known of the function of Nrx in redox control and regulation of cell signaling. However, recent studies have documented the involvement of Nrx in the redox regulation of transcription factors and the suppression of the Wnt-catenin signaling pathway through its oxidation-sensitive association with disheveled (Dvl) [Funato et. al., 2006].

Protein disulfide isomerase (PDI). PDI is a member of the Trx family that is located in the ER (Figure 2). PDI possesses two Trx-like active sites that are involved in the oxidation of cysteine residues from nascent proteins to form intramolecular disulfide bonds. While the exact mechanism is not well defined, the re-oxidation of PDI through the addition of GSH redox buffer (GSH+GSSG) is the rate-limiting step in PDI-catalyzed disulfide bond formation. To date, fourteen PDI-family members are known in human ER which probably exhibit overlapping functions in disulfide bond formation [Gorlach et. al., 2006].

II.3. Cysteine and Cystine (Cys/CySS) Redox

Cysteine and cystine (Cys/CySS) comprise the major extracellular low-molecular-weight thiol pool. The E_h for this redox system is approximately -80 mV, which is more oxidized than the GSH/GSSG and Trx/TrxSS redox couples. The oxidation of the extracellular cysteine pool is a determinant of elevated ROS production and the activation of the p44/p42 MAPK pathway in Caco2 cells [Nkabyo et. al., 2005]. In neuronal cells, cysteine oxidation through the formation of S-nitrosothiols (section III.3) has been demonstrated to have an important role in signal transduction pathway that governs neuronal survival [Hess et. al., 2005]. Together with the other redox couples, Cys/CySS participates in maintaining the intracellular redox environment. Interestingly, recent studies have shown that Cys/CySS as well as Trx/TrSS redox pools do not vary in a quantitative fashion as the GSH/GSSG pool; Jones et. al. [2004; 2006] have described that these redox couples, in fact, function independently in the control of the intracellular redox environment. A redox circuitry model with three nodes of redox modulation by differential variations in cellular Cys/CySS, GSH/GSSG, and Trx/TrxSS has been proposed. In this model, the three redox switches function in parallel circuits with electrons being donated from NADPH. Since the proteins that are reduced by one redox pair do not serve as adequate substrate for another redox pair, and because these redox systems have different rates of reaction with different proteins, their respective redox-dependent reactions exist in discrete redox circuitry within cells [Hansen et. al., 2006]. Given that Cys/CySS is the main extracellular thiol redox couple in plasma, its redox variation may be more important in extracellular or receptor mediated cell signaling such as in the control of cell proliferation [Jonas et. al., 2002; Noda et. al., 2002], cell susceptibility to oxidant-induced apoptosis [Nkabyo et. al., 2005] or atherosclerosis development by promoting monocyte adhesion to vascular endothelial cells [Go and Jones, 2005]. GSH/GSSG is also present in plasma but its redox state is less oxidized (-137mV) than Cys/CySS (-80 mV). The finding that these two plasma redox couples are not in equilibrium is consistent with independent mechanisms of transport/export and/or oxidation/reduction of each redox system [Jones, 2006b].

III. Redox Regulation of Cysteine Sulfhydryl Groups and Cell Signaling

Many cellular proteins contain cysteine residues that provide a redox-sensitive switch for regulating protein function. The ROS-induced oxidation of cysteine-SH group can result in the formation of intra- and/or inter-chain disulfide bonds. The direct addition of GSH disulfide could lead to the S-glutathionylation of the thiol while sequential thiol oxidation results in the formation of sulfenic, sulfinic and sulfonic acid (Figure 3). Nitric oxide (NO) has also been shown to S-nitrosylate specific cysteine thiols in proteins [Biswas et. al., 2006]. Protein-thiol oxidation products can be reversed by the GSH, Trx and Grx systems.

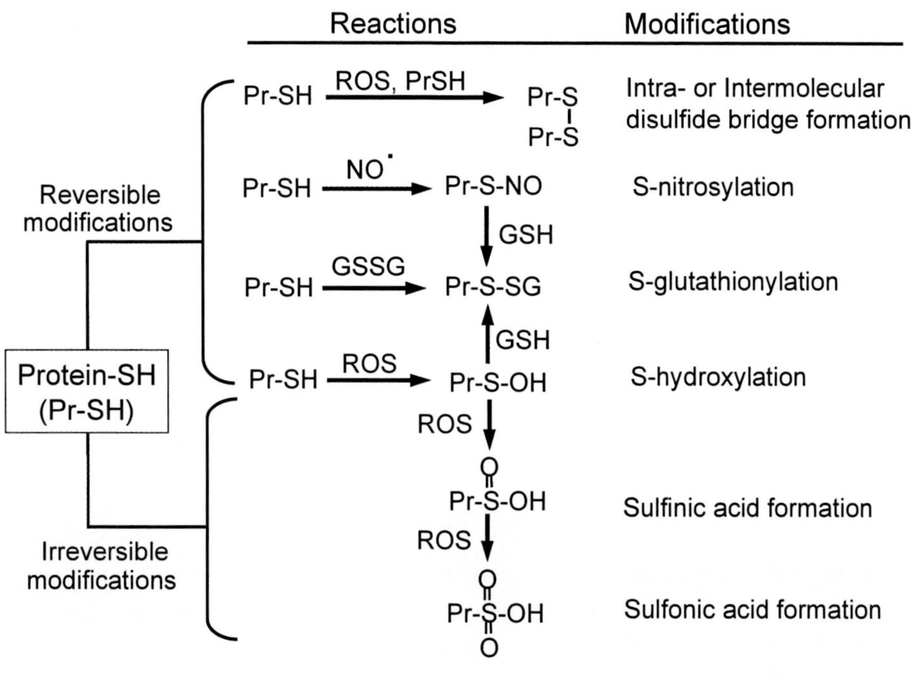

Figure 3. Redox modifications of protein thiols. Cysteine residues within proteins can undergo reversible and irreversible modifications by reactions with ROS and RNS. ROS can mediate intra-or intermolecular disulfide bridge formation to yield protein disulfide cross links or S-hydroxylation of protein thiols to form sulfenic acids (Pr-S-OH). NO, through its higher nitrogen species (e.g., N_2O_3) can cause S-nitrosylation of protein thiols to form nitrosothiols (Pr-S-NO). GSSG can induce S-glutathionylation of protein thiols through the exchange of disulfide bonds with protein thiols to form mixed disulfides (Pr-SSG). As secondary reactions, GSH reacts with nitrosothiols and/or sulfenic acids to induce S-glutathionylation.

III.1. Proteins with Redox Active Cys Active Sites

III.1.1. Peroxiredoxins

Peroxiredoxins (Prx) are a group of non-seleno thiol-specific peroxidases with an oxidizable cysteine active site and a role in antioxidant defense that involves the breakdown of organic hydroperoxides and H_2O_2 [Rhee et. al., 2005]. Prx are divided into three classes: typical 2-Cys Prx; atypical 2-Cys Prx and 1-Cys Prx, but all have the same catalytic mechanism with an active cysteine (the peroxidatic cysteine) being oxidized to a sulfenic acid (Cys-SOH) by the peroxide substrate. The regeneration of the thiol status of the active cysteine differs among the Prx classes. For the typical and atypical 2-Cys Prx classes, thiol regeneration is catalyzed by a disulfide reductase and the Trx/TrxR system while the reduction of 1-Cys Prx requires a thiol containing reductant [Wood et. al., 2003]. In mammalian cells six isoforms of Prx (PrxI to PrxVI) have been isolated. Different biological roles have been attributed to these Prx members including cell cycle arrest in response to oxidative stress, induced super-oxidation of the cysteine residue in the active site, and cell proliferation in response to the reduction of active site cysteine residues [Phalen et. al., 2006]. Prx are located mainly in the cytosol but these peroxidases have been found in the mitochondria and peroxisomes, as well as being associated with the nuclei and membranes [Hofmann et. al., 2002].

III.1.2.Sulfiredoxins

Sulfiredoxins (Srx) are enzymes which contain a highly conserved cysteine residue in the active site that serves as a phosphate carrier and function to reduce the sulfinic and sulfonic acid derivatives from oxidized proteins like peroxiredoxins [Biteau et. al., 2003]. Studies with human cells demonstrated that the reduction mechanism catalyzed by Srx involves ATP hydrolysis and requires Mg^{2+} and thiols as electron donors such as GSH or Trx [Chang et. al., 2004]. ATP can be efficiently replaced by GTP, dATP or dGTP. A novel role for human Srx in cellular redox control has been recently described to involve the de-glutathionylation of oxidized proteins. Findlay et. al. [2006], using two-dimensional gel electrophoresis have identified several targeted protein that undergoes Srx-catalyzed de-glutathionylation like actin and PTP1B, a tyrosine phosphatase that participates in the regulation of insulin signaling.

III.2. Redox regulation by S-Glutathionylation

Protein-containing accessible thiol groups can form mixed disulfides with low-molecular weight thiols, such as GSH, by a process termed S-glutathionylation (Figure 3). The S-glutathionylation reaction comprises the direct addition of GSH to a target cysteine residue within the protein structure. Formation of glutathionylated protein occurs through several mechanisms: a) thiol-disulfide exchange between free PrSH and GSSG formed under oxidative stress; b) via reaction of cysteine oxidation products: sulfenic (Protein-SOH) or sulfinic acid (Protein-SO_2H) with GSH; c) through formation of protein S-nitrosothiols (Protein-SNO) and subsequent reaction with GSH, and d) reaction of thiyl radical ($RS^{\bullet}$)

formed by oxidized proteins which could form mixed disulfide adduct with cellular GSH (Figure 3) [Ghezzi, 2005]. The disulfide bond formed between GSH and proteins can remain permanent under certain oxidative conditions or can be reversed by redox enzymes like Grx or Trx in a reduced cellular environment [Beer et. al., 2004]. Protein S-glutathionylation is a reversible process that has been associated with the regulation of protein structure and function, with protection of essential cysteine groups within protein structures that are highly prone to oxidizing conditions, and with stabilization of extracellular proteins. The reversibility of S-glutathionylation depends on the cellular GSH concentrations; GSH can directly de-glutathionylate proteins or serve as a substrate for Grx and Trx. An increasing body of evidence suggests that S-glutathionylation could act as an intracellular redox sensor, and functions as a signaling mechanism that is analogous to protein phosphorylation [Biswas et. al., 2006; Ghezzi, 2005]. For instance, a new mechanism of NO-induced vasodilatation that involves glutathionylation of the sarcoplasmic reticulum Ca^{2+} ATPase (SERCA) which is responsible for vascular smooth muscle relaxation has been proposed. SERCA contains a cysteine residue (Cys^{674}) which is S-glutathionylated in the process. Pathophysiological conditions like hyperlipidemia, hyperglycemia, and hypertension determine the oxidation of Cys^{674} that results in impairment of vasodilatation regulation and consequent cardiovascular complication [Cohen and Adachi, 2006].

III.3. Redox Regulation by S-Nitrosylation

The current discussion will focus specifically on the function of nitric oxide (NO) in redox signaling; details of NO chemistry and biology are covered in other chapters in this book. Like ROS, NO plays an important role in the regulation of cell function through S-nitrosylation signaling [Gaston et. al., 2003; Sun et. al., 2006]. Importantly, loss of normal NO signaling associated with oxidative stress has been implicated in disease pathogenesis. S-nitrosylation refers to the covalent binding of an NO residue to the sulfhydryl moieties of proteins in the presence of an electron acceptor to form an S-NO bond that results in the formation of S-nitrosothiols (SNOs, Figure 3). This interaction of NO with redox-active motifs within proteins can occur through transnitrosation/transfer reactions or metalloproteinase-catalyzed reactions. Transfer between thiol residues is favored over transfers to N- or C-containing species. Increasingly, the formation of SNOs is emerging to be a crucial post-translational protein modification in redox-based cellular signaling, ranging from the control of ion channels to nuclear regulatory proteins to cell apoptosis. Protein SNOs within the cytosol are unstable and are readily reduced by GSH (Figure 3) or Trx through transnitrosation and subsequent denitrosylation [Freedman et. al., 1995]. To protect against reductive or transnitrosative degradation, SNOs are sequestered in membranes, lipophilic protein folds, vesicles and interstitial spaces. This compartmentation of SNOs has a role in apoptotic signaling as S-nitrosylated inactive caspases are localized within the mitochondrial intermembrane space; during apoptosis, they undergo de-nitrosylation and activation in the cytosol [Mannick et. al., 2001]. Protein activity can be stimulated (e.g., p21ras or Trx) or inhibited (e.g., caspases) by S-nitrosylation.Two features of SNO function are noteworthy. First, specific protein thiols are targeted for S-nitrosylation. For instance, among 50 redox-active cysteine residues, only Cys^{3635} is selectively S-nitrosylated in NO-mediated calcium channel activity in skeletal muscle [Sun et. al., 2001]. Similarly, NO-mediated Trx activity selectively targets Cys^{69} on Trx [Haendeler et. al., 2002]. Second, SNO function has been shown to exhibit stereoselectivity in that the L-isomer, but not D-isomer is bioactive. Some key nuclear regulatory proteins that are controlled by S-nitrosylation include hypoxia-inducible factor 1 [HIF-1, Palmer et. al., 2000], stimulating proteins 1 and 3 [Sp1, Sp3, Zaman et. al., 2002] and nuclear factor–κB [NF-κB, Marshall and Stamler, 2001]. The S-nitrosylation mediated nuclear regulatory function is dose-dependent as physiological levels of SNOs sustain the transcription of physiological genes [Zaman et. al., 2002] while supraphysiological levels of SNOs induce S-nitrosative stress and increased expression of stress response genes [Marshall and Stamler, 2001]. The cystic fibrosis transmembrane regulatory gene, CTFR, HIF-1, and NF-κB are some examples whose expression subscribes to dose-dependent control by nitroso-glutathione (GSNO) [Zaman et. al., 2002]. The mechanism by which S-nitrosylation regulates cytosolic, mitochondrial, or extracellular proteins has been a subject of intense interest. Notable examples include regulation of caspases and more intriguingly, hemoglobin. S-nitrosylated inactive caspases are localized within the mitochondrial intermembrane space; during apoptosis, they undergo de-nitrosylation and activation within the cytosol resulting in initiation of apoptosis [Lluis et. al., 2005]. The chemistry between hemoglobin and NO is complex. To date, the general consensus is that hemoglobin is endogenously S-nitrosylated at the β93 Cys [Jia et. al., 1996], but the physiological significance of nitroso-hemoglobin is unresolved and remains a subject of active debate. Possible universality of S-nitrosylation signaling warrants mention as SNO signaling has been shown to be relevant in non-mammalian systems as well such as in viral replication [Gaston et. al., 2003]. An understanding of redox regulation of cell signaling in viral biology could have important implications for infectious diseases.

III.4. Redox Modulation of Protein Tyrosine Phosphorylation Signaling

Signaling by protein tyrosine phosphorylation controls many biological processes like cell motility and proliferation and involves the action of two classes of enzymes: protein tyrosine kinases (PTK) and protein tyrosine phosphatases (PTP). PTK belong to the transmembrane receptor family or to the cytosolic non-receptor PTK with a functional role in regulating cell metabolism, growth, migration and differentiation [Chiarugi and Buricchi, 2007]. PTP reverses the action of PTK. A characteristic feature of PTP is the presence of a highly reactive cysteine group in the catalytic site that is susceptible to oxidation. A model of redox control of tyrosine phosphorylation signaling comprises of ligand stimulation of PTK activity and down-regulation of PTP activity [Chiarugi and Cirri, 2003; Denu and Tanner, 1998]. ROS induces the reversible oxidation of the PTP active-site cysteine to cysteine sulfenic derivative, a process that avoids irreversible damage to PTP while inducing a loss of enzyme function [Salmee and Barford, 2005; Tonks and Neel, 2001]. Five reversible oxidation states of the sulfhydryl moiety in the catalytic site of PTP can be induced by low ROS concentrations produced at the level of the tyrosine receptor, namely, thiolate, sulfenic acid, glutathionyl disulfide, cyclic sulfonamide and sulfinic acid. The reversible and transient inactivation of PTP therefore represents an indirect redox mechanism in the regulation of PTK activity and is a strategy by which the cell can enhance or terminate receptor tyrosine kinase signaling. Redox-inactivation of PTP results in hyperphosphorylation of PTK and enhanced enzyme activity [Minetti et. al., 2002] while thiol reduction of the cysteine residue in the PTP catalytic site promoted by GSH or Trx, results in attenuated PTK activity and termination of receptor signaling. This mechanism of indirect redox regulation has been described for platelet-derived growth factor receptor (PDGFR); the two PTPs involved in regulation are the low molecular weight PTP (LMW-PTP) and SHP2 [Chiarugi et. al., 2001; Meng et. al., 2002]. The PTPs that participate in redox regulation of insulin receptor signaling include PTP1B and TC45 [Meng et. al., 2004]. The active site cysteine of the phosphatase and tensin homolog (PTEN) can form a disulfide bond with a neighboring cysteine when oxidized by H_2O_2 [Caselli et. al., 1998]. Analyses of cysteine mutants revealed that the critical Cys^{124} in the active site and Cys^{71} are involved in this disulfide bond formation [Lee et. al., 2002]. PTEN activity and suppression of PTEN recruitment into the PTEN-associated complex is negatively regulated by the phosphorylation of Ser^{380}, Thr^{382} and Thr^{383} of the PTEN tail [Vasquez et.al., 2000; 2001]. The redox control of PTEN has been suggested to involve Trx, but the precise mechanism is incompletely understood, and the biological impact of this interaction for PTEN activity is unresolved. For instance, Lee et. al. [2002] proposed that the binding of Trx reduces oxidized PTEN and restores its activity. However, Meuillet et. al. [2004] demonstrated that reduced Trx1 binding inhibits PTEN activity through intermolecular disulfide bond formation between Trx and PTEN. Unlike most members of the PTPs, the biological function of PTEN as a tumor suppressor is mediated through the phosphatidylinositol 3,4,5-triphosphate (PIP_3), its biological substrate, which reverses the action of PI3K, a regulator of PI3K/Akt signaling and cell survival [Cho et. al., 2004].

Other intracellular kinases such as Src tyrosine kinase, focal adhesion kinase (FAK) as well as certain serine/threonine kinases or dual specific tyrosine/threonine kinases like MAPK or Akt and apoptosis signal-regulating kinases (ASK, see III.5 below) have also been demonstrated to be redox-modulated through oxidation of redox active cysteine in their catalytic sites [Chiarugi et. al., 2003; Giannoni et. al., 2005]. Collectively, these findings support the growing evidence that redox-dependent modification of proteins is a key event in cellular signaling. This mode of protein S-glutathionylation complements protein phosphorylation in transducing cellular signals; greater effectiveness in signal transduction could further result from the cross talk between these two signaling processes [Chiarugi and Buricchi, 2007; Kamata and Hirata, 1999].

III.5. Redox Modulation of Mitogen-Activated Protein Kinase (MAPK) Signaling: Relationship to Apoptosis Signal-Regulating Kinase 1 (ASK-1)

Mitogen activated protein kinases (MAPK) represent an important class of signaling enzymes that phosphorylate specific serine and threonine of target proteins. In response to different stimuli including oxidative stress, MAPK elicit a variety of cellular processes such as gene expression, mitosis or apoptosis by interacting with specific transcription factors. The transduction of signal involves a cascade of phosphorylation in which upstream MAPK kinase kinase (MAPKKK) activates MAPK kinase (MAPKK) that in turn activates MAPK [Kyriakis and Avruch, 2001].

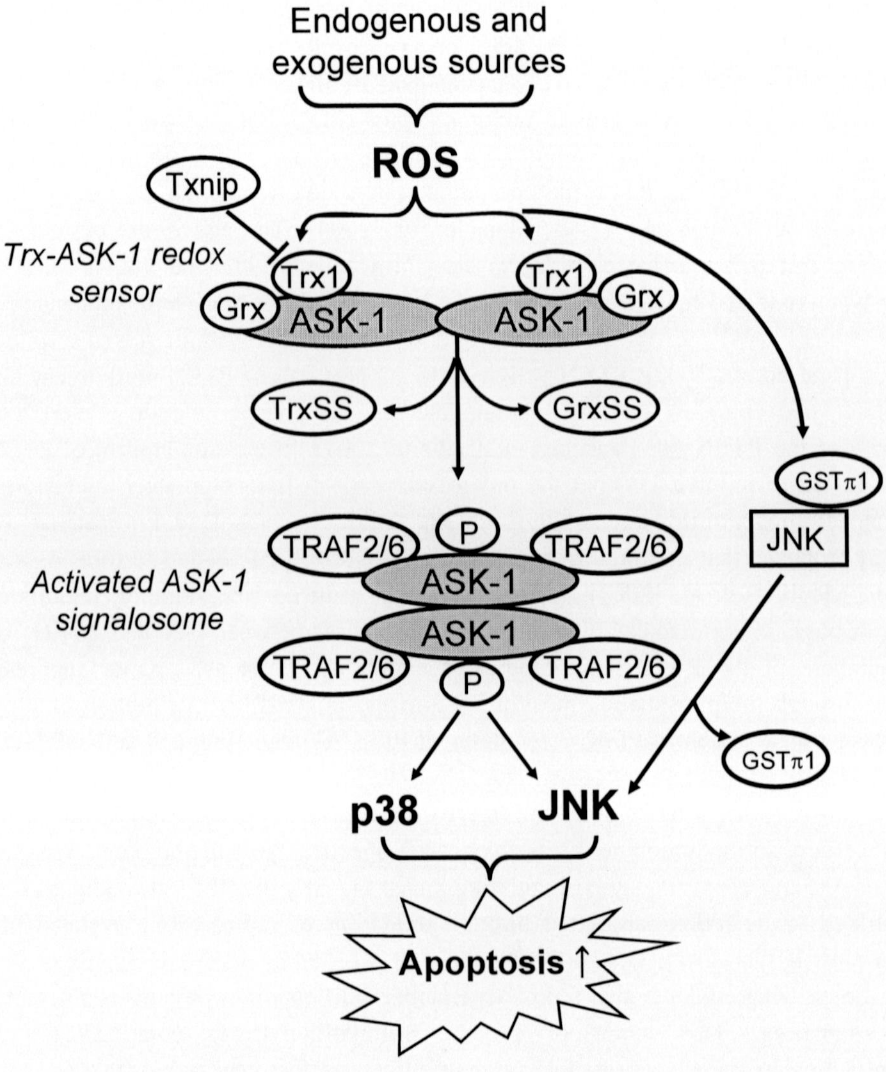

Figure 4. ROS-mediated ASK-1 signaling. Reactive oxygen species (ROS) generated from different sources oxidizes sulphydryl groups from the redox active site of thioredoxin (Trx) and glutaredoxin (Grx) and cause their dissociation from apoptosis signaling-regulating kinase 1 (ASK-1). Oxidized Trx (TrxSS) is sequestered in the cytosol by its inhibitor, thioredoxin-interacting protein (Txnip). Since the ASK-1-Trx complex functions as a redox sensor in ROS-dependent activation, it has been named "ASK-1 signalosome". ASK-1 release from this complex determines its auto-phosphorylation and the formation of "activated signalosome" where proteins such as tumor necrosis factor receptor-associated factor 2 and 6 (TRAF 2/6) are recruited. The signalosome signals the activation of p38 and JNK resulting in cell apoptosis. In addition, ROS can directly induce JNK activation by promoting the release from its inhibitor, the glutathione S-transferase π1 (GST-P1).

In mammalian cells there are 3 classes of MAPK: extracellular signal–regulated protein kinase (ERK); c-Jun NH2-terminal kinase (JNK) and p38 MAPK. MAPK cascades are involved in ROS-induced cellular response and therefore play a key role in redox signaling. A mechanism that results in JNK and p38 activation has been identified and involves the redox modulation of apoptosis signal-regulating kinase 1 (ASK-1) activity [Nishitoh et. al., 2002]. Figure 4 summarizes the mechanism of ROS/redox modulation of ASK-1 activity and its activation of p38 kinase and JNK signaling. ASK-1 is a MAPKKK that is activated by ROS [Saitoh et. al., 1998], TNFα [Liu et. al., 2002], ER stress and calcium influx [Nishitoh et. al., 2002]. Given the broad spectrum of biological responses mediated by ASK-1 signaling, oxidative stress-induced ASK-1 activation has received particular attention in recent years. Trx1 is an inhibitor of ASK-1 that binds ASK-1 through the cysteine residue in the catalytic site and prevents the downstream propagation of the apoptotic signal [Liu and Min, 2002]. It has been shown that only the reduced form of Trx1 promotes ASK-1 degradation; this Trx1/ASK-1 couple could be regarded as an intracellular redox switch designated "ASK-1 signalosome" that can be turned on or off under oxidative conditions [Fujino et. al., 2006]. The signalosome comprises the homo-dimers of ASK-1 that form higher molecular mass complex upon Trx1 dissociation following ROS stimulation [Noguchi et. al., 2005]. The redox state of Trx1 is an important player in ASK-1 redox signaling [Patwari et. al., 2006]. Under oxidizing conditions, oxidized thioredoxin-interacting protein (Txnip) also called vitamin D3 upregulated protein 1 (VDUP1) or Trx binding protein-2 (TPB-2), binds to catalytic cysteines of reduced Trx1 and forms mixed disulfide bonds that result in Trx1 loss of function.

The formation of stable mixed disulfide bonds within Trx1 indicates that the redox status of Trx1 itself is under redox control. Apart from Trx1, ASK-1 activity has been shown to be modulated by other factors. The redox protein Grx1 possesses an additional cysteine residue which can bind to ASK-1 and regulate its kinase activity in response to the cellular GSH redox state [Song and Lee, 2003]. Tumor necrosis factor receptor-associated factors 2 and 6 (TRAF2/6) were reported to be recruited to the ASK-1 signalosome and forms an activated ASK-1 signalosome complex that results in JNK-induced apoptosis [Fujino et. al., 2006; Nishitoh et. al., 1998]. Additionally, JNK has been shown to interact directly with GSTπ that inhibits JNK signaling independent of ASK-1 function. An increase in ROS results in GSTπ dissociation and JNK activation [Townsend and Tew, 2003]. Negative regulators of ASK-1 function have been described. During oxidative stress, protein serine-threonine phosphatase 5 (PP5) has been demonstrated to down regulate ASK-1 activity [Morita et. al., 2001]. Additionally, GSPM1 is a redox sensor that binds to the C-terminal domain of ASK-1 and inhibits its activity [Townsend and Tew, 2003].

III.6. Redox Control of Transcription Factors

As it is with activities of protein phosphatases and protein kinases, the intracellular redox environment also influences the activation of transcription factors and gene expression. In this regard, the cellular redox status can regulate both the upstream and downstream events of signal transduction pathways. A variety of redox sensitive transcription factors have been described, some of which are directly activated by upstream redox-dependent signaling events.

III.6.1. Insulin Signaling and Redox Activation of Nuclear Factor Erythroid 2-Related Factor (Nrf2)

Insulin is generally known for its peripheral effects on glucose metabolism, but in liver and endothelial cells, insulin has been demonstrated to confer cytoprotection that involves GSH increases [Kim et al., 2004; Okouchi et. al., 2006]. An important pathway by which insulin promotes endothelial cell survival is the suppression of mitochondrial apoptotic signaling through phosphatidylinositol 3-kinase (PI3K)-Akt signaling and the phosphorylation of Bad and procaspase-9 [Herman et. al., 2000; Kim and Tannenbaum, 2004]. More recently, a novel signaling pathway of insulin protection has been described in liver [Kim et al., 2004] and brain endothelial cells [Okouchi et. al., 2006] wherein insulin mediates cytosolic GSH increases through PI3K/Akt/mTOR signaling and expression of glutamate-L-cysteine ligase (GCL), the rate limiting enzyme in GSH synthesis (Figure 5).

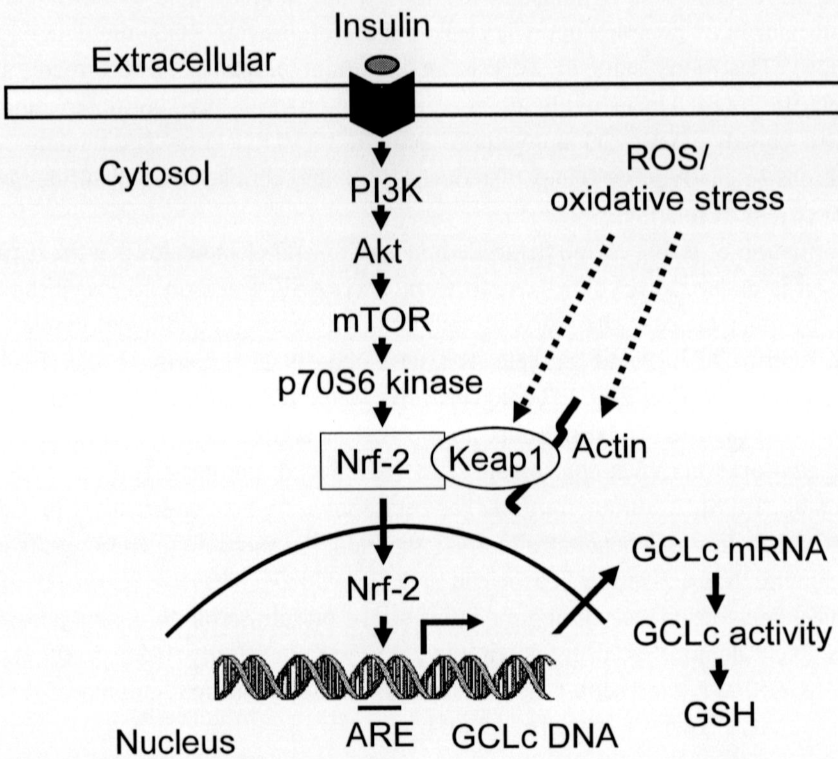

Figure 5. Insulin signaling and Nrf2 activation and nuclear translocation. Insulin confers protection against oxidative stress through a sequence of signaling events that include (a) activation of the PI3K/Akt/mTOR/p70S6K pathway, (b) Nrf2 phosphorylation and nuclear translocation, (c) transcriptional upregulation of GCLc and increased GCL activity, and (d) maintenance of cellular GSH. ROS/oxidative stress induced activation of Nrf2 is mediated by S-glutathionylation of actin and Keap1, which induces Nrf2 dissociation and nuclear translocation.

The fact that Akt is redox sensitive [Chiarugi et. al., 2003; Giannoni et. al., 2005; Murata et. al., 2003], and that redox sensitive PTPs participate in regulation of insulin receptor signaling [Meng et. al., 2004] implies that the insulin receptor/PI3K/Akt/mTOR signaling pathway will subscribe to redox control. Our recent studies further demonstrated that downstream signaling of PI3K/Akt/mTOR mediates the transcription up-regulation of the catalytic subunit of GCL through activation of Nrf2 (Figure 5) [Okouchi et. al., 2006] which is integral to transactivation of GCL [Zipper and Mulcahy, 2000]. Nrf2 is a redox-sensitive transcription factor that has been implicated in cellular response to oxidative stress by induction of gene expression of many antioxidant enzymes. In its unstimulated state, Nrf-2 is sequestered in the cytosol through its association with the actin binding protein, Kelch-like ECH associating protein 1 (Keap1).

During oxidative challenge, Keap1 is subjected to proteasomal degradation which results in Nrf-2 dissociation and nuclear translocation. The GSH and Trx1 systems function in the redox activation of Nrf-2 in the cytosol and the nucleus. Oxidation of cytosolic GSH/GSSG and S-glutathionylation of actin are two cytosolic redox mechanisms by which Keap1 can be induced to dissociate from Nrf-2 [Okouchi et. al., 2006]. Within the nucleus, nuclear Trx1 participates in the redox regulation of the critical Cys^{508} of Nrf-2 that is essential for DNA binding [Hansen et. al., 2004]. Thus, the Nrf-2/Keap1 complex functions as a crucial cellular redox sensor of oxidative/electrophilic stress [Itoh et. al., 2003; Zipper and Mulcahy, 2000].

III.6.2. Activator Protein 1 (AP-1)

Activator protein-1 (AP-1) is a transcription factor that regulates the expression of genes that are involved in cell growth. The oncogene products of Jun and Fos form AP-1 by hetero-/homo-dimerization through their leucine-zipper structures that are located at the C-terminal region [Angel and Karin, 1991]. Studies have demonstrated that AP-1 is redox regulated through the conserved cysteine residue located in the DNA-binding domain of each monomer, and it has been suggested that redox modification of this cysteine is a part of the mechanism that controls the transcriptional activity of AP-1 [Abate et. al, 1990]. Elevated oxidants induce the oxidation of the cysteine which results in decreased AP-1 binding to DNA and inhibition of gene expression. It has been shown that apurinic/apyrimidinic endonuclease (APE) is responsible for the redox regulation of AP-1, and that Trx1 is critical in the redox regulation of APE [Hirota et. al., 1997].

III.6.3. Nuclear Factor Kappa B (NF-κB)

NF-κB is a transcription factor that regulates the expression of genes that govern cell survival during oxidative challenge. In non-stimulated cells, NF-κB is bound to its endogenous inhibitor IκB, and sequestered in the cytosol. Activation of NF-κB by stimulus such tumor necrosis factor-α, ionizing radiation, and ROS results in its dissociation from IκB and translocation into the nucleus. The DNA-binding activity of NF-κB is redox sensitive; binding is decreased by oxidants (e.g., diamide) and increased by thiol-reducing compounds (e.g., dithiothreitol and β-mercaptoethanol) [Hayashi et. al., 1993]. NF-κB activity has been demonstrated to be differentially modulated by nuclear or cytosolic localized Trx1. Within the nucleus, Trx1 promotes NF-κB binding to DNA through the reduction of a disulfide bond involving the cysteine residue Cys^{62} of the p50 subunit [Matthews at; al., 1992]. Within the cytosol, Trx1 inhibits NF-κB activation by blocking the dissociation and degradation of IκB [Hirota et. al., 1999]. More recently, a mechanism for redox-induced inhibition of NF-κB binding to DNA has been described and involves glutathionylation of the p50 subunit catalyzed by Grx1 [Pineda-Molima et. al., 1999]. Grx1 and the nuclear oxidoreductase Nrx, can also participate in redox regulation of NF-κB, but the precise mechanism of their involvement is not known; overlapping roles of Trx1 in the cytosol and nucleus have been suggested [Hirota et. al., 2000].

III.6.4. Tumor Suppressor Protein, p53

The tumor suppressor protein p53 is induced by oxidative stress and controls the expression of cell cycle-related genes after DNA damage [Harris and Levine, 2005]. Trx1 has been shown to mediate redox-dependent p53/DNA-binding affinity and potentiates APE/Ref-1-induced p53 activity [Ueno et. al., 1999]. Seemann and Hainaut [2005] recently proposed a role for the APE/Ref-1 and Trx/TrxR cooperative regulation of p53 activity at the level of DNA binding, nuclear translocation and protein stability. In this system, oxidized Trx acts as a redox sensor that promotes p53 activation and its subsequent accumulation in the nucleus. APE/Ref-1 acts as a critical switch in this process, under normal conditions APE/Ref-1 functions in base excision repair of oxidized DNA, and under oxidative stress, APE/Ref-1 facilitates p53 activation and growth suppression [Seemann and Hainaut, 2005].

IV. Cell Responses to Redox Signaling and Redox Imbalance

IV.1. Concept of Redox Control of Cell Phenotypic Transitions

Three thiol/disulfide redox control nodes have been described to be involved in control of cellular function and the determination of cell transition from a quiescent to a proliferative, growth arrested, differentiated or apoptotic phenotype. Figure 6 illustrates the influence of redox environment on different cell transitions. Prevailing evidence shows that in normal cells, a more reducing environment favors cell proliferation, whereas a more oxidized environment is associated with cell differentiation or apoptosis.

For instance, cells that are rapidly proliferating exhibit an approximately 50 mV more reduced redox potential than cells that are differentiated or growth-arrested. An additional ~50 mV more oxidized redox potential is characteristic of apoptotic cells [Jones, 2002; Nkabyo et.al., 2002]. The relationship between cell phenotypes and cellular redox potential is best illustrated from studies of cell transitions in response to changes in the GSH/GSSG redox system. During proliferation, E_h values (for GSH/GSSG) vary from -260 mV to -230 mV (Figure 6A). A variation from -200 mV to -190mV induces cell transition to differentiation and growth arrest, while an E_h = -150mV is associated with apoptotic cell death [Jones, 2002; Siobhan and Jones, 2004]. The extracellular Cys/CySS redox potential is more oxidized than the intracellular GSH/GSSG system, but a change from reduced (> -80mV) to oxidized status (0 to less than -80mV) elicits a similar parallel pattern of cell transition from proliferation to growth arrest and apoptosis as that elicited by the GSH/GSSG redox system [Siobhan and Jones, 2004].

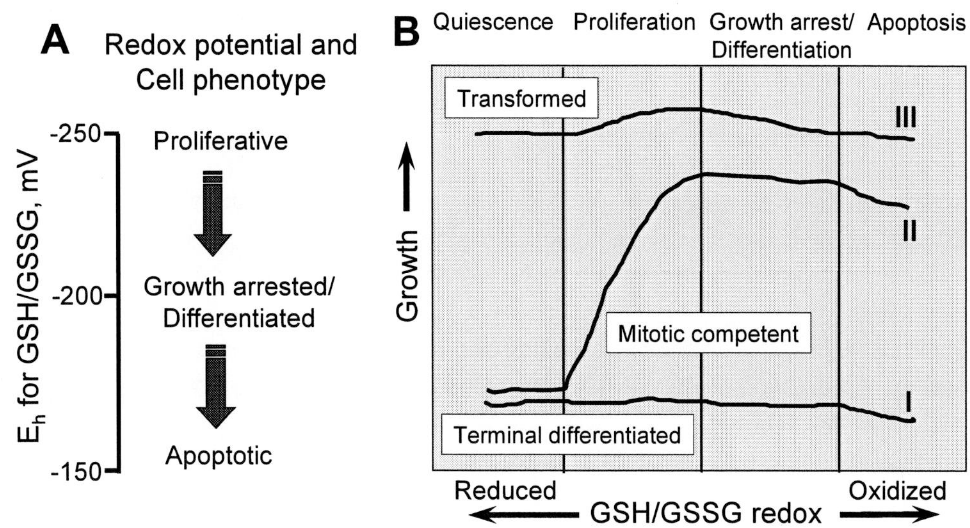

Figure 6. Relationship between cell phenotypes and GSH/GSSG redox potential (Eh) and cell responses to changes in GSH/GSSG redox. A, As cells transition from the proliferative to growth arrested/differentiated to apoptotic states, they undergo a progression from a reduced to oxidized redox potential (i.e., from a more negative to a more positive Eh). Apoptotic cells exhibit ~50 mV more oxidized Eh relative to growth arrested cells, while differentiated and growth arrested cells exhibit ~50mV more oxidized potential than the Eh in proliferating cells. B, the quiescent state is exhibited by cells that are fully differentiated. The magnitude of change in the cellular GSH/GSSG redox state can initiate cell proliferation, growth arrest/differentiation, or apoptosis depending on the initial phenotypic state of the cell. Curves I, II, and III, respectively, represents terminally differentiated, mitotically competent, and transformed cell types.

Studies from our laboratory further demonstrate that the promotion of cell proliferation by exogenous cysteine is independent of changes in intracellular GSH [Noda et. al., 2002], consistent with the notion that the two redox systems are not in equilibrium and each functions independently in oxidation/reduction reactions (sections II.1 and II.3 above) [Jones, 2006b]. As for the Trx/TrSS redox couple, the magnitude of change in its redox potential during cell transitions is small. For instance, in intestinal Caco-2 cells, the transition from a proliferative to a differentiated phenotype elicited little change in the Trx1 redox status despite a significant decrease in cell GSH that equates with a 40 mV oxidation of the GSH/GSSG redox [Watson and Jones, 2003]. It is noteworthy that the experimentally measured E_h for Trx is in the range of -330mV in proliferating cells is highly reduced and is between -70 to -100 mV more reduced than that of the GSH/GSSG and -250 mV more reduced than that of the Cys/CySS redox systems. However, as cells transition from proliferation to growth arrest and apoptosis the Trx redox does become more oxidized, indicating that the Trx system responds in a similar manner as the GSH/GSSG couple [Jones, 2002]. Collectively, the three nodes of redox systems of Cys/CySS, GSH/GSSG, and Trx/TrSS, while far from equilibrium with one another, functions independently to influence cell phenotypic transitions.

IV.2. Cellular Redox Sate and Signaling Governs Cell Proliferation and Apoptosis

Fully differentiated tissues including liver, kidney, brain and intestine are characterized by an arresting of cells in the quiescent state. While severe oxidative challenge results in cytotoxicity, milder oxidative stress elicits cell transitions from a quiescent state to that of the proliferative, growth arrested or apoptotic states [Aw, 1999; 2003]. Central to the concept of oxidant (e.g., ROS) regulation of cell transition is the shift of cellular redox in the direction of reduction or oxidation that results in a cell that favors quiescence, proliferation or death. A cell typically exhibits one of three characteristic responses to increased cellular oxidation (Figure 6B). Cells that are terminally differentiated such as hepatocytes exhibit mitotic block and biological constraint to proliferative, and therefore remain quiescent (curve I). Elevated oxidative stress induces cell apoptosis. In certain differentiated cells such as enterocytes, a genetic program is maintained that allows the cells to proliferate in response to antigenic, mitogenic or oxidative stimuli (curve II). This initial stimulus serves to lower regulatory checkpoint barriers (e.g., G_0 to G_1 transition) such that subsequent small increases in oxidation induce a heightened proliferative response. Indeed, an enhanced intestinal mitogenic response to hydroperoxide challenge has been shown to be associated with an oxidation of the mucosal GSH/GSSG redox status [Tsunada et. al., 2003 a and b]. At higher oxidative stress, cells die by apoptosis. Little barrier to proliferation are exhibited by transformed or tumor cells; subsequent oxidative challenge can further enhance cell proliferation or result in growth arrest or apoptosis when it is beyond the point of maximal cell division (curve III). Collectively, these examples illustrate the fundamental concept that redox status governs characteristic cell responses to oxidative events and that the phenotypic endpoint depends on the initial phenotypic state of the cell and the degree and duration of the oxidative stress. These considerations also underscore the notion that cell necrosis is not an obligatory biological endpoint of all oxidative challenge.

We have tested the concept of differential cell responses to oxidative stress experimentally in intestinal CaCo2 cells. We found that lipid hydroperoxide (LOOH) exposure elicited differential CaCo-2 cell transition into the proliferative or apoptotic states as a function of the degree of perturbation of the cellular GSH/GSSG redox. A mild redox imbalance induced by low LOOH levels (1-5 μM) favors cell proliferation as evidenced by increases in DNA synthesis and ornithine decarboxylase activity, and expression of cyclin D1 and cdk4 [Gotoh et. al., 2002], consistent with G_1 exit and S phase progression. One suggested mechanism of redox-induced cell proliferation involves the redox modulation of protein tyrosine kinases/phosphatases (PTK/PTP) which is necessary in the adhesion-dependent cell proliferation response [Chiarugi and Giannoni, 2005]. By extending the duration of the redox imbalance, CaCo2 cells were growth arrested at the G_2M phase which was reversible depending on the severity and duration of the sustained redox shift [Noda et. al., 2001]. At 10-20μM, LOOH caused significant cellular GSH/GSSG redox disruption, mitochondrial dysfunction, caspase-3 activation, and DNA fragmentation [Wang et. al., 2000], consistent with induction of cell apoptosis. Apoptosis initiation was rapid (within 30 min post LOOH-induced GSH/GSSG imbalance); interestingly, subsequent GSH/GSSG recovery did not influence the apoptotic outcome, indicating that redox-induced apoptotic initiation occurred within an early and narrow window of redox shift. This early redox imbalance in peroxide-mediated apoptosis is not cell type specific. An early loss of GSH/GSSG redox balance similarly mediated apoptosis caused by *tert*-butyl hydroperoxide (*t*BH) in mitotic competent undifferentiated PC12 cells [Pias and Aw, 2002a]. Notably, this redox-dependent induction of apoptosis in naïve PC12 cells occurs independently of ROS production [Pias and Aw, 2002b] and was prevented by pretreatment with the thiol antioxidant, N-acetylcysteine (NAC). However NAC was ineffective when administered at a time after *t*BH-induced GSH/GSSG loss or ROS formation [Pias et. al., 2003], consistent with early redox signaling mediating oxidative cell death. The result that early redox shifts control apoptosis initiation in intestinal and neuronal cells suggests that GSH/GSSG redox signaling is likely to be a generalized mechanism in oxidative cell killing in various cell types. Apart from intracellular GSH/GSSG redox, the extracellular redox potential also influences cell transition states; we discovered that exogenous cysteine and cystine promoted cell proliferation in CaCo-2 cells that was unrelated to the *de novo* synthesis of cellular GSH [Noda et. al., 2002]. These results highlight a role for redox control of cell apoptosis and proliferation by intra- and extracellular redox status. Taken together, our findings validated the hypothesis that cellular redox is an important fundamental determinant of cell responses; the phenotypic endpoint of proliferation, growth arrest, or apoptosis directly tracks with the extent of GSH/GSSG redox alteration.

The conclusion that cellular redox status controls cell transition is also corroborated by our studies in rat intestine *in vivo.* Under physiological conditions, mucosal proliferation and apoptosis subscribe to a circadian rhythm that was coordinated with the feeding and postprandial patterns of the animal wherein peak proliferative and apoptotic responses were separated by 12h during a 24h cycle [Iwakiri et. al., 2001]. Interestingly, however, chronic administration of subtoxic levels of LOOH (2 - 4 weeks) suppressed both intestinal proliferative and apoptotic activities in accordance with significant loss of mucosal GSH/GSSG redox balance, indicating that sustained redox imbalance elicits cell cytostasis [Tsunada et. al., 2003 a and b]. This disruptive effect of chronic LOOH exposure on normal enterocyte turnover was reversed by exogenous GSH in conjunction with preserved mucosal GSH and restored tissue GSH/GSSG redox status [Tsunada et. al., 2003b]. Moreover, the suppressed enterocyte proliferative activity was alleviated by EGF which induced a 30-fold increase in colonic ornithine decarboxylase activity [Tsunada et. al., 2003a], indicating a highly elevated proliferative response. These results show that mucosal redox governs differential intestinal cell responses to oxidative challenge *in vivo* and that chronic redox imbalance induces a cytostatic state that primes the cell for a heightened response to exogenous mitogenic stimuli.

V. Conclusion

The recognition that oxidants, apart from their widely accepted roles as agents of cytotoxicity, can serve as important mediators of cellular metabolism and gene expression, and that oxidant effects are mediated by changes in the cellular redox state are two major conceptual breakthroughs that have advanced our current understanding of redox signaling and regulation of cell function. Indeed, the notion that redox play a central role in signaling cell transitions and that aberrant proliferation and apoptosis have a wide range of clinical implications in disease processes, have spurred much of the current research investigations into identifying redox sensitive molecular components of various signaling pathways and the redox mechanisms of its regulation. While substantial progress has been made in uncovering the roles of reactive oxygen and nitrogen species in redox signaling as well as the molecular targets for ROS and RNS modification, the challenge remains of linking these specific redox protein modifications to individual biological endpoint. An area of continued interest and challenge for future research will be the establishment of a mechanistic link between cellular redox signaling and disease etiology. The success of such an investigation will provide the crucial knowledge base for future design of targeted and therapeutic interventions and treatment modalities for a variety of disease states that includes inflammation, cancer, cardiovascular and degenerative disorders.

Acknowledgements

Research in the authors' laboratory is supported by grants from the National Institutes of Health, DK44510 and DK43785.

References

Abate, C., Patel, L., Rauscherll, F.J., Curran, T. (1990). Redox regulation of fos and jun DNA-binding activity. *Science,* 249, 1157-1161.

Angel, P. and Karin, M. (1991). The role of Jun, Fos and AP-1 complex in cell-proliferation and transformation. *Biochim. Biophys.Acta*, 1072, 129-157.

Aw, T.Y. (1999). Cellular and molecular responses to oxidative stress and redox imbalance. *Am. J. Clin. Nutr.*, 70, 557-565.

Aw, T.Y. (2003). Cellular redox: a modulator of intestinal epithelial cell proliferation. *News Physiol. Sci.*, 18, 201-204.

Bass, R., Ruddock, L.W., Klappa, P., Freedman, R.B. (2004). A major fraction of endoplasmic reticulum-located glutathione is present as mixed disulfides with proteins. *J. Biol. Chem.,* 297, 5257-5262.

Beer, S.M., Taylor, E.R., Brown, S.E., Dahm, C.C., Costa, N.J., Runswick, M.J., Murphy, M.P. (2004). Glutaredoxin 2 catalyzes the reversible oxidation and glutathionylation of mitochondrial membrane thiol proteins. *J. Biol. Chem.*, 279, 47939-47951.

Biswas, S., Seema, C., Rahman, I. (2006). Redox modifications of protein-thiols: emerging roles in cell signaling. *Biochem. Pharmacol.*, 71, 551-564.

Biteau, B., Labarre, J., Toledano, M.B. (2003). ATP-dependent reduction of cysteine-sulphinic acid by S. cerevisiae sulphiredoxin. *Nature,* 425, 980-984.

Caselli, A., Marzocchini, R., Camici, G., Manao, G., Moneti, G., Pieraccini, G., Ramponi, G. (1998). The inactivation mechanism of low molecular weight phosphotyrosine-protein phosphatase by H_2O_2. *J. Biol. Chem.*, 273, 32554-32560.

Chang, T.S., Jeong, W., Woo, H.A., Lee, S.M., Park, S., Rhee, S.G. (2004). Characterization of mammalian sulfiredoxin and its reactivation of hyperoxidized peroxiredoxin through reduction of cysteine sulfinic acid in the active site to cysteine. *J. Biol. Chem.*, 279, 50994-51001.

Chen, Z. and Lash, L.H. (1998). Evidence for mitochondrial uptake of glutathione by dicarboxylate and 2-oxoglutarate carriers. *J. Pharmacol. Exp. Ther.*, 285, 608-618.

Chiarugi, P. and Buricchi, F. (2007). Protein tyrosine phosphorylation and reversible oxidation: two cross-talking posttranslational modifications. *Antiox. Redox Signal.*, 9, 1-24.

Chiarugi, P. and Cirri, P. (2003). Redox regulation of protein tyrosine phosphatases during receptor tyrosine kinase signal transduction. *Trends Biochem. Sci.*, 28, 509-514.

Chiarugi, P., Fiaschi, T., Taddei, M.L., Talini, D., Giannoni, E., Raugei, G., Ramponi, G. (2001). Two vicinal cysteines confer a peculiar redox regulation to low molecular weight protein tyrosine phosphatase in response to platelet-derived growth factor receptor stimulation. *J. Biol. Chem.,* 276, 33478-33487.

Chiarugi, P. and Giannoni, E. (2005). Anchorage-dependent cell growth: tyrosine kinases and phosphatases meet redox regulation. *Antiox. Redox Signal.,* 7, 578-592.

Chiarugi, P., Panni, G., Giannoni, E., Taddei, L., Colavitti, L., Raugei, G., Symons, M., Borello, S., Galleoti, T., Ramponi, G. (2003). Reactive oxygen species as essential mediators of cell adhesion: the oxidative inhibition of a FAK tyrosine phosphatase is required for cell adhesion. *J. Cell Biol.,* 161, 933-944.

Cho, S.H., Lee, C.H., Ahn, Y., Kin, H., Kin, H., Ahn, C., Yang, K., Lee, S. (2004). Redox regulation of PTEN and protein tyrosine phosphatases in H_2O_2-mediated cell signaling. *FEBS Lett.,* 560, 7-13.

Cohen, R.A. and Adachi, T. (2006). Nitric oxide-induced vasodilatation: regulation by physiologic S-glutathionylation and pathologic oxidation of the sarcoplasmic endoplasmic reticulum calcium ATPase. *Trends. Cardiovasc. Med.*, 16, 109-114.

Cotgreave, I.A. (2003). Analytical developments in the assay of intra- and extra-cellular GSH homeostasis: specific protein S-glutathionylation, cellular GSH and mixed disulfide compartmentalization and interstitial GSH redox balance. *Biofactors,* 17, 269-277.

Damdimopoulos, A.E., Miranda-Vizuete, A., Pelto-Huikko, M., Gustafsson, J.A., Spyrou, G. (2002). Human mitochondrial thioredoxin. *J. Biol.Chem.*, 36, 33249-33257.

Denu, J.M. and Tanner, K.G. (1998). Specific and reversible inactivation of protein tyrosine phosphatases by hydrogen peroxide: evidence for a sulfenic acid intermediate and implication for redox regulation. *Biochemistry,* 37, 5633-5642.

Fernandez-Checa, J.C., Garcia-Ruiz, C., Ookhtens, M., Kaplowitz, N. (1991). Impaired uptake of glutathione by hepatic mitochondrial from chronic ethanol-fed rats. Tracer kinetic studies in vitro and in vivo and susceptibility to oxidant stress. *J. Clin. Invest.*, 87, 397-405.

Findlay, V.J., Townsend, D.M., Morris, T.E, Fraser, J.P., He, L., Tew, K.D. (2006). A novel role for human sulfiredoxin in the reversal of glutathionylation. *Cancer Res.* 66, 6800-6806.

Frand, A.R. and Kaiser, C.A. (2000). Two pairs of conserved cysteines are required for the oxidative activity of Ero1p in protein disulfide bond formation in the endoplasmic reticulum. *Mol. Biol. Cell*, 11, 2833-2843.

Freedman, J., Frei, B., Welch, G., Loscalzo, J. (1995). Glutathione peroxidase potentiates the inhibition of platelet function by S-nitrosothiols. *J. Clin. Invest.*, 96, 394-400.

Fujino, G., Noguchi, T., Takeda, K., Ichijo, H. (2006). Thioredoxin and protien kinases in redox signaling. *Seminars Cancer Biol.,* 16, 427-435.

Funato, Y., Michiue, T., Asashima, M., Miki, H. (2006). The thioredoxin-related redox-regulating protein nucleoredoxin inhibits Wnt-beta-catenin signalling through dishevelled. *Nat. Cell Biol.*, 8, 501-508.

Gaston, B.M., Carver, J., Doctor, A., Palmer, L.A. (2003). S-nitrosylation signaling in cell biology. Mol. Interventions, 3, 253-263.

Ghezzi, P. (2005). Regulation of protein functions by glutathionylation. *Free Rad. Research*, 39, 573-580.

Giannoni, E., Buricchi, F., Raugei, G., Ramponi, G., Chiarugi, P. (2005). Intracellular reactive oxygen species activate Src tyrosine kinase during cell adhesion and anchorage-dependent cell growth. *Moll. Cell Biol.*, 25, 6391-6403.

Gladyshev, V.N., Liu, A., Novoselov, S.V., Krysan, K., Sun, Q.A., Kryukov, V.M., Kryukov, G.V., Lou, M.F. (2001). Identification and characterization of a new mammalian glutaredoxin (thioltransferase), Grx2. *J. Biol. Chem.,* 276, 30374-30380.

Go, Y. and Jones, D.P. (2005). Intracellular proatherogenic and cell adhesion modulated by extracellular thiol/disulfide redox state. *Circulation*, 111, 2973-2980.

Gorlach, A., Klappa, P., Kietzmann T. (2006). The endoplasmic reticulum: folding, calcium homeostasis, signaling and redox control. *Antioxid. Redox Signal,* 8, 1391-1418.

Gotoh, Y., Noda, T., Iwakiri, R., Fujimoto, K., Rhoads, C.A., Aw, T.Y. (2002). Lipid peroxide-induced redox imbalance differentially mediates CaCo-2 cell proliferation and growth arrest. *Cell Prolif.,* 35, 221-235.

Haendeler, J., Hoffmann, J., Tischler, V., Berk, B.C., Zeither, A.M., Dimmler, S. (2002). Redox regulatory and anti-apoptotic function of thioredoxin depends on S-nitrosylation at Cysteine 69. *Nat. Cell Biol.*, 4, 743-749.

Hallberg, E. and Rydstrom, J. (1989). Selective oxidation of mitochondrial glutathione in cultured rat adrenal cells and its relation to polycyclic aromatic hydrocarbon-induced cytotoxicity. *Arch. Biochem. Biophys.*, 270, 662-671.

Hansen, J.M., Go, Y.M., Jones, D.P. (2006). Nuclear and mitochondrial compartmentation of oxidative stress and redox signaling. *Ann. Rev. Pharmacol. Toxicol.,* 46, 215-234.

Hansen, J.M., Watson, W.H., Jones, D.P. (2004). Compartmentation of Nrf-2 redox control: regulation of cytoplasmic activation by glutathione and DNA binding by thioredoxin-1. *Toxicol. Sci.,* 82, 308-317.

Harris, S. and Levine, A.J. (2005). The p53 pathway: positive and negative feedback loops. *Oncogene,* 24, 2899-2908.

Hayashi, T., Ueno, Y., Okamoto, T. (1993). Oxidoreductive regulation of nuclear factor kappa B. Involvement of cellular reducing catalyst thioredoxin. *J. Biol. Chem.*, 268, 11380-11388.

Hermann, C., Assmus, B., Urbich, C., Zeiher, A.M., Dimmeler, S. (2000). Insulin-mediated stimulation of protein kinase Akt: A potent survival signaling cascade for endothelial cells. *Arterioscler. Thromb. Vasc Biol.*, 20, 402-409.

Hess, D.T., Matsumoto A., Kim, S.O., Marshall, H., Stamler, J.S. (2005). Protein S-nitrosylation: purview and parameters. *Nat. Rev. Moll. Cell Biol.*, 6, 150-166.

Hirota, K., Matsui, M., Iwata, S., Nishiyama, A., Mori, K., Yodoi, J. (1997). AP-1 transcriptional activity is regulated by a direct association between thioredoxin and Ref-1. *Proc. Natl. Acad. Sci. USA.*, 94, 3633-3638.

Hirota, K., Matsui, M., Murata, M., Takashima, Y., Cheng, F.S., Itoh, T., Fukuda, K., Yodoi, J. (2000). Nucleoredoxin, glutaredoxin, and thioredoxin differentially regulate NF-κB, AP-1, and CREB activation in HEK293 cells. *Biochem. Biophys. Res. Commun.*, 274, 177-182.

Hirota, K., Murata, M., Sachi, Y., Nakamura, H., Takeuchi, J., Mori, K., Yodoi, J. (1999). Distinct roles of thioredoxin in the cytoplasm and in the nucleus. A two step mechanism of redox regulation of transcription factor NF-kB. *J. Biol. Chem.*, 274, 27891-27897.

Hofmann, B., Hecht, H.J., Flohe, L. (2002). Peroxiredoxins. *Biol. Chem.*, 383, 347-364.

Hwang, C., Sinskey, A.J., Lodish, H.F. (1992). Oxidized redox state of glutathione in the endoplasmic reticulum. *Science,* 257, 1496-1502.

Itoh, K., Wakabayashi, N., Katoh, Y., Ishii, T., O'Connor, T., Yamamoto, M. (2003). Keap1 regulates both cytoplasmic-nuclear shuttling and degradation of Nrf2 in response to electrophiles. *Genes Cells*, 8, 379-391.

Iwakiri, R., Gotoh, Y., Noda, T., Sugihara, H., Fujimoto, K., Fuseler, J.W., Aw, T.Y. (2001). Programmed cell death in rat intestine: effect of feeding and fasting. *Scan. J. Gastroenterol.,* 36, 39-47.

Jia, L., Bonaventura, C., Bonaventura, J., Stamler, J.S. (1996). Nitrosohaemoglobin: A dynamic activity of blood involved in vascular control. *Nature*, 380, 221-226.

Jocelyn, P.C. and Kamminga, A. (1974). The non-protein thiol of rat liver mitochondria. *Biochem. Biophys. Acta*, 343, 356-362.

Johansson, C., Lilig, C.H., Holmgren, A. (2004). Human mitochondrial glutaredoxin reduces S-glutathionylated proteins with high affinity accepting electrons from either glutathione or thioredoxin reductase. *J. Biol. Chem.*, 279, 7537-7543.

Jonas, C.R., Ziegler, T.R., Gu, L.H., Jones, D.P. (2002). Extracellular thiol/disulfide redox state affects proliferation rate in a human colon carcinoma (Caco2) cell line. *Free Radic. Biol. Med.,* 33, 1499-506.

Jones, D.P. (2002). Redox potential of GSH/GSSG couple: assay and biological significance. *Methods Enzymol.*, 348, 93-112.

Jones, D.P. (2006). Redefining oxidative stress. *Antiox. Redox Signal.,* 8, 1865-1879.

Jones, D.P. (2006b). Extracellular redox state: refining the definition of oxidative stress in aging. *Rejuvenation Res.,* 9, 169-181.

Jones, D.P., Go, C.L., Anderson, T.R., Ziegler, J.M., Kincade, J., Kirlin, W.G. (2004). Cysteine/cystine couple is a newly recognized node in the circuitry for biologic redox signaling and control. *FASEB J.,* 18, 1246-1248.

Kamata, H. and Hirata, H. (1999). Redox regulating of cellular signaling. *Cell.Signal.,* 11, 1-14.

Kim, J.E. and Tannenbaum, S.R. (2004). Insulin regulates cleavage of procaspase-9 via binding of X chromosome-linked inhibitor of apoptosis protein in HT-29 cells. *Cancer Res.* 64, 9070-9075.

Kim, S.K., Woodcroft, K.J., Khodadadeh, S.S., Novak, R.F. (2004). Insulin signaling regulates gamma-glutamylcysteine ligase catalytic subunit expression in primary cultured rat hepatocytes. *J. Pharmacol. Exp. Ther.,* 311, 99-108.

Kyriakis, J.M. and Avruch, J. (2001). Mammalian mitogen-activated protein kinase signal transduction pathways activated by stress and inflammation. *Physiol. Reviews*, 81, 807-860.

Lash, L.H. (2005). Role of glutathione transport process in kidney function. *Toxicol. Appl. Pharmacol.,* 204, 329-342.

Lee, S.R., Yang, K.S., Kwon, J., Lee, C., Jeong, W., Rhee, S.G. (2002). Reversible inactivation of the tumor suppressor PTEN by H_2O_2. *J. Biol. Chem.*, 277, 20336-20342.

Lillig, C.H. and Holmgren, A. (2007). Thioredoxin and related molecules-from biology to health and disease. *Antioxid Redox Signal*., 9, 25-47.

Lillig, C.H., Berndt, C., Vergnolle, O., Lonn, M.E., Hudemann, C., Bill, E., Holmgren, A. (2005). Characterization of human glutaredoxin 2 as iron-sulfur protein: a possible role as redox sensor. *Proc.Natl. Acad. Sci. USA.,* 102, 8168-8173.

Liu, H., Nishitoh, H., Ichijo, H., Kyriakis, J.M. (2000). Activation of apoptosis signal-regulating kinase 1 (ASK 1) by tumor necrosis factor receptor-associated factor 2 requires prior dissociation of the ASK1 inhibitor thioredoxin. *Mol. Cell Biol*., 20, 2198-2208.

Liu, Y. and Min, W. (2002). Thioredoxin promotes ASK1 ubiquination and degradation to inhibit ASK1-mediated apoptosis in a redox activity-independent manner. *Circ.Res*., 90, 1259-1266.

Lluis, J.M., Morales, A., Blacol, C., Corell, A., Mari, M., Garcia-Ruiz, C., Fernandez-Checa, J.C. (2005). Critical role of mitochondrial glutathione in the survival of hepatocytes during hypoxia. *J. Biol. Chem*., 280, 3324-3232.

Mannick, J., Schonhoff, C., Papeta, N, Ghafourifar, P., Szibor, M., Fang, K., Gaston, B. (2001). S-nitrosylation of mitochondrial caspases. *J. Cell. Biol*., 154, 1111-1116.

Marshall, H. and Stamler, J. (2001). Inhibition of NF-kappa B by S-nitrosylation. *Biochemistry*, 40, 1688-1693.

Matthews, J.R., Wakasugi, N., Verelizier, J.R., Yodoi, J., Hay, R.T. (1992). Thioredoxin regulates the DNA binding activity of NF-kappaB by reduction of a disulfide bond involving cysteine 62. *Nucleic Acid Res*., 20, 3821-3830.

Meister, A. and Anderson, M.E. (1983). Glutathione. *Ann. Rev. Biochem*., 52, 711-760.

Meng, T.C., Buckley, D.A., Galic, S., Tiganis, T., Tongs, N.K. (2004). Regulation of insulin signaling through reversible oxidation of the protein-tyrosine phosphatases TC45 and PTP1B. *J. Biol. Chem*., 279, 37716-37725.

Meng, T.C., Fukada, K., Tongs, N.K. (2002). Reversible oxidation and inactivation of protein-tyrosine phosphatases *in vivo*. *Mol. Cell*, 9, 383-399.

Meredith, M.J. and Reed, D.J. (1982). Status of the mitochondrial pool of glutathione in the isolated hepatocytes. *J. Biol. Chem*., 257, 3747-3753.

Meuillet, E.J., Mahadevan, D., Berggren, M., Coon, A., Powis, G. (2004). Thioredoxin-1 binds to the C2 domain of PTEN inhibiting PTEN's lipid phosphatase activity and membrane binding: a mechanism for the functional loss of PTEN's tumor suppressor activity. *Arch. Biochem. Biophys.,* 429, 123-133.

Minetti, M., Mallozzi, C., Stasi, A.M. (2002). Peroxynitrite activates kinases of src family and upregulates tyrosine phosphorylation signaling. *Free Rad. Biol. Med.*, 33, 744-754.

Moldovan, L. and Moldovan, N.L. (2004). Oxygen free radicals and redox biology of organelles. Histochem. *Cell Biol.*, 122, 395-412.

Morita, K., Saitoh, M., Tobiume, K., Matsuura, H., Enomoto, S., Nishitoh, H., Ichijo, H. (2001). Negative feedback regulation of ASK1 by protein phosphatase 5 (PP5) in response to oxidative stress. *EMBO J.,* 20, 6028-6036.

Muratam, H., Ihara, Y., Nakamura, H., Yodoi, J., Sumikawa, K., Kondo, T. (2003). Glutaredoxin exerts an antiapoptotic effect by regulating the redox state of Akt. *J. Biol. Chem.*, 278, 50226-50233.

Nakamura, H. (2005). Thioredoxin and its related molecules: update. *Antiox. Redox Signaling,* 7, 823-828.

Nakamura, H., Vaage, J., Valen G., Padilla, C.A, Bjornstedt, M., Holmgren, A. (1998). Measurements of plasma glutaredoxin and thioredoxin in healthy volunteers and during open-heart surgery. *Free Radic Biol Med.*, 24, 1176-1186.

Nishitoh, H., Matsuzawa, A., Tobiume, K., Saegusa, K., Takeda, K., Inoue, K., Hori, S., Kakizuka, A., Ichijo, H. (2002). ASK1 is essential for endoplasmic reticulum stress-induced neuronal cell death triggered by expanded polyglutamine repeats. *Genes Dev.*, 16, 1345-1355.

Nishitoh, H., Saitoh, M., Mochida, Y., Takeda, K., Nakano, H., Rothe, M., Miyazono, K., Ichijo, H. (1998). ASK1 is essential for JNK/SAPK activation by TRAF2. *Mol. Cell*, 2, 389-395.

Nkabyo, Y.S., Go, Y.M., Ziegler, T.M., Jones, D.P. (2005). Extracellular cysteine/cystine redox regulates the p44/p42 MAPK pathway by metalloproteinase-dependent epidermal growth factor receptor signaling. *Am. J. Physiol.*, 289, G70- G78.

Nkabyo, Y.S., Siegler, T.R., Li, H., Watson, W.H., Jones, D.P. (2002). Glutathione and thioredoxin redox during differentiation in a human colon carcinoma (Caco-2) cell line. *Am. J. Physiol. Gastrointest. Liver*, 283, G1352-G1359.

Noda, T., Iwakiri, R., Fujimoto, K., Aw, T.Y. (2001). Induction of mild intracellular redox imbalance inhibits proliferation of CaCo-2 cells. *FASEB J.*, 15, 2131-2139.

Noda, T., Iwakiri, R., Fujimoto, K., Rhoads, C.A., Aw, T.Y. (2002). Exogenous cysteine and cystine promote cell proliferation in Caco-2 cells. *Cell Prolif.*, 35, 117-129.

Noguchi, T., Takeda, K., Matsuzawa, A., Saegusa, K., Nakano, H., Gohda, J., Inoue, J., Ichijo, H. (2005). Recruitment of tumor necrosis factor receptor-associated factor family proteins to the ADK1 signalosome is essential for oxidative stress-induced cell death. *J. Biol. Chem.,* 280, 37033-37040.

Okouchi, M., Okayama, N., Alexander, J.S., Aw, T.Y. (2006). Nrf2-dependent glutamate-L-cysteine ligase catalytic subunit expression mediates insulin protection against hyperglycemia-induced brain endothelial cell apoptosis. *Current Neurovascular Research*, 3, 249-261.

Olafsdottir, K. and Reed, D.J. (1988). Retention of oxidized glutathione by isolated rat liver mitochondria during hydroperoxide treatment. *Biochim. Biophys. Acta*, 964, 377-382.

Palmer, L.A., Gaston, B., Johns, R.A. (2000). Normoxic stabilization of hypoxia inducible factor 1 expression and activity: Redox-dependent effect of nitrogen oxides. *Mol. Pharmaol.*, 58, 1197-1203.

Patwari, P., Higgins, L.J., Chutkow, W.A., Yoshioka, J., Lee, R.T. (2006). The interaction of thioredoxin with Txnip, evidence for formation of a mixed disulfide by disulfide exchange. *J. Biol. Chem.*, 281, 21884-21891.

Pekkari, K. and Holmgren, A. (2004). Truncated thioredoxin: physiological functions and mechanism. *Antioxid. Redox Signal*, 6, 53-61.

Phalen, T.J., Weirather, K., Denning, P.B., Anathy, V., Howe, A.K., van der Vliet, A., Jonsson, T.J., Poole, L.B., Heintz, N.H. (2006). Oxidation state govens structural transitions in peroxiredoxin II that correlate with cell cycle arrest and recovery. *J. Cell Biol.*, 175, 779-789.

Pias, E.K. and Aw, T.Y. (2002a). Early redox imbalance mediates hydroperoxide-induced apoptosis in mitotic competent undifferentiated cells. *Cell Death and Differ.*, 9, 1007-1016.

Pias, E.K. and Aw, T.Y. (2002b). Apoptosis in mitotic competent undifferentiated cells is induced by redox imbalance independently of reactive oxygen species production. *FASEB J.*, 16, 781-790.

Pias, E.K., Ekshyyan, O.Y., Rhoads, C.A., Fuseler, J.W., Harrison, L., Aw, T.Y. (2003). Differential effects of superoxide dismutase isoform expression on hydroperoxide-induced apoptosis in PC12 cells. *J. Biol. Chem.*, 278. 13294-13301.

Pineda-Molima, E., Klatt, P., Vasquez, J., Marina, A., Garcia de la Coba, M., Perez-Sala, D., Lamas, S. (1999). Glutathionylation of the p50 subunit of NF-kB: a mechanism for redox-induced inhibition of DNA binding. *Biochemistry*, 40, 14134-14142.

Rhee, S.G., Chae, H.Z., Kim, K. (2005). Peroxiredoxins: a historical overview and speculative preview of novel mechanisms and emerging concepts in cell signaling. *Free Rad. Biol. Med.*, 38, 1543-1552.

Rubartelli, A., Bajetto, A., Allavena, G., Wollman, E., Sitia, R. (1992). Secretion of thioredoxin by normal and neoplastic cells through a leaderless secretory pathway. *J. Biol. Chem.*, 267, 24161-24164.

Saitoh, M., Nishitoh, H., Fujii, M., Takeda, K., Tobiume, K., Sawada, Y., Kawabata, M., Miyazono, K., Ichijo, H. (1998). Mammalian thiredoxin is a direct inhibitor of apoptosis signal-regulating kinase (ASK). *EMBO J.*, 17, 2596-2606.

Salmee, A. and Barford, D. (2005). Functions and mechanisms of redox regulation of cysteine-based phosphatases. *Antiox. Redox Signal.*, 7, 560-577.

Schafer, F.Q. and Buetner, G.R. (2001). Redox environment of the cell as viewed through the redox state of the glutathione disulfide/glutathione couple. *Free Rad. Biol. Med.*, 30, 1191-1212.

Seemann, S. and Hainaut, P. (2005). Roles of thioredoxin reductase in APE/Ref-1 in the control of basal p53 stability and activity. *Oncogene*, 24, 3853-3863.

Sies, H. (1999). Glutathione and its cellular functions. *Free Rad. Biol. Med.* 27, 916-921.

Siobhan, E. and Jones, D.P. (2004). Extracellular thiols and thiol/disulfide redox in metabolism. *Annu. Rev. Nutr.*, 24, 481-509.

Soderdahl, T., Enoksson, A., Lundberg, M., Ottersen, O.P., Orrenius, S.., Bolfoldi, G., Cotgrave, I.A. (2002). Visualization of the compartmentalization of glutathione and protein-glutathione mixed disulfide in cultured cells. *FASEB J.*, 17, 124-129.

Song, J.J. and Lee, Y.D. (2003). Differential role of glutaredoxin and thioredoxin in metabolic oxidative stress-induced activation of apoptosis signal-regulated kinase. *Biochem. J.*, 373, 844-853.

Sun, J., Steenbergen, C., Murphy, E. (2006). S-nitrosylation: NO-related redox signaling to protect against oxidative stress. *Antioxid. Redox Signal,* 8, 1693-17075.

Sun, J., Xin, C., Eu, J., Stamler, J., Meissner, G. (2001). Cysteine-3635 is responsible for skeletal muscle ryanodine receptor modulation by NO. *Proc. Natl. Acad. Sci. USA.*, 98, 11158-11162.

Tonks, N.K. and Neel, B.G. (2001). Combinatorial control of the specificity of protein tyrosine phosphatases. *Curr. Opin. Cell Biol.*, 13, 182-195.

Townsend, D.M. and Tew, K.D. (2003). The role of glutathione-S-transferase in anti-cancer drug resistance. *Oncogene*, 22, 7369-7375.

Tsunada, S., Iwakiri, R., Fujimoto, K., Aw, T.Y. (2003a). Chronic lipid hydroperoxide stress suppresses mucosal proliferation in rat intestine: potentiation of ornithine decarboxylase activity by epidermal growth factor. *Dig. Dis. Sci.*, 48, 2333-2341.

Tsunada, S., Iwakiri, R., Noda, T., Fujimoto, K., Fuseler, J.W., Aw, T.Y. (2003b). Chronic exposure to subtoxic levels of peroxidized lipids suppresses mucosal cell turnover in rat small intestine and the reversal by glutathione. *Dig. Dis. Sci.*, 48, 210-222.

Ueno, M., Masutami, H., Arai, R.J., Yamauchi, A., Hirota, K., Sakai, T., Inamoto, T., Yamaoka, Y., Yodoi, J., Nikaido, T. (1999). Thioredoxin-dependent redox regulation of p53-mediated p21 activation. *J. Biol. Chem.*, 50, 35809-35815.

Vazquez, F., Grossman, S.R., Takahashi, Y., Rokas, M.V., Nakamura, N., Sellers, W.R. (2001). Phosphorylation of the PTEN tail acts as an inhibitory switch by preventing its recruitment into a protein complex. *J. Biol. Chem.*, 276, 48627-48630.

Vazquez, F., Ramaswamy, S., Nakamura, N., Sellers, W.R. (2000). Phosphorylation of the PTEN tail regulates protein stability and function. *Mol. Cell Biol.*, 20, 5010-5018.

Wang, T.G., Gotoh, Y., Jennings, M.H., Rhoads, C.A., Aw, T.Y. (2000). Cellular redox imbalance induced by lipid hydroperoxide promotes apoptosis in human colonic CaCo-2 cells. *FASEB J.,* 14, 1567-1576.

Watson, W.H. and Jones, D.P. (2003). Oxidation of thioredoxin during oxidative stress. *FEBS Lett.*, 543, 144-147.

Wood, Z.A., Schroder, J., Harris, R., Poole, L.B. (2003). Structure, mechanism and regulation of peroxiredoxins. *Trends Biochem Sci.,* 28, 32-40.

Zaman, J., Vaughan, J., Hunt, J., Palmer, L., Gaston, B. (2002). Transcription factor Sp1 is involved in the regulation of cystic fibrosis transmembrane conductance regulator by endogenous S-nitroglutathione. *Am. J. Respir. Crit. Care Med.,* 165, A279-A286.

Zhong, L., Arner, E.S., Holmgren, A. (2000). Structure and mechanism of mammalian thioredoxin reductase: the active site is a redox-active selenothiol/selenylsulfide formed from the conserved cysteine-selenocysteine sequence. *Proc. Natl. Acd. Sci. USA.*, 97, 5854-5859.

Zipper, L.M. and Mulcahy, R.T. (2000). Inhibition of ERK and p38 MAP kinases inhibits binding of Nrf2 and induction of GCS genes. *Biochem. Biophys Res. Commun.*, 278, 484-492.

In: Oxidative Stress: Clinical and Biomedical Implications ISBN: 978-1-60021-888-0
Editors: B. M. Matata, M. M. Elahi, pp. 133-144

Chapter VIII

Oxidative Stress: Dysregulation or Bad Setpoint? Interactions with Basal NO

David R. Whitlock*
Nitroceutic LLC
40 Nash Street
Watertown, MA 02472

Abstract

This chapter discusses whether a state of oxidative stress is due to dysregulation or to a bad setpoint and presents literature data supporting the bad setpoint hypothesis, how NO is involved in oxidative stress as a control signal transmitting information on a cell's state of oxidative stress to neighboring cells, a few examples of how cells use oxidative stress as a control parameter to modulate O_2 metabolism in part by removing NO by generating superoxide to disinhibit heme containing enzymes. Because a state of oxidative stress and a low NO state are essentially the same thing, NO physiology is intimately associated with oxidative stress. The chapter closes by emphasizing the importance of basal NO on NO mediated pathways, and how changes in basal NO will affect every feedback pathway utilizing NO with no threshold, including those pathways communicating oxidative stress.

Introduction

One view of oxidative stress is because it can damage cells, oxidative stress is "bad", and occurs due to dysregulation brought about by hypoxia, hyperoxia, xenobiotic toxins, infection, cytokines, a deficiency of antioxidant nutrients, ionizing radiation, or other insults to normal metabolism. I suggest oxidative stress is a powerful modulator of cell metabolism, and cells use it as a control parameter to effect global changes in cell and organ physiology to

* 617-924-4368 (h); 781-972-2317 (w);dwhitlock@nitroceutic.com

deal with those insults. With this view, "correcting" a state of oxidative stress can *only* be done by changing the oxidative stress setpoint, because there is no dysregulation.

Under the bad setpoint hypothesis, during a state of oxidative stress the redox potential of the cell remains as fully regulated as under non-oxidatively stressed conditions. The implications of this change in perspective are significant. If oxidative stress is due to dysregulation, improving regulation should augment normal physiology and rapidly restore a "normal" state of oxidative stress. If oxidative stress is due to a "bad" setpoint, facilitating regulation will be ineffective. In other words, nutritive levels of antioxidants should be sufficient to remedy dysregulation, but pharmacological levels of antioxidants would be necessary to overwhelm a "bad" setpoint (and perhaps unsuccessful even then). I am not suggesting that the "setpoint" hypothesis is incompatible with a "dysregulation" hypothesis. There could be some of both going on. It should be recognized that the concept of oxidative stress "dysregulation" necessitates a "setpoint" about which oxidative stress is regulated. Attempting to overwhelm the oxidative stress setpoint through pharmacological doses of antioxidants is likely to have adverse side effects. Cells evolved hundreds of pathways to control and utilize oxidative stress parameters. It is unlikely crude external control by systemic pharmacological agents will improve on 3+ billion years of evolution in each tissue and cellular compartment which may have idiosyncratic needs and regulatory schemes. Many of the oxidative stress protective and repair pathways are induced by free radicals themselves. Expression and activation of the P450 enzymes are part of the Phase I system of drug detoxification. The Phase II detoxification pathways [1] are induced by Nrf2 [2], the loss of which increases sensitivity to carcinogenesis [3]. Dietary agents cause induction of Phase II compounds [4], as do electrophiles produced by Phase I pathways. Large doses of antioxidants may prevent the induction of systems necessary to deal with the levels of free radicals that remain.

Epidemiology: Diet and Oxidative Stress

Numerous studies show oxidative stress associated with disease states [5], including coronary heart disease, atherosclerosis, dilative cardiomyopathy [6], hypertension [7], diabetes, obesity, end stage kidney failure[8], primary biliary cirrhosis, Alzheimer's disease, Lewy body neuropathies, Parkinson's disease, ischemic brain damage [9], amyotrophic lateral sclerosis, cancer, and autism [10]. There is accumulation of oxidatively damaged proteins and lipids. There is good evidence of lower levels of antioxidant nutrients such as vitamin C, E, selenium, and carotenes.

Ingestion of fruits and vegetables (significant dietary sources of these nutrients), in a self-selected diet, is robustly associated with reduced cardiovascular disease with protection most strongly associated with green leafy vegetables [11, 12]. A large prospective study found the same thing in female health care workers for diabetes type 2 [13]. Consumption of a self-selected diet rich in vegetables was strongly associated with reduced risk for lung cancer in women [14].

A number of large clinical trials have shown antioxidant dietary supplementation is completely ineffective in preventing cardiovascular disease [15]. Supplementation with vitamins E, C and beta carotene "did not produce any significant reductions in the 5-year mortality from, or incidence of, any type of vascular disease, cancer, or other major outcome" [16]. In this very large data set (20,536 starting patients), the supplement group faired slightly worse (though the 0.05 confidence on the odds ratio did overlap with 1). Similarly, in another large trial (9,541 patients), those given vitamin E also faired slightly worse, (though the 0.05 confidence on the odds ratio did overlap with 1) [17]. There was no reduction in disease despite significant increases in blood levels of vitamin E. Modest doses of combinations of antioxidants (vitamin C, E, beta carotene, Se, Zn) for ~7 years had no positive effect on artery health; again, the placebo group had slightly better measures of vascular health [18]. Experimentally induced high intake of fruits and berries in otherwise healthy individuals does not reduce markers for oxidative stress [19].

A recent very large scale meta-analysis of antioxidant supplementation shows increased mortality from dietary supplementation; with all cause mortality increased with beta-carotene, vitamin A, and vitamin E [20].

Supplementation with antioxidants could worsen disease if there is an oxidative stress setpoint and it remains unchanged. The body will simply make more superoxide to overwhelm the increased levels of antioxidants. Free radicals are required components of many signaling pathways [21]. Antioxidant levels beyond the normal physiological range compatible with a specific oxidative stress setpoint may disrupt those signals and interfere with normal pathways regulated via free radicals. The disparity observed between self-selected dietary and supplemental antioxidants is consistent with a "setpoint" not a "dysregulation" hypothesis. The composition of a self-selected diet may be part of the oxidative stress regulatory pathway. A person with a high oxidative stress setpoint may self-select a diet low in antioxidants simply to reduce the need for his body to generate superoxide to overwhelm dietary antioxidants.

That there is essentially no effect on oxidative stress from increases in antioxidants is extremely strong evidence that the level of oxidative stress is not regulated *at all* by levels of dietary antioxidants. If antioxidants were regulating the level of oxidative stress, then increasing antioxidants would reduce oxidative stress. A reduction is not observed; therefore *dietary antioxidants are not regulating oxidative stress*. Pharmacological doses of antioxidants may overwhelm what ever is regulating the level of oxidative stress, but that is a pharmacological effect, not a regulatory one.

Nitric Oxide and Diet: Protective Effects of Nitrate in Leafy Vegetables?

A dietary component active in oxidative stress pathways usually not accounted for in dietary studies is nitrate [22]. Nitrate is particularly abundant in green leafy vegetables, precisely the vegetables that exhibit the greatest "protection" in self-selected diets against diseases associated with oxidative stress. Lettuce and spinach are on the order of 0.1 to 0.3% nitrate. Nitrate is well absorbed, and concentrates in the saliva where bacteria on the tongue reduce nitrate to nitrite. When the nitrite is swallowed, the low pH of the stomach causes reduction of the nitrite to nitric oxide (NO) where a level of 90 ppm was measured [23].

Considerable quantities can be produced, and it is believed that this NO and acidified nitrite are significant components of the normal antimicrobial activity of the stomach [24]. In particular Helicobacter pylori is killed (>5 log CFU reduction) [25] by acidification (pH = 2) of nitrite at levels found in saliva after a high nitrate meal (~200 mg nitrate, ~ 120 g lettuce) [26]. This level of nitrate ingestion has also been shown to generate SNO-thiols in the blood and inhibit platelet function [27]. Nitrite in saliva increases mucosal blood flow and the production of gastric mucus, which is important in protecting the stomach from stomach contents [28]. Dietary nitrate does protect rats from experimental gastritis [29]. Unfortunately, there have been no published trials of dietary nitrate supplementation.

Oxidative Stress Regulated in Each Cell Compartment Independently

Superoxide is produced by a number of different pathways. Two main pathways are mitochondria and the cytochrome P450 enzymes. These enzymes are confined to particular cellular compartments, mitochondria and peroxisomes. In both cases superoxide is vectorally produced to the inside of the organelle. In the case of mitochondria there are 2 membranes between the source of superoxide and the cytoplasm. Superoxide is an anion, and does not pass through cell membranes except through anion channels. In contrast NO is uncharged, and freely diffuses to all cellular compartments. Superoxide confined to the mitochondria or the peroxisome, can thus act as a "sink" for cytoplasm NO, without superoxide entering the cytoplasm. Similarly extracellular superoxide (as from xanthine oxidoreductase or myeloperoxidase) can destroy extracellular NO and affect intracellular NO levels without superoxide entering the cytoplasm. Because superoxide is confined, "oxidative stress" is a local phenomenon, that is, each cell, each mitochondria, each peroxisome determines its own state of oxidative stress. NO is an uncharged radical that has sufficient mobility and sufficiently low reactivity with non-radicals to have a significant lifetime and diffusion distance, so NO can act to directly "communicate" the state of oxidative stress between cell compartments. The solubility of NO in lipid membranes is ~9x higher than in aqueous phase [30], so lipid membranes act as "antennas" and distribution channels to collect NO.

Because regulation of oxidative stress is local, there must be signals so all cellular compartments can be regulated in concert. NO may be one of the species that communicates and thus determines the "setpoint" for oxidative stress. A "high" NO level is a state of low superoxide and low oxidative stress. "High" is a relative term. For example, O_2 is required by all cells and within quite a narrow range. Reduction in O_2 level of 50% (equivalent to ~5,000 m altitude) is not universally tolerable. Mice exposed to 100% O_2 (5 times greater) die after 3 or 4 days [31]. Thus, the tolerable O_2 range (that is the tolerable range, not the optimal range) is less than an order of magnitude. O_2 and NO are similar in a number of respects but NO is more reactive and more complicated in its chemistry. They both bind to heme, but the binding constant for NO is many times greater. The association binding constants of O_2, CO and NO to sperm whale myoglobin are 1.2, 27 and 220,000 respectively [32]. It would be unreasonable to assume the tolerable NO range would be greater than that for O_2. Measuring

NO on the length, time and concentrations scales that must be important is extremely challenging, and not yet possible.

NO is a signaling molecule, so its concentration is not constant in either time or space. The best characterized target for NO is soluble guanylyl cyclase (sGC), which has an EC50 of ~ 20 nM/L. The response curve is quite non-linear in that range, and drops off rapidly below 10 nM/L [33]. The basal level of NO in the body therefore cannot exceed ~20 nM/L because activation of sGC would cause vasodilatation and local hyperemia. Hemoglobin is a sink for NO, so hyperemia reduces local NO levels. It must be appreciated that if iNOS is highly expressed, as in septic shock, the basal NO level doesn't get very high. For the most part, sGC while activated is not fully activated, so the NO concentration remains below that point. The NO production rate becomes very high, but there is still minimal NO accumulation. As fast as NO is produced, it is degraded to nitrite and nitrate. Accumulation of nitrite and nitrate reflect NO production rates, *not* NO concentrations. NO effects are mediated by NO concentrations, not NO production rates. The NO production rate equals the NO destruction rate (no accumulation).

Nitric Oxide: Influences on Oxidative Stress

Essentially all free radicals react very rapidly with NO. As a small molecule, NO can diffuse to stericly hindered sites that more bulky antioxidants cannot reach. NO rapidly reacts with peroxidized lipids, quenching the free radical mediated chain reaction. NO reacts with peroxidized lipid radicals considerably faster than does vitamin E (~2E9 M^{-1} s^{-1} vs. 5E5), and considerably faster than the propagation reaction (1.3E3) [34]. In vitro, 7 nM/min NO was as effective as 50 μM/L α- tocopherol (50% inhibition of 2,2'-azobis(2-amidinopropane) dihydrochloride-catalyzed linoleic acid oxidation). Oxidative consumption of α- tocopherol did not occur until NO concentrations fell below ~10 nM/L. The combination of NO and α-tocopherol prevented lipid oxidation better than either alone.

NO is more reactive than vitamin E toward radical species, therefore modest increases in NO will have greater antioxidant effect than modest increases in vitamin E. NO inhibits metal induced peroxidation of lipids by deactivating Fenton active metals. NO per se, is not a strong enough oxidant to peroxidize lipids. Virtually all NO derived species that are active enough to peroxidize lipids are derived from compounds that are themselves capable of oxidizing lipids [35]. It is not NO that causes these peroxidation reactions; rather these reactions only proceed when external oxidants are in excess of NO [36]. When there is an excess of NO, the free radical chain reaction oxidation of lipids is terminated by NO quenching lipid radicals. When there is insufficient NO, the radicals are unquenched and chain reaction oxidation proceeds.

Superoxide reacts with NO, at near diffusion limited reaction rates (k = 4.3E9 (L mol^{-1} s^{-1})) [37]. One cannot have both superoxide and NO present simultaneously. Which ever one is in excess will destroy the one that isn't [38]. The nitration of tyrosine by fluxes of superoxide and NO occurs in vitro only at near equimolar fluxes [39]. The actually nitrating species is thought to be NO_2. NO_2 rapidly reacts with NO to form N_2O_3 (k=1.1E9). When NO or superoxide is in excess, nitration does not occur [40]. With a 2 to 3 fold excess of NO over superoxide, the oxidative effects of peroxynitrite are not observed..

The diffusion limited reaction of NO and superoxide has particular relevance as to the location where peroxynitrite is generated, and so what species is "attacked" by the peroxynitrite generated. If the level of NO is high, when superoxide is generated, peroxynitrite will be created near the superoxide source, and inhibit the production of superoxide. If the level of NO is low, then superoxide will dominate, and peroxynitrite will be generated at the source of NO, likely the NOS enzymes. Peroxynitrite has the effect of oxidizing the zinc thiolate couple in NOS and causing it to operate in the uncoupled mode, where it generates superoxide rather than NO [41]. Thus a state dominated by superoxide exhibits hysteresis because under conditions of superoxide dominance, NOS becomes a source of superoxide and not NO. To switch from a superoxide dominated state to a NO dominated state requires an external source of NO, that is, NO not from NOS.

The presence of NO metabolites following oxidative stress (i.e. nitrotyrosine or nitrate) is not proof that "too much" NO caused or exacerbated that oxidative stress, any more than O_2 metabolites, (i.e. peroxidized compounds or CO_2) is proof that "too much" O_2 caused it either. Just as too little or too much O_2 can cause "oxidative stress", so to can too little or too much NO cause "nitrosative stress". In the previous example, it is not the absolute quantity of NO or NO metabolites that is important; rather it is whether NO is in excess over superoxide in the particular cellular compartment under consideration. Conceivably, one could have 3 different states, high superoxide and low NO (the oxidative stress state), low superoxide and high NO, or low levels of both. This last state is not stable, there is O_2 in all tissues, there are free electrons from normal and abnormal metabolic activities, so there are always sources of superoxide. If a state of low superoxide and low NO could be induced, it would rapidly decay to one dominated by superoxide which then exhibits hysteresis. A state of oxidative stress is a low NO state. Whether adverse effects are due to high superoxide or low NO is important in elucidating mechanisms, but treatment is likely the same because both states only occur simultaneously.

Oxidative Stress: A Feature Increasing Aerobic Metabolism via NO Destruction

Oxidative stress is produced in response to a number of different metabolic states. I suggest that oxidative stress is adaptive, in that it provides the cell with greater resources to deal with what ever metabolic insult is being dealt with, and is in fact purposefully generated to deal with that insult. But it is a state that while beneficial in the short term is harmful in the long term, primarily due to the different time constants of different physiological processes regulated by NO, for example inhibition of cytochrome oxidase (seconds) vs. mitochondria biogenesis (weeks).

Mammalian cells are aerobic. Organic compounds (glucose and fatty acids) are conveyed via the blood stream, actively ported to cells, fed into the citric acid cycle, oxidized to CO_2 and water in mitochondria, producing reducing equivalents and ATP. Oxygen is absorbed in the lung, transferred to hemoglobin in erythrocytes, carried by the blood stream, where it diffuses from the terminal capillaries to the mitochondria. The transport of O_2 is by purely passive diffusion down a concentration (actually chemical potential) gradient. There is no active O_2 transport. The chemical potential of O_2 (often measured as a partial pressure) at the mitochondria is the lowest point in the body. It must be, because O_2 diffuses to mitochondria and is consumed.

ATP production is extremely well regulated [42]. Some organs, such as the heart, can change their ATP production by an order of magnitude. O_2 consumption changes by an order of magnitude also. The geometry of the vasculature does not change appreciably during this change (although there is some increased recruitment of blood vessels). With a constant O_2 partial pressure in the blood, and a constant mass transfer area, and a constant diffusion length, the only way 10x more O_2 can be delivered to the mitochondria, is if the concentration gradient increases. The only way for the concentration gradient to increase is for the O_2 level at the mitochondria to go down (the level in the capillary is nearly constant and is fixed by the O_2 content of the atmosphere). If the mitochondria O_2 level goes down an order of magnitude, and the mitochondria O_2 consumption goes up an order of magnitude, the specific O_2 consumption (O_2 consumed per cytochrome oxidase per Torr O_2) must go up 2 orders of magnitude.

O_2 consumption occurs at cytochrome oxidase and is tonally inhibited by NO [43]. To remove NO inhibition, the NO must be removed. One way is to generate superoxide, which destroys NO, disinhibits cytochrome oxidase, the mitochondria now consume O_2 at a higher rate, the O_2 level local to the mitochondria drops, the concentration gradient of O_2 from the vessel to the mitochondria increases, and more O_2 can diffuse to the now more active mitochondria. Thus generation of superoxide is a "feature" that increases local metabolic capacity by disinhibiting cytochrome oxidase. However, this only occurs if the cytochrome oxidase is already inhibited by NO. If cytochrome oxidase is not inhibited by NO (i.e. under conditions of low basal NO), adding superoxide does not increase metabolism, it simply causes oxidative damage.

Production of superoxide is observed in hypoxia and in reperfusion, and is a major cause of the damage. A little superoxide might be good, if it increased O_2 availability by increasing O_2 diffusion, but this can *only* occur when there is sufficient NO present initially to have inhibited cytochrome oxidase in the first place.

Similarly, the cytochrome P450 enzymes produce superoxide, most of the P450 enzymes are associated with membranes of peroxisomes, and the superoxide as well as H_2O_2 is released into the inside of the peroxisome [44]. Peroxisomes contain SOD, and also catalase. SOD dismutates superoxide to H_2O_2, and catalase decomposes H_2O_2 to O_2. Nitric oxide inhibits catalase [45], and superoxide reactivates catalase that has been inhibited by NO [46]. The coupling of electron flow of the P450 enzymes is less efficient than the coupling of the electron flow in mitochondria. The P450 enzymes collect electrons from NADPH even under basal conditions, and so there is some production of ROS even under basal conditions. This basal production of ROS may be one of the major determinants of oxidative stress of the cell.

The P450 enzymes are major pathways for the synthesis of steroids and also for the metabolizing of xenobiotic chemicals. NO inhibition of the P450 enzymes [47]. involved in steroid synthesis is part of the normal pathway by which steroid synthesis is regulated [48]. Production of ROS by the P450 enzymes does regulate their expression and post translation function. I suspect that the production of superoxide by the P450 enzymes serves to disinhibit them by locally destroying NO, similar to how production of superoxide by mitochondria releases the inhibition of cytochrome oxidase by NO.

Many xenobiotic chemicals cause the release of superoxide, for example many azo dyes generate superoxide when metabolized by P450 enzymes in microsomes [49]. By releasing the inhibition of P450 enzymes, the capacity of a cell to metabolize xenobiotic compounds in increased. This may be an adaptive rationale for a state of oxidative stress under conditions of infection or toxin ingestion. The cell's capacity to deal with xenobiotic insults is increased, and H_2O_2 is generated to diffuse to the extracellular space and interact with myeloperoxidase. But again, just as in the mitochondria, peroxynitrite oxidizes the Zn thiolate center of NOS and converts it from a source of NO to a source of superoxide. Perhaps an appropriate response to rapidly mobilize metabolic resources to deal with a toxin, but not a state that can not easily be reversed without an external source of NO.

The sensitivity of some individuals to azo dyes and other xenobiotic chemicals might reflect a low NO state. When such compounds are metabolized, they will generate superoxide at P450 enzymes, which will draw down NO levels. The extreme sensitivity that some individuals have may reflect a pathologically low NO level to start with. Any metabolic state that increases superoxide production will cause a similar exacerbation of a low NO state. For example hyperglycemia is well known to increase superoxide formation [50]. The exacerbation of some symptoms upon consuming high glycemic index foods may result from a similar effect.

No Threshold for Basal NO Changes on NO Mediated Feedback Pathways

NO is involved in many pathways as a regulatory species. NO sensors, such as sGC only sense the sum of NO from all NO sources. NO is generated, diffuses some distance, and then activates the NO sensor when a threshold is exceeded. Because NO is already in the active range; for all pathways that NO is actively regulating there is no threshold for changes in basal NO levels affecting NO regulated pathways. Any change in basal NO will affect the range, onset time, and duration of every NO signal with *no threshold.* As a species inside the feedback loop which controls radical production (mostly superoxide), NO can be thought of as a part of the "oxidative stress setpoint". Because NO is inside the feedback loop involved in superoxide regulation, any changes to the NO level by conditions outside that feedback loop will affect the feedback loop with *no threshold.* That is, because the NO level is in the "active range", where changes in NO reflect changes in superoxide output, *any change in NO level* will affect all pathways that involve NO as a signaling molecule. Because of the positive feedback and hysteresis associated with switching from a high NO state to the oxidative stress state, a low basal NO level will cause that transition to happen.

Competing Financial Interest

D. R. Whitlock is working on commercializing products using skin resident commensal autotrophic ammonia oxidizing bacteria to naturally supply basal NO under normal physiological control via sweating to prevent and treat a number of disorders caused or exacerbated by low NO, including all of those associated with oxidative stress.

References

[1] Wolf CR. Chemoprevention: increased potential to bear fruit. *Proc. Natl. Acad. Sci. USA.* 2001 Mar 13;98(6):2941-3.

[2] Lee JS, Surh YJ. Nrf2 as a novel molecular target for chemoprevention. *Cancer Lett.* 2005 Jun 28;224(2):171-84.

[3] Ramos-Gomez M, Kwak MK, Dolan PM, Itoh K, Yamamoto M, Talalay P, Kensler TW. Sensitivity to carcinogenesis is increased and chemoprotective efficacy of enzyme inducers is lost in nrf2 transcription factor-deficient mice. *Proc. Natl. Acad. Sci. USA.* 2001 Mar 13;98(6):3410-5.

[4] Chen C, Kong AN. Dietary chemopreventive compounds and ARE/EpRE signaling. *Free Radic. Biol. Med.* 2004 Jun 15;36(12):1505-16.

[5] Mayne ST. Antioxidant nutrients and chronic disease: use of biomarkers of exposure and oxidative stress status in epidemiologic research. *J. Nutr.* 2003 Mar;133 Suppl 3:933S-940S.

[6] Keith M, Geranmayegan A, Sole MJ, Kurian R, Robinson A, Omran AS, Jeejeebhoy KN. Increased Oxidative Stress in Patients With Congestive Heart Failure. *J. Am. Coll Cardiol.* 1998 May;31(6):1352-6.

[7] Kedziora-Kornatowska K, Czuczejko J, Pawluk H, Kornatowski T, Motyl J, Szadujkis-Szadurski L, Szewczyk-Golec K, Kedziora J. The markers of oxidative stress and activity of the antioxidant system in the blood of elderly patients with essential arterial hypertension. *Cell Mol. Biol. Lett.* 2004;9(4A):635-41.

[8] Himmelfarb J, Stenvinkel P, Ikizler TA, Hakim RM. The elephant in uremia: Oxidant stress as a unifying concept of cardiovascular disease in uremia. *Kidney Int.* 2002 Nov;62(5):1524-38.

[9] Warner DS, Sheng H, Batinic-Haberle I. Oxidants, antioxidants and the ischemic brain. *J. Exp. Biol.* 2004 Aug;207(Pt 18):3221-31.

[10] McGinnis WR. Oxidative stress in autism. *Altern. Ther. Health Med.* 2005 Jan-Feb;11(1):19.

[11] Joshipura KJ, Hu FB, Manson JE, Stampfer MJ, Rimm EB, Speizer FE, Colditz G, Ascherio A, Rosner B, Spiegelman D, Willett WC. The effect of fruit and vegetable intake on risk for coronary heart disease. *Ann. Intern. Med.* 2001 Jun 19;134(12):1106-14.

[12] Bazzano LA, He J, Ogden LG, Loria CM, Vupputuri S, Myers L, Whelton PK. Fruit and vegetable intake and risk of cardiovascular disease in US adults: the first National

Health and Nutrition Examination Survey Epidemiologic Follow-up Study. *Am. J. Clin. Nutr*. 2002 Jul;76(1):93-9.

[13] Liu S, Serdula M, Janket SJ, Cook NR, Sesso HD, Willett WC, Manson JE, Buring JE. A prospective study of fruit and vegetable intake and the risk of type 2 diabetes in women. *Diabetes Care*. 2004 Dec;27(12):2993-6.

[14] Wright ME, Mayne ST, Swanson CA, Sinha R, Alavanja MC. Dietary carotenoids, vegetables, and lung cancer risk in women: the Missouri women's health study (United States). *Cancer Causes Control*. 2003 Feb;14(1):85-96.

[15] Morris CD, Carson S. Routine Vitamin Supplementation To Prevent Cardiovascular Disease: A Summary of the Evidence for the U.S. Preventive Services Task Force. *Ann. Intern. Med.* 2003 Jul 1;139(1):56-70.

[16] Heart Protection Study Collaborative Group. MRC/BHF Heart Protection Study of antioxidant vitamin supplementation in 20,536 high-risk individuals: a randomised placebo-controlled trial. *Lancet*. 2002 Jul 6;360(9326):23-33.

[17] Yusuf S, Dagenais G, Pogue J, Bosch J, Sleight P. Vitamin E supplementation and cardiovascular events in high-risk patients. The Heart Outcomes Prevention Evaluation Study Investigators. *N. Engl. J. Med.* 2000 Jan 20;342(3):154-60.

[18] Zureik M, Galan P, Bertrais S, Mennen L, Czernichow S, Blacher J, Ducimetiere P, Hercberg S. Effects of long-term daily low-dose supplementation with antioxidant vitamins and minerals on structure and function of large arteries. *Arterioscler Thromb. Vasc. Biol.* 2004 Aug;24(8):1485-91.

[19] Freese R, Alfthan G, Jauhiainen M, Basu S, Erlund I, Salminen I, Aro A, Mutanen M. High intakes of vegetables, berries, and apples combined with a high intake of linoleic or oleic acid only slightly affect markers of lipid peroxidation and lipoprotein metabolism in healthy subjects. *Am. J. Clin. Nutr*. 2002 Nov;76(5):950-60.

[20] Bjelakovic G, Nikolova D, Gluud LL, Simonetti RG, Gluud C. Mortality in Randomized Trials of Antioxidant Supplements for Primary and Secondary Prevention: Systematic Review and Meta-analysis. *JAMA*. 2007 Feb 28;297(8):842-57.

[21] Droge W. Free radicals in the physiological control of cell function. *Physiol. Rev*. 2002 Jan;82(1):47-95.

[22] Vallance P. Dietary nitrate: poison or panacea? *Gut*. 1997 Feb;40(2):288.

[23] McKnight GM, Smith LM, Drummond RS, Duncan CW, Golden M, Benjamin N. Chemical synthesis of nitric oxide in the stomach from dietary nitrate in humans. *Gut*. 1997 Feb;40(2):211-4.

[24] Dykhuizen RS, Frazer R, Duncan C, Smith CC, Golden M, Benjamin N, Leifert C. Antimicrobial effect of acidified nitrite on gut pathogens: importance of dietary nitrate in host defense. *Antimicrob Agents Chemother*. 1996 Jun;40(6):1422-5.

[25] Dykhuizen RS, Fraser A, McKenzie H, Golden M, Leifert C, Benjamin N. Helicobacter pylori is killed by nitrite under acidic conditions. *Gut*. 1998 Mar;42(3):334-7.

[26] Pannala AS, Mani AR, Spencer JP, Skinner V, Bruckdorfer KR, Moore KP, Rice-Evans CA. The effect of dietary nitrate on salivary, plasma, and urinary nitrate metabolism in humans. *Free Radic. Biol. Med*. 2003 Mar 1;34(5):576-84.

[27] Richardson G, Hicks SL, O'Byrne S, Frost MT, Moore K, Benjamin N, McKnight GM. The ingestion of inorganic nitrate increases gastric S-nitrosothiol levels and inhibits platelet function in humans. *Nitric Oxide*. 2002 Aug;7(1):24-9.

[28] Bjorne H H, Petersson J, Phillipson M, Weitzberg E, Holm L, Lundberg JO. Nitrite in saliva increases gastric mucosal blood flow and mucus thickness. J Clin Invest. 2004 Jan;113(1):106-14. *Erratum in: J. Clin. Invest.* 2004 Feb;113(3):490.

[29] Larauche M, Anton PM, Garcia-Villar R, Theodorou V, Frexinos J, Bueno L, Fioramonti J. Protective effect of dietary nitrate on experimental gastritis in rats. *Br. J. Nutr.* 2003 Jun;89(6):777-86.

[30] Liu X, Miller MJ, Joshi MS, Thomas DD, Lancaster JR Jr. Accelerated reaction of nitric oxide with O2 within the hydrophobic interior of biological membranes. *Proc. Natl. Acad. Sci. USA.* 1998 Mar 3;95(5):2175-9.

[31] Barazzone C, Horowitz S, Donati YR, Rodriguez I, Piguet PF. 1 Oxygen toxicity in mouse lung: pathways to cell death. *Am. J. Respir. Cell Mol. Biol.* 1998 Oct;19(4):573-81.

[32] Gardner PR, Gardner AM, Martin LA, Dou Y, Li T, Olson JS, Zhu H, Riggs AF. Nitric-oxide dioxygenase activity and function of flavohemoglobins. sensitivity to nitric oxide and carbon monoxide inhibition. *J. Biol. Chem.* 2000 Oct 13;275(41):31581-7.

[33] Gibb BJ, Wykes V, Garthwaite J. Properties of NO-activated guanylyl cyclases expressed in cells. *Br. J. Pharmacol.* 2003 Jul;139(5):1032-40.

[34] Rubbo H, Radi R, Anselmi D, Kirk M, Barnes S, Butler J, Eiserich JP, Freeman BA. Nitric oxide reaction with lipid peroxyl radicals spares alpha-tocopherol during lipid peroxidation. Greater oxidant protection from the pair nitric oxide/alpha-tocopherol than alpha-tocopherol/ascorbate. *J. Biol. Chem.* 2000 Apr 14;275(15):10812-8.

[35] Hogg N, Kalyanaraman B. Nitric oxide and lipid peroxidation. *Biochim. Biophys. Acta.* 1999 May 5;1411(2-3):378-84.

[36] Bloodsworth A, O'Donnell VB, Freeman BA. Nitric oxide regulation of free radical- and enzyme-mediated lipid and lipoprotein oxidation. *Arterioscler Thromb. Vasc. Biol.* 2000 Jul;20(7):1707-15.

[37] Goldstein S, Czapski G. The reaction of NO. with O2.- and HO2.: a pulse radiolysis study. Free Radic Biol Med. 1995 Oct;19(4):505-10. Erratum in: *Free Radic. Biol. Med.* 1995 Dec;19(6):953.

[38] Daiber A, Frein D, Namgaladze D, Ullrich V.. Oxidation and nitrosation in the nitrogen monoxide/superoxide system. *J. Biol. Chem.* 2002 Apr 5;277(14):11882-8.

[39] Espey MG, Xavier S, Thomas DD, Miranda KM, Wink DA. Direct real-time evaluation of nitration with green fluorescent protein in solution and within human cells reveals the impact of nitrogen dioxide vs. peroxynitrite mechanisms. *Proc. Natl. Acad. Sci. USA*. 2002 Mar 19;99(6):3481-6.

[40] Espey MG, Thomas DD, Miranda KM, Wink DA. Focusing of nitric oxide mediated nitrosation and oxidative nitrosylation as a consequence of reaction with superoxide. *Proc. Natl. Acad. Sci. USA*. 2002 Aug 20;99(17):11127-32.

[41] Zou MH, Shi C, Cohen RA. Oxidation of the zinc-thiolate complex and uncoupling of endothelial nitric oxide synthase by peroxynitrite. *J. Clin. Invest*. 2002 Mar;109(6):817-26.

[42] Dzeja PP, Terzic A. P Phosphotransfer networks and cellular energetics. *J. Exp. Biol.* 2003 Jun;206(Pt 12):2039-47.

[43] Brunori M, Giuffre A, Forte E, Mastronicola D, Barone MC, Sarti P. Control of cytochrome c oxidase activity by nitric oxide. *Biochim. Biophys. Acta*. 2004 Apr 12;1655(1-3):365-71.

[44] Zangar RC, Davydov DR, Verma S. Mechanisms that regulate production of reactive oxygen species by cytochrome P450. *Toxicol Appl. Pharmacol.* 2004 Sep 15;199(3):316-31.

[45] Mohazzab-H KM, Fayngersh RP, Wolin MS. Nitric oxide inhibits pulmonary artery catalase and H2O2-associated relaxation. *Am. J. Physiol.* 1996 Nov;271(5 Pt 2):H1900-6.

[46] Kim YS, Han S. Superoxide reactivates nitric oxide-inhibited catalase. *Biol. Chem.* 2000 Dec;381(12):1269-71.

[47] Morgan ET, Ullrich V, Daiber A, Schmidt P, Takaya N, Shoun H, McGiff JC, Oyekan A, Hanke CJ, Campbell WB, Park CS, Kang JS, Yi HG, Cha YN, Mansuy D, Boucher JL. Cytochromes P450 and flavin monooxygenases--targets and sources of nitric oxide. *Drug Metab. Dispos*. 2001 Nov;29(11):1366-76.

[48] Drewett JG, Adams-Hays RL, Ho BY, Hegge DJ. Nitric oxide potently inhibits the rate-limiting enzymatic step in steroidogenesis. *Mol. Cell Endocrinol.* 2002 Aug 30;194(1-2):39-50.

[49] Docampo R, Moreno SN, Mason RP. G Generation of free radical metabolites and superoxide anion by the calcium indicators arsenazo III, antipyrylazo III, and murexide in rat liver microsomes. *J. Biol. Chem*. 1983 Dec 25;258(24):14920-5.

[50] Brodsky SV, Gao S, Li H, Goligorsky MS. Hyperglycemic switch from mitochondrial nitric oxide to superoxide production in endothelial cells. *Am. J. Physiol. Heart Circ Physiol*. 2002 Nov;283(5):H2130-9.

In: Oxidative Stress: Clinical and Biomedical Implications ISBN: 978-1-60021-888-0
Editors: B. M. Matata, M. M. Elahi, pp. 145-155

Chapter IX

The Biological Significance of the Nitric Oxide-Dependent Mitochondrial Biogenesis

Sestina Falcone and Emilio Clementi*
E. Medea Scientific Institute, 23842 Bosisio Parini, Italy and
Department of Preclinical Sciences,
University of Milan, 20157 Milan, Italy

Abstract

The function of mitochondria, the site of oxidative energy metabolism in eukaryotic cells, is regulated in complex ways by several agonists and environmental conditions, through activation of specific transcription factors and signalling pathways. A key player in this scenario is the gaseous messenger nitric oxide (NO). Its binding to cytochrome c oxidase in the mitochondrial respiratory chain, which is reversible and in competition with oxygen, plays a role in acute oxygen sensing and in the cell response to hypoxia. Evidence of the last two years showed that NO has also long-term effects, leading to biogenesis of functionally active mitochondria that complement its oxygen sensing function. Mitochondrial biogenesis is triggered by NO through activation of guanylate cyclase and generation of cyclic GMP. The combined action of NO at its two known intracellular receptors, cytochrome c oxidase and guanylate cyclase, plays a role in coupling energy generation with energy demand. This may explain why de-regulation of the NO signalling pathway is often associated with the pathogenesis of metabolic disorders.

Keywords: mitochondria; nitric oxide; cyclic GMP; mitochondrial biogenesis; thermogenesis; respiration; gene transcription; hypoxia; metabolism.

---

* Author for correspondence: Prof. Emilio Clementi, Department of Preclinical Sciences, Via G.B. Grassi 74, 20157 Milan, Italy. Tel: +390226434814. Fax: +390226434815. E-mail: emilio.clementi@unimi.it

Introduction

Mitochondria first captured the attention of cell physiologists some 50 years ago. The role of these organelles as the producers of most of the energy in animal cells was soon discovered. The steps involved in such process, i.e. the passing of electrons along the series of respiratory enzyme complexes located in the inner mitochondrial membrane, and the ensuing build up of a trans-membrane electrochemical proton gradient, enabling ATP synthase, to synthesize the energy carrier ATP are now characterized [1]. Recently, the gaseous messenger NO has been shown to play a crucial role in regulating the ability of mitochondria to generate ATP. In addition it stimulates their biogenetic process. This review article will outline our understanding and the biological significance of this complex regulation of mitochondrial structure and function by NO.

Mitochondria, NO, and the Sensor for Acute Changes in Cellular Oxygen Concentrations

NO is synthesised from L-arginine and O_2 by NO synthases (NOS) in almost all mammalian cells [2, 3]. Three distinct cellular isoforms of NOS have been identified: the endothelial (eNOS) and neuronal (nNOS) isoforms are regulated by second messengers; whilst inducible NOS (iNOS) is induced by cytokines and bacterial products. All three isoforms are regulated by transcriptional and post-transcriptional mechanisms and are constitutively expressed in various tissues [2, 3]. Moreover, a mitochondrial NOS isoform (mtNOS) has been described as a constitutive protein of the mitochondrial inner membrane that generates NO in a Ca^{2+}-dependent reaction [4], although its existence has been doubted [5]. It is notable that eNOS is attached to the outer mitochondrial membrane in neurons and endothelial cells [6, 7], indicating that mitochondria might regulate NOS activity and, conversely, that eNOS might regulate mitochondrial function. Recent evidence by our group shows that also nNOS may play a regulatory role in mitochondria. The enzyme contains a unique protein sequence that allows its physical interaction with cytochrome *c* oxidase, the terminal enzyme in the electron-transport chain, and such physical interaction has been shown to occur both in brain tissues and neuronal cell lines [8]. NO binds at physiological, nanomolar concentrations to cytochrome *c* oxidase reversibly and in competition with O_2, inhibiting the activity of the enzyme [9-11] and thus negatively regulating mitochondrial oxidative phosphorylation – particularly at the low O_2 concentrations usually found in tissues [12]. This also leads to redistribution of O_2 to neighbouring cells [13]. Moreover, the NO-dependent inhibition of cell respiration can change as part of the adaptive response to stress – for example, in response to alcohol toxicity [14] and cardiac failure [15]. Together with hypoxia inducible factor 1 α, the NO-cytochrome *c* oxidase system helps fine-tuning cell metabolism [16]. Such a sensing mechanism might, for example, allow cardiac myocytes to adapt their metabolic function to hypoxia [17].

NO and Mitochondrial Biogenesis

Recent evidence suggests that ATP generation is not only regulated at the level of oxidative phosphorylation but through dynamic changes in mitochondrial structure. Mitochondria are in constant movement within the cells, with several fusion/fission events taking place. This high degree of plasticity is accompanied by variations of mitochondrial size, number and mass, in complex processes triggered by a variety of physiological stimuli and differentiation states, involving approximately 1000 genes and producing about 20% of cellular proteins. Such complexity underlies the existence of a complex network connecting many different regulatory pathways coordinated and regulated tightly [18, 19]. Among the factors involved in the regulation and coordination of mitochondrial gene expression identified thus far, peroxisome proliferator-activated receptor-γ coactivator 1 α (PGC-1α) and PGC-1β play crucial roles. In particular, both in brown adipocytes and myocytes, PGC-1α enforces the expression of nuclear respiration factors (NRF-1 and NRF-2), i.e. Transcription factors that trigger the expression of genes coding for both nuclear subunits of the respiratory chain and proteins involved in mitochondrial DNA transcription and replication [20]. The hypothesis of an involvement of NO in the regulation of mitochondrial biogenesis came from observations by several groups about the role of NO, either exogenous or generated by eNOS, in the differentiation programme of the brown adipose tissue (BAT) [21, 22]. It was found that treatment with NO-releasing drugs increases the mtDNA content in a way that is sensitive to NO removal by the NO scavenger oxyhaemoglobin, indicating that it was mediated by NO generation [23]. The NO-stimulated mitochondrial biogenesis occurs through activation of PGC-1α, NRF-1 and the mitochondrial transcription factor A (Tfam) i.e. the pathway responsible for the process of mitochondrial biogenesis triggered by most mitochondriogenic stimuli in BAT, cardiac and skeletal muscle. Using a cyclic GMP (cGMP) analogue and a guanylate cyclase inhibitor, it was shown that mitochondrial biogenesis depends on cyclic GMP. Activation of guanylate cyclase and generation of cGMP occur at physiological NO concentrations [2] and this indicates that mitochondrial biogenesis is a physiological process regulated by NO. The mechanisms by which NO and/or cGMP increase expression of PGC-1α and genes encoding mitochondrial proteins have not been analyzed in detail. Interestingly, in brown adipocytes, which exhibit several responses to NO (proliferation, differentiation, respiration and mitochondrial biogenesis), eNOS, NO and cGMP are located in both the cytoplasm and the nucleus. Their levels and activity are dynamically modulated by noradrenaline (NA) [24]. In animals acclimatised to cold, in which high levels of NA are released from sympathetic nerve terminals that innervate richly brown adipocytes, elevations of eNOS, NO and cGMP levels in both the cytosol and nucleus of these cells correlates with an increase in the level of uncoupling protein 1 (UCP1). UCP1 is an inner mitochondrial membrane proton channel that dissipates the inner trans-membrane potential to produce heat [25, 26]. Interestingly, NA induces expression of PGC-1 α, which binds to the *UCP1* promoter and increases UCP1 expression [27]. Thus, eNOS expression and NO and/or cGMP levels in brown adipocytes are linked to cell activation, mitochondrial biogenesis and heat production.

Experiments in mouse white-fat 3T3-L1, human U937 monocytic and HeLa epithelial cells, as well as rat L6 myoblasts and PC12 neurosecretory cells, yielded results similar to those obtained in brown adipocytes, revealing that the NO-dependent mitochondrial biogenesis is restricted neither to brown adipocytes and their differentiation processes nor to a specific cell lineage or species [23, 28]. An important aspect of mitochondrial biogenesis by NO is that it leads to formation of functionally active mitochondria. In particular, the newly generated mitochondria are capable of sustaining coupled respiration, and generate ATP, through the oxidative phosphorylation pathway [28]. Interestingly, this increase in ATP generation by mitochondria is not accompanied by a decreased generation of ATP through the glycolytic pathway, i.e. the Warburg effect [29], yielding steady state concentrations of ATP higher than those observed in cells not exposed to NO. The significance of this profound change in energy metabolism remains to be investigated. However, mitochondrial activity plays critical roles in various processes, such as the switch of skeletal muscle fibres from glycolytic to oxidative metabolism [30], and the regeneration of cardiac and skeletal muscles [31, 32]. NO regulates skeletal and cardiac muscle energy demand [31, 32] and the change in energy metabolism induced by NO may explain its role in these processes.

The obligatory role of eNOS in mitochondrial biogenesis was proved by using $eNOS^{-/-}$ mice. Histological analysis indicated that BAT of $eNOS^{-/-}$ is functionally inactive, and indeed exposure of these animals to cold did not result in the mitochondriogenetic process that is commonly observed in wild-type animals [23]. In addition, deletion of eNOS was sufficient to reduce the mitochondrial mass even in tissues that have a basal expression of nNOS, and possibly iNOS, such as the brain, liver, muscle and heart. The reduced mitochondrial content in tissues from $eNOS^{-/-}$ mice was accompanied by reduction of both basal oxygen consumption and steady state ATP levels, associated with a significant reduction in the mitochondrial size. This occurred in tissues dependent on oxidative metabolism, such as the brain, and in glycolytic tissues, such as the gastrocnemius muscle, indicating that the effect of NO on mitochondrial biogenesis is a general phenomenon. Interestingly, in the gastrocnemius muscle mitochondrial biogenesis is important in the conversion of type II (glycolytic) fibres into type I (oxidative) fibres [33]. The mitochondrial volume and network extension [34] have been shown to correlate positively with the efficiency of the mitochondrial respiratory function [35]. Thus, it is likely that the decrease in mitochondrial size observed in tissues from $eNOS^{-/-}$ mice is a morphological correlate of the reduced mitochondrial function.

Biological Role of NO-Dependent Mitochondrial Biogenesis

In muscle cells a multilayer of mitochondria is found below the outer cell membrane (sarcolemma) where the oxygen supply is high. These organelles are connected by mitochondrial filaments to mitochondria located in the oxygen-poor core part of the muscle fibre. It has been proposed that the mitochondrial network in muscle cells serves to transmit the membrane potential generated by subsarcolemmal mitochondria to intermyofibrillar organelles, which can then use oxygen from the cell periphery to produce ATP in the core of the muscle fibre [36]. Morphological observations [37] are in agreement with the view that mitochondrial fusion is required for the establishment of the intermyofibrillar network and maintenance of muscle function. The mechanisms by which NO is beneficial in different myopathies are only in part linked to increased muscle perfusion and reduced functional ischemia. In view of the critical effects of NO on mitochondria and the importance of these organelles in muscle physiology in terms of dynamics, morphology and activity it is conceivable that the connection NO/mitochondria, both in terms of regulation of oxygen sensing and mitochondrial biogenesis, is an important component of the mechanisms leading to functional improvement in these diseases.

Other pathologies associated to an impairment of mitochondrial function include also neurodegenerative diseases, neuromuscular disorders, liver and heart failure, and type 2 diabetes [31, 38-41]. The possibility of generating new, metabolically active mitochondria might therefore improve the outcome of these pathologies. In this respect particularly relevant appears the role of NO in diabetes and obesity. In $eNOS^{-/-}$ mice, oxygen consumption, an indicator of metabolic rate, is decreased, indicating an impaired BAT-dependent thermogenesis. In genetic models of obesity, defective energy expenditure is involved in increased food intake and body-weight gain; eight week-old $eNOS^{-/-}$ mice showed similar food consumption to, but weighed more than, wild-type mice. Thus, the increased body weight of $eNOS^{-/-}$ mice could be accounted for by higher feed efficiency (i.e. weight gain/food intake) caused by defective energy expenditure [23]. The features shown by $eNOS^{-/-}$ mice, i.e. reduced mitochondrial number and energy expenditure, weight gain, insulin resistance and hypertension, are all typical signs encountered in the metabolic syndrome.

NO-induced mitochondrial biogenesis also has a role in the physiological phases of the rat ovarian cycle [42]. A moderately increased rate of NO production in the proliferative phase is associated with mitochondrial biogenesis, whereas a high rate of NO generation (probably produced by mtNOS) at pro-oestrus phase triggers mitochondria-dependent apoptosis [42]. Thus, the roles of mitochondria as ATP provider, and as a source of NO to signal for mitochondrion proliferation and mitochondria-dependent apoptosis, appear well adapted to serve the proliferation/apoptosis sequence of the ovarian cycle.

Increase in mitochondrial biogenesis is linked also to pathological condition: NO mediates mitochondrial hyperpolarisation (MHP) and production of reactive oxygen species (ROS), as well as sustained Ca^{2+} fluxing in human T lymphocytes in which the two receptors CD3 and CD28 are both activated [43]. Persistent MHP is associated with increased mitochondrial mass and Ca^{2+} content in T cells of patients with systemic lupus erythematosus (SLE) which exhibit both enhanced spontaneous apoptosis and defective activation-induced cell death. Moreover, monocytes of patients with SLE exhibit significantly higher amounts of NO than normal monocytes [44].

Another disorder associated to abnormal proliferation of mitochondria is oncocytoma, a large-cell tumour in which numerous nuclear genes encoding proteins involved in mitochondrial biogenesis, such as NRF-1 and eNOS, and components of the respiratory chain are up-regulated [45]. Remarkably, primary B-cell chronic lymphocytic leukaemia (CLL) cells contain significantly more mitochondria than do normal lymphocytes and their mitochondrial mass correlates with endogenous NO levels [46]. Expression of NRF-1 and Tfam is elevated in most CLL specimens examined and appears to be related to cellular NO levels [46]. Thus, NO-induced mitochondrial biogenesis might cause primary alterations of the cell cycle and/or apoptotic mechanisms that lead to malignant cell development.

Alternatively, NO-induced mitochondrial biogenesis might compensate for a primary defect in mitochondrial ATP production by a feedback mechanism in tumour cells in which oxidative phosphorylation is decreased and substituted for by high rates of aerobic glycolysis [29].

NO-Dependent Mitochondrial Biogenesis and Aging

The hypothesis that mitochondrial dysfunctions play a role in aging has been proposed in 1972 [47]. This hypothesis postulates that the rate of aging and the onset of degenerative diseases are determined by the rate of free radical leakage and attacks on the adjacent mitochondrial DNA (mtDNA), which causes mutations that undermine mitochondrial function. As mitochondria decay the performance of the cell as a whole declines and this leads to the traits observed in aging. The inability of antioxidants to prolong life has weakened support for the above theory [48, 49], and the accumulation of mitochondrial mutations with age has not been generally observed, also because mitochondria are reasonably efficient at repairing damage to their genes [50, 51]. Unexpectedly, a homoplasmic C150T transition near an origin of mtDNA replication has even been associated with greater longevity in an Italian population [52]. Mutations that impair electron transport and consequently lead to ROS production might be able to signal the synthesis of new respiratory chain components encoded in the nucleus. Such a 'retrograde response' would enable the cell to compensate for a defect. In particular, a retrograde signalling mechanism involving increased cytosolic free Ca^{2+} concentration exists in skeletal myoblasts, human lung carcinoma cells and rat pheochromocytoma cells [53]. Ca^{2+} released by mitochondria leads to the activation of calcineurin, which in turn activates the transcription factors NFAT and NF-κB. One target gene activated by NFAT and NF-κB is eNOS. NO is known to induce the production of ROS and trigger redox signalling [54] and thus might itself enhance the retrograde response further; in addition, by generating cGMP it might trigger mitochondrial biogenesis.

Numerous studies have demonstrated that calorie restriction (CR) extends lifespan in organisms from yeast to rodents and possibly primates [55]. In mammals, CR delays the onset of age-associated diseases including cancer, atherosclerosis and diabetes [55]. The molecular mechanisms underlying this effect are not known. Recent evidence indicates that CR induces eNOS and that the resulting generation of NO activates the synthesis of a broad array of mitochondrial proteins and increases production of mtDNA, respiration and ATP levels in several different tissues and organs, including white and brown fat, brain, liver and heart [56]. Intriguingly, NO also activates expression of the mammalian Sir2 ortholog SIRT1, which is induced in mouse tissues by CR in wild-type mice but not in $eNOS^{-/-}$ mice [56]. Because SIRT1 is known to mediate resistance to cellular stress by a variety of mechanisms [57], increases in its levels might be essential for greater longevity of the organism. These findings lead to a model in which CR induces eNOS, which results in mitochondrial biogenesis through increased PGC-1α expression and upregulation of SIRT1 and perhaps other longevity-promoting agents. Recent observations confirm that CR induces mitochondrial biogenesis and enhances bioenergetic efficiency both in vitro and in vivo [58]. The increase in both PGC-1α and SIRT1 levels after CR is also relevant for fat metabolism [59]. PGC-1α co-ordinately regulates genes involved in mitochondrial biogenesis and in downregulation of adipogenesis [60]. This should reduce fat accumulation, which is known to have an impact on lifespan [61, 62]. Thus, the NO-mediated mitochondrial biogenesis seems to play a role in slowing aging.

Conclusion

An intriguing question is how to reconcile the apparent paradox of the two actions of NO on mitochondria, i.e. short-term inhibition of respiration and long-term generation of new mitochondria. A possibility resides in the structure of the energy generating system itself. Oxidative phosphorylation is never fully coupled and the leak of electrons is enhanced when mitochondria are working full speed [63]. Thus, increasing the number of oxidative phosphorylation units by increasing the number of mitochondria while placing a "brake" on their function may allow increasing generation of energy, when needed by changed metabolic conditions, without enhancing the threat posed by concomitant ROS generation. To fully understand this issue, however, we need to learn more about the mechanisms presiding over the mitochondriogenic action of NO. In particular, we need to elucidate the precise mechanism(s) by which NO/cGMP activates PGC- 1α NRF-1 and Tfam to trigger mitochondrial biogenesis, and the exact nature of the NO effects on mitochondrial dynamic properties. In addition, dietary factors (i.e. proteins formed by certain amino acids, carbohydrates and lipids, or minerals and vitamins) that are able to modulate NO production in different tissues specifically activating the diverse NOS isoforms [64], may cooperate with NO to modulate mitochondrial properties. Elucidation of the relationship of these factors with NO signalling may be relevant also to understand their possible use in disease prevention or treatment.

References

[1] Kennedy EP and Lehninger AL. (1949). Oxidation of fatty acids and tricarboxylic acid cycle intermediates by isolated rat liver mitochondria. *J. Biol. Chem.*, 179, 957-972.

[2] Moncada, S., Palmer, R.M. and Higgs, E.A., (1991). Nitric oxide: physiology, pathophysiology, and pharmacology. *Pharmacol. Rev.* 43, 109–142.

[3] Alderton,W.K., Cooper, C.E. and Knowles, R.G., (2001). Nitric oxide synthases: structure, function and inhibition. *Biochem. J.*, 357, 593– 615.

[4] Ghafourifar, P. and Cadenas, E. (2005). Mitochondrial nitric oxide synthase. *Trends Pharmacol. Sci.* 26, 190-195.

[5] Lacza, Z., Pankotai, E., Csordas, A., Gero, D., Kiss, L., Horvath, E. M., Kollai, M., Busija, D. W. and Szabo, C. (2006). Mitochondrial NO and reactive nitrogen species production: does mtNOS exist? *Nitric Oxide* 14, 162-168.

[6] Henrich, M., Hoffmann, K., Konig, P., Gruss, M., Fischbach, T., Godecke, A., Hempelmann, G. and Kummer, W. (2002). Sensory neurons respond to hypoxia with NO production associated with mitochondria. *Mol. Cell. Neurosci.,* 20, 307-322.

[7] Gao, S., Chen, J., Brodsky, S. V., Huang, H., Adler, S., Lee, J. H., Dhadwal, N., Cohen-Gould, L., Gross, S. S. and Goligorsky, M. S. (2004). Docking of endothelial nitric oxide synthase (eNOS) to the mitochondrial outer membrane: a pentabasic amino acid sequence in the autoinhibitory domain of eNOS targets a proteinase K cleavable peptide on the cytoplasmic face of mitochondria. *J. Biol. Chem.*, 279, 15968-15974.

[8] Persichini T, Mazzone V, Polticelli F, Moreno S, Venturini G, Clementi E and Colasanti M. (2005). Mitochondrial type I nitric oxide synthase physically interacts with cytochrome c oxidase. *Neurosci Lett.*, Aug 26, 384 (3), 254-9.

[9] Cleeter, M. W., Cooper, J. M., Darley-Usmar, V. M., Moncada, S. and Schapira, A. H. (1994). Reversible inhibition of cytochrome c oxidase, the terminal enzyme of the mitochondrial respiratory chain, by nitric oxide. Implications for neurodegenerative diseases. *FEBS Lett.*, 345, 50-54.

[10] Brown, G. C. and Cooper, C. E. (1994). Nanomolar concentrations of nitric oxide reversibly inhibit synaptosomal respiration by competing with oxygen at cytochrome oxidase. *FEBS Lett.,* 356, 295-298.

[11] Clementi, E., Brown, G. C., Feelisch, M. and Moncada, S. (1998). Persistent inhibition of cell respiration by nitric oxide: crucial role of S-nitrosylation of mitochondrial complex I and protective action of glutathione. *Proc. Natl. Acad. Sci. USA*, 95, 7631-7636.

[12] Clementi, E., Brown, G. C., Foxwell, N. and Moncada, S. (1999). On the mechanism by which vascular endothelial cells regulate their oxygen consumption. *Proc. Natl. Acad. Sci. USA*, 96, 1559-1562.

[13] Hagen, T., Taylor, C. T., Lam, F. and Moncada, S. (2003). Redistribution of intracellular oxygen in hypoxia by nitric oxide: effect on HIF1α. *Science*, 302, 1975-1978.

[14] Venkatraman, A., Shiva, S., Wigley, A., Ulasova, E., Chieng, D., Bailey, S. M. and Darley-Usmar, V. M. (2004). The role of iNOS in alcohol-dependent hepatotoxicity and mitochondrial dysfunction in mice. *Hepatology*, 40, 565-573.

[15] Brookes, P. S., Zhang, J., Dai, L., Zhou, F., Parks, D. A., Darley-Usmar, V. M. and Anderson, P. G. (2001). Increased sensitivity of mitochondrial respiration to inhibition by nitric oxide in cardiac hypertrophy. *J. Mol. Cell. Cardiol.,* 33, 69-82.

[16] Semenza, G. L. (1999). Perspectives on oxygen sensing. Cell, 98, 281-284.

[17] Budinger, G. R., Chandel, N., Shao, Z. H., Li, C. Q., Melmed, A., Becker, L. B. and Schumacker, P. T. (1996). Cellular energy utilization and supply during hypoxia in embryonic cardiac myocytes. *Am. J. Physiol.*, 270, L44-L53.

[18] Attardi G and Schatz G. (1988). Biogenesis of mitochondria. *Annu. Rev. Cell Biol.*, 4, 289–333.

[19] Neupert, W., (1997). Protein import into mitochondria. *Annu. Rev. Biochem*. 66, 863–917.

[20] Scarpulla, R.C., (2002). Nuclear activators and coactivators in mammalian mitochondrial biogenesis. *Biochim. Biophys. Acta* 1576, 1– 14.

[21] Puigserver, P., Wu, Z., Park, C.W., Graves, R., Wright, M. and Spiegelman, B.M., (1998). A cold inducible coactivator of nuclear receptors linked to adaptive thermogenesis. *Cell,* 92, 829– 839.

[22] Nisoli, E., Clementi, E., Tonello, C., Sciorati, C., Briscini, L. and Carruba, M.O., (1998). Effects of nitric oxide on proliferation and differentiation of rat brown adipocytes in primary cultures. *Br. J. Pharmacol.*, 125, 888– 894.

[23] Nisoli, E., Clementi, E., Paolucci, C., Cozzi, V., Tonello, C., Sciorati, C., Bracale, R., Valerio, A., Francolini, M., Moncada, S. and Carruba, M.O., (2003). Mitochondrial biogenesis in mammals: the role of endogenous nitric oxide. *Science*, 299, 896–899.

[24] Giordano, A., Tonello, C., Bulbarelli, A., Cozzi, V., Cinti, S., Carruba, M. O. and Nisoli, E. (2002). Evidence for a functional nitric oxide synthase system in brown adipocyte nucleus. *FEBS Lett.*, 514, 135-140.

[25] Matthias, A., Jacobsson, A., Cannon, B. and Nedergaard, J. (1999). The bioenergetics of brown fat mitochondria from UCP1-ablated mice. Ucp1 is not involved in fatty acid induced de-energization ("uncoupling"). *J. Biol. Chem.*, 274, 28150-28160.

[26] Nedergaard, J., Golozoubova, V., Matthias, A., Asadi, A., Jacobsson, A. and Cannon, B. (2001). UCP1: the only protein able to mediate adaptive non-shivering thermogenesis and metabolic inefficiency. *Biochim. Biophys. Acta,* 1504, 82-106.

[27] Lowell, B. B. and Spiegelman, B. M. (2000). Towards a molecular understanding of adaptive thermogenesis. *Nature,* 404, 652-660.

[28] Nisoli, E., Falcone, S., Tonello, C., Cozzi, V., Palombo, L., Fiorani, M., Pisconti, A., Brunelli, S., Cardile, A., Francolini, M., Cantoni, O., Carruba, M.O., Moncada, S. and Clementi, E. (2004). Mitochondrial biogenesis by NO yields functionally active mitochondria in mammals. *Proc. Natl. Acad. Sci. USA.*, 101, 16507–16512.

[29] Warburg, O. (1956). On the origin of cancer cells. Science, 123, 309-314.

[30] Lin J, Wu H, Tarr PT, Zhang CY, Wu Z, Boss O, Michael LF,Puigserver P, Isotani E, Olson EN, Lowell BB, Bassel-Duby R and Spiegelman BM. (2002). Transcriptional co-activator PGC-1α drives the formation of slow-twitch muscle fibres. *Nature*, 418, 797-801.

[31] Lehman, J. J., Barger, P. M., Kovacs, A., Saffitz, J. E., Medeiros, D. M. and Kelly, D. P. (2000). Peroxisome proliferator-activated receptor γ coactivator-1α promotes cardiac mitochondrial biogenesis. *J. Clin. Invest.,* 106, 847-856.

[32] Stamler, J. S. and Meissner, G. (2001). Physiology of nitric oxide in skeletal muscle. *Physiol. Rev,.* 81, 209-237.

[33] Baddeley, A.J., Gundersen, H.J. and Cruz-Orive, L.M., (1986). Estimation of surface area from vertical sections. *J. Microsc.*, 142, 259–276.

[34] Bach, D., Pich, S., Soriano, F.X., Vega, N., Baumgartner, B., Oriola, J., Daugaard, J.R., Lloberas, J., Camps, M., Zierath, J.R., Rabasa-Lhoret, R., Wallberg-Henriksson, H., Laville, M., Palacin, M., Vidal, H., Rivera, F., Brand and M., Zorzano, A., (2003). Mitofusin-2 determines mitochondrial network architecture and mitochondrial metabolism. A novel regulatory mechanism altered in obesity. *J. Biol. Chem.*, 278, 17190– 17197.

[35] Skulacev V.P. (2001). Mitochondrial filaments and clusters as intracellular power-transmitting cables. *Trends Biochem. Sci..*, Jan, 26 (1), 23-9.

[36] Bakeeva LE, Chentsov YuS and Skulachev VP. (1983). Intermitochondrial contacts in myocardiocytes. *J. Mol. Cell Cardiol.,* Jul, 15 (7), 413-20.

[37] Kopecky, J., Clarke, G., Enerback, S., Spiegelman, B. M. and Kozak, L. P. (1995). Expression of the mitochondrial uncoupling protein gene from the aP2 gene promoter prevents genetic obesity. *J. Clin. Invest*., 96, 2914-2923.

[38] Hansford, R. G., Tsuchiya, N. and Pepe, S. (1999). Mitochondria in heart ischaemia and aging. *Biochem. Soc. Symp*., 66, 141-147.

[39] Lehman, J. J., Barger, P. M., Kovacs, A., Saffitz, J. E., Medeiros, D. M. and Kelly, D. P. (2000). Peroxisome proliferator-activated receptor γ coactivator-1 promotes cardiac mitochondrial biogenesis. *J. Clin. Invest.,* 106, 847-856.

[40] Nardin, R.A. and Johns, D.R., (2001). Mitochondrial dysfunction and neuromuscular disease. *Muscle Nerve,* 24, 170– 191.

[41] Patti, M.E., Butte, A.J., Crunkhorn, S., Cusi, K., Berria, R., Kashyap, S., Miyazaki, Y., Kohane, I., Costello, M., Saccone, R., Landaker, E.J., Goldfine, A.B., Mun, E., DeFronzo, R., Finlayson, J., Kahn, C.R. and Mandarino, L.J., (2003). Coordinated reduction of genes of oxidative metabolism in humans with insulin resistance and diabetes: potential role of PGC1 and NRF1. *Proc. Natl. Acad. Sci. USA*., 100, 8466–8471.

[42] Navarro, A., Torrejon, R., Bandez, M. J., Lopez-Cepero, J. M. and Boveris, A. (2005). Mitochondrial function and mitochondria-induced apoptosis in an overstimulated rat ovarian cycle. *Am. J. Physiol. Endocrinol. Metab.,* 289, E1101-E1109.

[43] Nagy, G., Koncz, A. and Perl, A. (2003). T cell activation-induced mitochondrial hyperpolarization is mediated by Ca2+- and redox-dependent production of nitric oxide. *J. Immunol.,* 171, 5188-5197.

[44] Nagy, G., Barcza, M., Gonchoroff, N., Phillips, P. E. and Perl, A. (2004). Nitric oxide-dependent mitochondrial biogenesis generates Ca2+ signaling profile of lupus T cells. *J. Immunol*., 173, 3676-3683.

[45] Baris, O., Savagner, F., Nasser, V., Loriod, B., Granjeaud, S., Guyetant, S., Franc, B., Rodien, P., Rohmer, V., Bertucci, F., Birnbaum D., Malthièry Y., Reynier P. and Houlgatte R. (2004). Transcriptional profiling reveals coordinated up-regulation of oxidative metabolism genes in thyroid oncocytic tumours. *J. Clin. Endocrinol. Metab*., 89, 994-1005.

[46] Carew, J. S., Nawrocki, S. T., Xu, R. H., Dunner, K., McConkey, D. J., Wierda, W. G., Keating, M. J. and Huang, P. (2004). Increased mitochondrial biogenesis in primary leukemia cells: the role of endogenous nitric oxide and impact on sensitivity to fludarabine. *Leukemia,* 18, 1934-1940.

[47] Harman, D. (1972). The biological clock: the mitochondria? *J. Am. Geriatr*. Soc., 20, 145- 147.

[48] Lane, N. (2005). *Power, Sex, Suicide. Mitochondria and the Meaning of Life*. Oxford: Oxford University Press.

[49] Trifunovic, A., Hansson, A., Wredenberg, A., Rovio, A. T., Dufour, E., Khvorostov, I., Spelbrink, J. N., Wibom, R., Jacobs, H. T. and Larsson, N. G. (2005). Somatic mtDNA mutations cause aging phenotypes without affecting reactive oxygen species production. *Proc. Natl. Acad. Sci. USA*, 102, 17993-17998.

[50] Kraytsberg, Y., Schwartz, M., Brown, T. A., Ebralidse, K., Kunz, W. S., Clayton, D. A., Vissing, J. and Khrapko, K. (2004). Recombination of human mitochondrial DNA. *Science,* 304, 981.

[51] Spees, J. L., Olson, S. D., Whitney, M. J. and Prockop, D. J. (2006). Mitochondrial transfer between cells can rescue aerobic respiration. *Proc. Natl. Acad. Sci. USA*, 103, 1283-1288.

[52] Zhang, J., Asin-Cayuela, J., Fish, J., Michikawa, Y., Bonafe, M., Olivieri, F., Passarino, G., De Benedictis, G., Franceschi, C. and Attardi, G. (2003). Strikingly higher frequency in centenarians and twins of mtDNA mutation causing remodelling of replication origin in leukocytes. *Proc. Natl. Acad. Sci. USA,* 100, 1116-1121.

[53] Butow, R. A. and Avadhani, N. G. (2004). Mitochondrial signaling: the retrograde response. *Mol. Cell,* 14, 1-15.

[54] Brookes, P. S., Levonen, A. L., Shiva, S., Sarti, P. and Darley-Usmar, V. M. (2002). Mitochondria: regulators of signal transduction by reactive oxygen and nitrogen species. *Free Radic. Biol. Med.*, 33, 755-764.

[55] Masoro, E. J. (2003). Subfield history: caloric restriction, slowing aging, and extending life. *Sci. Aging Knowledge Environ,*. 8, RE2.

[56] Nisoli, E., Tonello, C., Cardile, A., Cozzi, V., Bracale, R., Tedesco, L., Falcone, S., Valerio, A., Cantoni, O., Clementi, E. Moncada S, Carruba M.O. (2005). Calorie restriction promotes mitochondrial biogenesis by inducing the expression of eNOS. *Science,* 310, 314-317.

[57] Luo J, Nikolaev AY, Imai S, Chen D, Su F, Shiloh A, Guarente L and Gu W. (2001). Negative control of p53 by Sir2alpha promotes cell survival under stress. *Cell.*, Oct 19, 107(2),137-48.

[58] López-Lluch, G., Hunt, N., Jones, B., Zhu, M., Jamieson, H., Hilmer, S., Cascajo, M. V., Allard, J., Ingram, D. K., Navas, P. and de Cabo R. (2006). Calorie restriction induces mitochondrial biogenesis and bioenergetic efficiency. *Proc. Natl. Acad. Sci. USA,* 103, 1768-1773.

[59] Bordone, L. and Guarente, L. (2005). Calorie restriction, SIRT1 and metabolism: understanding longevity. *Nat. Rev. Mol. Cell Biol,*., 6, 298-305.

[60] Picard, F., Kurtev, M., Chung, N., Topark-Ngarm, A., Senawong, T., Machado De Oliveira, R., Leid, M., McBurney, M. W. and Guarente, L. (2004). Sirt1 promotes fat mobilization in white adipocytes by repressing PPAR-gamma. *Nature,* 429, 771-776.

[61] Bluher, M., Kahn, B. B. and Kahn, C. R. (2003). Extended longevity in mice lacking the insulin receptor in adipose tissue. *Science*, 299, 572-574.

[62] Chiu, C. H., Lin, W. D., Huang, S. Y. and Lee, Y. H. (2004). Effect of a C/EBP gene replacement on mitochondrial biogenesis in fat cells. *Genes Dev.,* 18, 1970-1975.

[63] Brown GC. (1992). Control of respiration and ATP synthesis in mammalian mitochondria and cells. *Biochem J.*, May 15, 284 (Pt 1), 1-13.

[64] Wu G. and Meininger CJ. (2002). Regulation of nitric oxide synthesis by dietary factors. *Annu. Rev. Nutr.,* 22, 61–86.

In: Oxidative Stress: Clinical and Biomedical Implications ISBN: 978-1-60021-888-0
Editors: B. M. Matata, M. M. Elahi, pp. 157-170

Chapter X

S-Nitrosothiols in Disease

Emma L. Taylor* and Paul G. Winyard
Inflammation and Musculoskeletal Diseases Research Group,
Institute of Biomedical and Clinical Science,
Peninsula Medical School, St Luke's Campus,
Exeter, Devon EX1 2LU

Abstract

S-nitrosothiols (RSNO) are the result of the covalent modification of reduced thiol groups in peptides or proteins by nitric oxide (NO). Studies have demonstrated that RSNO have physiologically relevant functional roles in human biology, and may mediate many of the effects attributed to free NO. Therefore, it is not surprising that alterations in rates of RSNO production or breakdown have been proposed to contribute to the pathogenesis of a number of conditions. In this chapter, we consider the formation and breakdown of RSNO and proposed physiological roles of RSNO. We then discuss the evidence suggesting that lowered or elevated RSNO play a role in various diseases, and consider the potential uses of RSNO, either as therapeutic agents or as biomarkers of disease.

Introduction

Studies of the potential roles of S-nitrosothiols in disease have been hindered by difficulties in establishing baseline physiological levels in biological systems, and discrepancies between the different methods utilised to measure them [1]. Nonetheless, it is becoming clear that disturbances in NO homeostasis, and its consequent effects on S-nitrosothiol levels, may be involved in the pathogenesis of a number of diseases.

* Tel: 01392 262940; Fax: 01392 262926; Email of corresponding author: emma.taylor@pms.ac.uk

Formation and Breakdown of RSNO

There are three proposed mechanisms through which RSNO formation can occur. The method most described in the literature involves the conversion of free NO to a higher oxide of nitrogen (e.g. NO_2, N_2O_3, $ONOO^-$) by autooxidation, with NO^+ then being the species which is 'donated' to the reduced thiol to form RSNO [2]. Such a pathway is more likely to occur under inflammatory or hydrophobic conditions, due to this reaction requiring high local NO concentrations [2]. However, two first-order reactions of NO have also been suggested, which would require lower concentrations of NO, and may be more relevant under resting physiological conditions. It is thought that a direct interaction of NO with a reduced thiol could generate an intermediate species (RSNOH), from which removal of an electron by an acceptor molecule leads to RSNO formation [3]. Finally, RSNO formation can occur via interaction of NO with the metal centres of proteins (e.g. copper or iron) with subsequent RSNO formation via transnitrosation. Complexes of metals with NO may either naturally dissociate, with the NO moiety transferred directly onto a thiol, or the reaction may occur via generation of NO^+ and reduced metal [2].

Both chemical and biological means exist by which RSNO may be broken down [2]. The NO moiety can be directly released from the thiol-containing protein to which it is attached, thus leaving the two components unaltered, via a mechanism that is catalysed by light or Cu(I). Alternatively, cleavage or transnitrosation may result in the release of nitrosonium ion from the protein. Furthermore, oxidation of the RSNO produces disulphide bonds and causes release of nitroxyl anion (NO^-) [4]. However, it seems that enzymatic degradation of RSNO may also occur, via the action of a formaldehyde dehydrogenase, or GSNO reductase (GSNOR) [5]. Other enzyme systems, such as xanthine/xanthine oxidase, thioredoxin/thioredoxin reductase, γ-glutamyl transpeptidase, glutathione peroxidase and Cu/Zn superoxide dismutase may also play roles of varying physiological importance in the catabolism of RSNO, and have been reviewed by Gaston et al [6].

Physiological Roles of RSNO

S-nitrosothiols are formed as intermediates during the biotransformation of certain traditional nitrovasodilator drugs, such as the organic nitrites [7] and glyceryl trinitrate [8], and their formation is required for the NO-donating functions of these compounds. However, far from being mere artificial chemical intermediates in the exogenous supply of NO, it has been suggested that S-nitrosothiols have important protective roles in physiology and homeostasis, and act both as an endogenous 'reservoir' of NO [9] and as important means of post-translational modification [10] to regulate protein function and stability. Recently, however, an alternative theory has been put forward claiming that it is the nitrite anion that represents the physiologically relevant NO reservoir through which many biological actions of NO are mediated [11,12]. For example, it has been proposed that nitrite reductase activity of haemoglobin is responsible for NO generation and blood flow regulation under hypoxic conditions [13]. Furthermore, yet another theory proposes a role for nitrated fatty acids (e.g. nitrolinoleic acid) as endogenous stores of NO that are able to mediate NO biology, such as post-translational modification of proteins [14,15]. Much further work is required to examine these controversial theories and fully elucidate the respective roles of nitroso lipids, nitrite and RSNO in biological signalling.

Circulating Pools of NO in the Bloodstream

Reservoirs of NO, in the form of S-nitrosothiols, are important within the blood vessels for the maintenance of vascular tone. Studies have shown that human plasma contains S-nitrosothiols, the bulk of which are in the form of S-nitrosoalbumin (SNO-alb), and that inhibition of NO synthesis correlates with decreased RSNO and increased blood pressure [9]. SNO-alb is considered to be primarily a buffer of NO levels through which vascular tone is maintained [9]. The intracellular actions of NO are effected by transfer of NO to its sites of biological activity via transnitrosation of the NO moiety to low molecular weight thiols such as cysteine, through which the NO can gain intracellular access [16]. Thus NO-conjugated low molecular weight thiols are more effective vasodilators than NO-conjugated proteins [17].

Erythrocytes also carry NO within the bloodstream, via S-nitroso-haemoglobin (SNO-Hb). NO is 'scavenged' from the local environment by the haem moiety of haemoglobin via interaction with Fe^{2+}, but a significant portion of the NO is then transferred to a cysteine residue within haemoglobin, generating SNO-Hb [18]. Through this adduct, vascular tone may be locally regulated, with NO being released from haemoglobin where required, e.g. in areas of hypoxia, to cause vasodilation [19]. Again, the transfer of NO from its circulating form to free, active NO occurs via transnitrosation of NO, this time onto a cysteine thiol of another protein, e.g. an anion-exchanger protein, as an intermediate carrier [17, 20]. Overall, evidence suggests that systemic blood pressure is directly correlated with RSNO levels, and is therefore regulated by the rates of both synthesis and metabolism of RSNOs [17].

Physiological S-nitrosation can also occur via dinitrosyl-iron complexes (DNICs), which are non-haem iron (e.g. iron-sulphur proteins, ferritin) nitrosyls, which are constitutively present in the body [21]. These donate NO^+ and may either S-nitrosate cellular targets to affect biological activity or contribute to the endogenous circulating pools of RSNO [22].

Other Biological Effects of RSNO

Many of the cellular effects of NOS-derived NO are mimicked by RSNO, and in some cases they may represent the intermediate species through which NO acts. These effects may be due to stimulation of cGMP, or may occur via transnitrosation reactions [23]. For example, smooth muscle relaxation mediated by cGMP does not only mediate vasorelaxation as described above, but also causes bronchorelaxation in the presence of a number of RSNO, in particular GSNO [24], and relaxes urethral [25] and gut [26] smooth muscle, although other mechanisms may also be involved. In addition, synthetic GSNO are able to inhibit platelet aggregation as free NO does [27]. This has been shown to be mediated by both cGMP-dependent and independent mechanisms [28]. RSNO have also been demonstrated to have the ability to induce apoptotic cell death at high concentrations [29, 30, 31].

S-Nitrosation as a Post-Translational Modification of Proteins

Physiological post-translational modification pathways to regulate protein function must, by definition, alter activity either positively or negatively. S-nitrosation of proteins would appear to be a mechanism of protein regulation which is parallel to phosphorylation, in which addition of NO to form RSNOs may either increase or decrease protein activity, depending on the nature of the protein (reviewed by Mannick and Schonhoff [10]). In particular, this method of altering protein activity seems to be relevant within channel proteins, with the NMDA receptor being inhibited by RSNO formation, whereas the ryanidine receptor is activated by this modification [2].

Additionally, S-nitrosation may influence a number of proteins involved in the regulation of cell survival and apoptosis, which may go some way towards explaining the mechanism of the anti-apoptotic effects of low concentrations of NO observed in some systems [32]. For example, almost all proteins of the caspase family, which mediate the process of apoptosis, are susceptible to inactivation by S-nitrosation [33]. The activity of Ras is increased by S-nitrosation, which in turn activates the anti-apoptotic MAP kinase, ERK1/2 [34]. In contrast, S-nitrosation of the proapoptotic MAP kinase, JNK, leads to decreased activity [35].

Studies in endothelial cells have shown levels of S-nitrosated proteins to be correlated with the degree of NOS activity. A number of proteins were shown to be modified by S-nitrosation, including GAPDH, vimentin, β-galactosidase, peroxiredoxin 1, β-actin and ubiquitin-conjugating enzyme, making them potential candidates for regulation of their activity by NO [35].

S-Nitrosation and Regulation of Gene Transcription

A further physiological effect of S-nitrosation is the up- or down-regulation of gene transcription. This is an important mechanism in the lung, where GSNO increases the DNA binding ability of the transcription factor, SP1. This leads to enhanced expression of several lung housekeeping genes, such as CFTR [36]. In contrast, however, high concentrations of GSNO ($> 10\mu M$) inhibit SP1 binding entirely, thus switching off the expression of these proteins [37]. RSNOs that are formed in conjunction with the deoxygenation of Hb also regulate gene expression. These act to stabilise the HIF 1α, increasing its DNA binding, leading to increased transcription of genes such as VEGF [38]. In mouse lung endothelial cells, S-nitrosation of zinc-thiolate clusters in metallothionein was shown to induce nuclear translocation of the transcription factor, MTF-1, and upregulate gene expression [39].

NO affects the activity of the transcription factor, NF-κB, by restricting its DNA binding, promoter and gene transcription abilities [40]. It has further been shown that this is due to S-nitrosation of the kinase, IKK, thus preventing the phosphorylation and ultimate breakdown of the inhibitor of NF-κB, I-κB [41]. In macrophages, decreased IL-1β expression in response to LPS has been shown to be due to NO-mediated S-nitrosation of NF-κB [42].

In prostate cancer cell lines, S-nitrosation of the transcription repressor, YinYang1 is responsible for the inhibition of its activity by NO, causing increased expression of Fas, and increased sensitivity of the tumour cells to Fas-induced apoptosis [43].

RSNO in the Pathogenesis of Disease

S-nitrosothiols naturally play a part in the physiological functioning of several systems within humans. The numerous roles of S-nitrosothiols make it inevitable that disruption of normal NO biology may contribute to the pathogenesis of numerous disorders, as described in the following sections.

Lowered RSNO levels

The beneficial physiological functions of RSNO in a number of systems suggest that an impairment of RSNO formation or function would remove the protective effect exerted by these molecules, and thus be detrimental to the organism.

There are three potential mechanisms through which RSNO levels may be decreased in disease; decreased synthesis, increased enzymatic breakdown and increased chemical breakdown. Decreased synthesis of RSNO has been suggested to occur in the airways of hypoxaemic babies and cystic fibrosis patients [17].

Excessive breakdown appears to be a more prevalent mechanism for depletion of RSNO. In one model of asthma, it has been proposed that enzymatic degradation of GSNO by GSNOR released from damaged or dying cells causes the decrease in GSNO observed, which is in contrast to increased exhaled NO [44]. Accelerated metabolism is also proposed to account for depletion of RSNO in the airways of patients with cystic fibrosis [45]. Cu(I)-catalysed breakdown, on the other hand, may contribute to depletion of RSNO in familial amyotrophic lateral sclerosis (fAMS), through mutant forms of superoxide dismutases (SODs) [46,47]. Similarly, the amyloid precursor protein of Alzheimer's disease and the prion protein associated with Creutzfeld-Jacob disease both have the ability to reduce Cu(II) to Cu(I), thereby increasing the degree of Cu(I)-induced degradation of RSNO [48].

In other conditions, the cause of reduced RSNO levels is unclear, though the biological consequences are known. Studies have demonstrated decreased levels of SNO-alb in patients with pulmonary arterial hypertension, with associated impairment of the ability of red blood cells to release NO to induce vasorelaxation under hypoxic conditions [17]. Studies have also shown that RSNO levels are negatively correlated with endothelial function in individuals with cardiovascular risk factors, with a stronger correlation observed with increasing numbers of such factors [49].

Elevated RSNO Levels

Supraphysiological levels of RSNO have also been observed in several conditions. However, it is unclear in many cases whether these levels of RSNO directly contribute to the pathogenesis of the disease, or are artefacts which merely reflect increased iNOS activity from activated inflammatory cells. Indeed, in some cases it has been suggested that elevated RSNO serve a protective purpose which is overwhelmed in the pathogenesis of disease. For example, elevated NO and/or RSNO in tuberculosis [50] and pneumonia [51], plus SNO-alb in *E. coli*-derived LPS-induced septic shock [52] have all been suggested to have anti-bacterial functions. Additionally, increased SNO-Hb in malaria has been proposed to defend against the parasites [53].

As many of the biological effects of NO are now thought to be mediated via the formation of RSNO intermediates, it is perhaps reasonable to assume that they are also responsible for the damaging effects observed with excessive NO production. In LPS-induced sepsis, increased RSNO have been proposed to account for hypoperfusion and shunting via increased haemoglobin nitrosation at haem [54] or thiol residues [52]. However, an alternative theory is that these modifications improve oxygenation through a rightward shift of the Hb oxygen dissociation curve [55] and are therefore protective. Studies in a rat model of cirrhosis have also shown increased RSNO synthesis when liver function is impaired, which is markedly exacerbated in the presence of LPS [56].

In cerebral ischaemia-reperfusion, S-nitrosation of MMP-9 has been shown to increase its activity and cause apoptotic cell death of neurons, thus contributing to ischaemia-reperfusion injury in stroke [57]. Similarly, S-nitrosation of GAPDH is able to promote apoptosis [58]. Furthermore, decreased release of free NO from RSNO has been proposed to account for both elevated RSNO levels and decreased NO activity observed in diabetes, in which SNO-Hb becomes glycosylated, decreasing its propensity to release the NO moiety [59] and in pre-eclampsia [60]. Excessive NO generation in neurodegenerative disorders such as Parkinson's disease and Alzheimer's disease also leads to S-nitrosation of protein disulphide isomerase (PDI), thus initiating the unfolded protein response of ER stress and neuronal apoptosis [61]. S-nitrosated PDI has been detected in the brains of Parkinson's and Alzheimer's patients, but not healthy brain [61].

In other cases, a correlation between RSNO and severity of disease has been observed, but a clear causative role for these in the aetiology of the disease has not yet been defined. Antibodies against SNO-Cys have been detected in the serum of some patients suffering from multiple sclerosis, at levels which are correlated with the extent of demyelination and periods of relapse [62], and nitrosothiols are increased in the cerebrospinal fluid of MS patients [63]. Other conditions in which elevated RSNO have been measured include rheumatoid arthritis, in which RSNO proteins are increased in synovial fluid and plasma [64,65], and in lung diseases [66,67], and hypercholesterolaemia [68]. Antibodies against SNO-Cys are also detectable in a rat model of peritonitis [69].

RSNO as Potential Therapeutic Agents or Biomarkers of Disease

An increasing number of studies in animal models suggest that supplementation of RSNO levels in conditions where NO is protective may provide a beneficial therapeutic strategy. For example, in sickle cell disease, SNO-alb has been shown to decrease hypoxia induced injury and thrombus formation [70], and SNO-Cys induced haemoglobin production, thus improving tissue oxygen delivery [71].

RSNO therapy also seems to be promising for use in ischaemia-reperfusion. Inhaled NO was shown to increase SNO-alb levels, leading to improved blood flow. This suggests that inhalation of NO may represent an effective system of NO delivery in cardiovascular disease [72]. Furthermore, in a rat epigastric flap model, cell viability following ischaemia-reperfusion was increased in the presence of SNO-alb [73]. Partial S-nitrosation of thioredoxin (Trx) increased its protective effect in a mouse model of myocardial ischaemia-reperfusion, which may have been due both to a conformational change within Trx and to increased availability of NO for the inactivation of pro-apoptotic proteins such as caspases, via transnitrosation reactions [74].

Following coronary artery bypass surgery, the anti-thrombotic effects of GSNO may potentially decrease the chances of early vein graft failure [75]. Pretreatment with SNO-alb also restored heart and lung function, and decreased neutrophil priming, in a model of bacterial LPS-induced endotoxaemia [76]. GSNO has been demonstrated to be more effective than inhaled NO in cystic fibrosis [36, 77].

A novel S-nitrosating agent which fails to release free NO, ethyl nitrite, has been developed, and has been shown to be more beneficial than NO in pulmonary hypertension [78]. Such agents therefore appear to be promising therapeutic strategies for conditions in which RSNO are depleted, and it is anticipated that further potential clinical uses will be determined in future.

In contrast, in diseases in which elevation of RSNO is associated with pathology, RSNO, or antibodies against them, may prove to be effective biomarkers of disease in conditions in which they have been shown to be elevated. For example, it has been proposed that anti-SNO-Cys antibodies may be effective markers of the extent of disease and periods of relapse in multiple sclerosis [62]. Conversely, the depletion of RSNO in human subjects with cardiovascular risk factors could potentially be useful as an index of impaired endothelial function [49]. Further potential uses for RSNO measurements in determining disease status will no doubt be elucidated as the roles of RSNO in various conditions receive additional study.

Conclusion

RSNO are important molecules in mammalian biology, in both health and disease. They are formed by the reaction of metabolites of NO with low molecular weight thiols or accessible reduced thiols in proteins, and can be degraded by both enzymatic and non-enzymatic means, or NO can be transferred from one thiol to another by transnitrosation.

RSNO have roles in regulating vascular tone and improving perfusion in hypoxic areas of the circulation through providing an endogenous circulating pool of NO. It is now becoming clear that in addition they are able to regulate the activity of a number of proteins and signalling pathways, in a manner analogous to phosphorylation. As the importance of RSNO is becoming clear, it is no surprise that dysregulation of their formation or breakdown may contribute to the pathogenesis of disease. As such, diseases marked by both depletion and excess of RSNO have been identified, although much further work is required to fully characterise their contribution to both the maintenance of human health and the progression of disease. Additionally, alterations in RSNO levels in such diseases may provide an opportunity to utilise measurements of RSNO as biomarkers of either the risk or progression of certain conditions.

Acknowledgements

Emma Taylor is a postdoctoral research fellow funded by the Strategic Promotion of Ageing Research Capacity (SPARC) initiative.

References

[1] Giustarini, D.; Milzani, A.; le-Donne, I.; Rossi, R. Detection of S-Nitrosothiols in Biological Fluids: A Comparison Among the Most Widely Applied Methodologies. *J. Chromatogr. B Analyt. Technol. Biomed. Life Sci*. 2006.

[2] Gow, A. J.; Farkouh, C. R.; Munson, D. A.; Posencheg, M. A.; Ischiropoulos, H. Biological Significance of Nitric Oxide-Mediated Protein Modifications. *Am. J. Physiol Lung Cell Mol. Physiol*. 2004, 287, L262-L268.

[3] Gow, A. J.; Buerk, D. G.; Ischiropoulos, H. A Novel Reaction Mechanism for the Formation of S-Nitrosothiol in Vivo. *J. Biol. Chem*. 1997, 272, 2841-2845.

[4] Arnelle, D. R.; Stamler, J. S. NO+, NO, and NO- Donation by S-Nitrosothiols: Implications for Regulation of Physiological Functions by S-Nitrosylation and Acceleration of Disulfide Formation. *Arch. Biochem. Biophys*. 1995, 318, 279-285.

[5] Liu, L.; Hausladen, A.; Zeng, M.; Que, L.; Heitman, J.; Stamler, J. S. A Metabolic Enzyme for S-Nitrosothiol Conserved From Bacteria to Humans. *Nature* 2001, 410, 490-494.

[6] Gaston, B. M.; Carver, J.; Doctor, A.; Palmer, L. A. S-Nitrosylation Signaling in Cell Biology. *Mol. Interv*. 2003, 3, 253-263.

[7] Gow, A. J.; Chen, Q.; Hess, D. T.; Day, B. J.; Ischiropoulos, H.; Stamler, J. S. Basal and Stimulated Protein S-Nitrosylation in Multiple Cell Types and Tissues. *J. Biol. Chem*. 2002, 277, 9637-9640.

[8] Janero, D. R.; Bryan, N. S.; Saijo, F.; Dhawan, V.; Schwalb, D. J.; Warren, M. C.; Feelisch, M. Differential Nitros(Yl) Ation of Blood and Tissue Constituents During Glyceryl Trinitrate Biotransformation in Vivo. *Proc. Natl. Acad. Sci. USA* 2004, 101, 16958-16963.

[9] Stamler, J. S.; Jaraki, O.; Osborne, J.; Simon, D. I.; Keaney, J.; Vita, J.; Singel, D.; Valeri, C. R.; Loscalzo, J. Nitric Oxide Circulates in Mammalian Plasma Primarily As an S-Nitroso Adduct of Serum Albumin. *Proc. Natl. Acad. Sci. USA* 1992, 89, 7674-7677.

[10] Mannick, J. B.; Schonhoff, C. M. Nitrosylation: the Next Phosphorylation? *Arch. Biochem. Biophys.* 2002, 408, 1-6.

[11] Gladwin, M. T.; Schechter, A. N.; Kim-Shapiro, D. B.; Patel, R. P.; Hogg, N.; Shiva, S.; Cannon, R. O., III; Kelm, M.; Wink, D. A.; Espey, M. G.; Oldfield, E. H.; Pluta, R. M.; Freeman, B. A.; Lancaster, J. R., Jr.; Feelisch, M.; Lundberg, J. O. The Emerging Biology of the Nitrite Anion. *Nat. Chem. Biol.* 2005, 1, 308-314.

[12] Gladwin, M. T.; Raat, N. J.; Shiva, S.; Dezfulian, C.; Hogg, N.; Kim-Shapiro, D. B.; Patel, R. P. Nitrite As a Vascular Endocrine Nitric Oxide Reservoir That Contributes to Hypoxic Signaling, Cytoprotection, and Vasodilation. *Am. J. Physiol Heart Circ. Physiol.* 2006, 291, H2026-H2035.

[13] Crawford, J. H.; Isbell, T. S.; Huang, Z.; Shiva, S.; Chacko, B. K.; Schechter, A. N.; rley-Usmar, V. M.; Kerby, J. D.; Lang, J. D., Jr.; Kraus, D.; Ho, C.; Gladwin, M. T.; Patel, R. P. Hypoxia, Red Blood Cells, and Nitrite Regulate NO-Dependent Hypoxic Vasodilation. *Blood* 2006, 107, 566-574.

[14] Schopfer, F. J.; Baker, P. R.; Giles, G.; Chumley, P.; Batthyany, C.; Crawford, J.; Patel, R. P.; Hogg, N.; Branchaud, B. P.; Lancaster, J. R., Jr.; Freeman, B. A. Fatty Acid Transduction of Nitric Oxide Signaling. Nitrolinoleic Acid Is a Hydrophobically Stabilized Nitric Oxide Donor. *J. Biol. Chem.* 2005, 280, 19289-19297.

[15] Batthyany, C.; Schopfer, F. J.; Baker, P. R.; Duran, R.; Baker, L. M.; Huang, Y.; Cervenansky, C.; Branchaud, B. P.; Freeman, B. A. Reversible Post-Translational Modification of Proteins by Nitrated Fatty Acids in Vivo. *J. Biol. Chem.* 2006, 281, 20450-20463.

[16] Scharfstein, J. S.; Keaney, J. F., Jr.; Slivka, A.; Welch, G. N.; Vita, J. A.; Stamler, J. S.; Loscalzo, J. In Vivo Transfer of Nitric Oxide Between a Plasma Protein-Bound Reservoir and Low Molecular Weight Thiols. *J. Clin. Invest* 1994, 94, 1432-1439.

[17] Foster, M. W.; McMahon, T. J.; Stamler, J. S. S-Nitrosylation in Health and Disease. *Trends Mol. Med.* 2003, 9, 160-168.

[18] Gross, S. S. Vascular Biology. Targeted Delivery of Nitric Oxide. *Nature* 2001, 409, 577-578.

[19] Allen, B. W.; Piantadosi, C. A. How Do Red Blood Cells Cause Hypoxic Vasodilation? The SNO-Hemoglobin Paradigm. *Am. J. Physiol Heart Circ. Physiol.* 2006, 291, H1507-H1512.

[20] Pawloski, J. R.; Hess, D. T.; Stamler, J. S. Export by Red Blood Cells of Nitric Oxide Bioactivity. *Nature* 2001, 409, 622-626.

[21] Boese, M.; Mordvintcev, P. I.; Vanin, A. F.; Busse, R.; Mulsch, A. S-Nitrosation of Serum Albumin by Dinitrosyl-Iron Complex. J. Biol. Chem. 1995, 270, 29244-29249.

[22] Mulsch, A.; Mordvintcev, P. I.; Vanin, A. F.; Busse, R. Formation and Release of Dinitrosyl Iron Complexes by Endothelial Cells. *Biochem. Biophys. Res. Commun.* 1993, 196, 1303-1308.

[23] Liu, Z.; Rudd, M. A.; Freedman, J. E.; Loscalzo, J. S-Transnitrosation Reactions Are Involved in the Metabolic Fate and Biological Actions of Nitric Oxide. *J. Pharmacol. Exp. Ther*. 1998, 284, 526-534.

[24] Jansen, A.; Drazen, J.; Osborne, J. A.; Brown, R.; Loscalzo, J.; Stamler, J. S. The Relaxant Properties in Guinea Pig Airways of S-Nitrosothiols. *J. Pharmacol. Exp. Ther*. 1992, 261, 154-160.

[25] Costa, G.; Labadia, A.; Triguero, D.; Jimenez, E.; Garcia-Pascual, A. Nitrergic Relaxation in Urethral Smooth Muscle: Involvement of Potassium Channels and Alternative Redox Forms of NO. Naunyn Schmiedebergs. *Arch. Pharmacol*. 2001, 364, 516-523.

[26] Slivka, A.; Chuttani, R.; Carr-Locke, D. L.; Kobzik, L.; Bredt, D. S.; Loscalzo, J.; Stamler, J. S. Inhibition of Sphincter of Oddi Function by the Nitric Oxide Carrier S-Nitroso-N-Acetylcysteine in Rabbits and Humans. *J. Clin. Invest* 1994, 94, 1792-1798.

[27] Mellion, B. T.; Ignarro, L. J.; Myers, C. B.; Ohlstein, E. H.; Ballot, B. A.; Hyman, A. L.; Kadowitz, P. J. Inhibition of Human Platelet Aggregation by S-Nitrosothiols. Heme-Dependent Activation of Soluble Guanylate Cyclase and Stimulation of Cyclic GMP Accumulation. *Mol. Pharmacol.* 1983, 23, 653-664.

[28] Crane, M. S.; Ollosson, R.; Moore, K. P.; Rossi, A. G.; Megson, I. L. Novel Role for Low Molecular Weight Plasma Thiols in Nitric Oxide-Mediated Control of Platelet Function. *J. Biol. Chem.* 2002, 277, 46858-46863.

[29] Pfeilschifter, J.; Huwiler, A. Nitric Oxide Stimulates Stress-Activated Protein Kinases in Glomerular Endothelial and Mesangial Cells. *FEBS Lett.* 1996, 396, 67-70.

[30] Brockhaus, F.; Brune, B. P53 Accumulation in Apoptotic Macrophages Is an Energy Demanding Process That Precedes Cytochrome c Release in Response to Nitric Oxide. *Oncogene* 1999, 18, 6403-6410.

[31] Jekabsone, A.; Dapkunas, Z.; Brown, G. C.; Borutaite, V. S-Nitrosothiol-Induced Rapid Cytochrome c Release, Caspase Activation and Mitochondrial Permeability Transition in Perfused Heart. *Biochem. Pharmacol.* 2003, 66, 1513-1519.

[32] Taylor, E. L.; Megson, I. L.; Haslett, C.; Rossi, A. G. Nitric Oxide: a Key Regulator of Myeloid Inflammatory Cell Apoptosis. *Cell Death. Differ.* 2003, 10, 418-430.

[33] Kim, P. K.; Kwon, Y. G.; Chung, H. T.; Kim, Y. M. Regulation of Caspases by Nitric Oxide. *Ann. N. Y. Acad. Sci*. 2002, 962, 42-52.

[34] Lin, Y. F.; Raab-Graham, K.; Jan, Y. N.; Jan, L. Y. NO Stimulation of ATP-Sensitive Potassium Channels: Involvement of Ras/Mitogen-Activated Protein Kinase Pathway and Contribution to Neuroprotection. *Proc. Natl. Acad. Sci. USA* 2004, 101, 7799-7804.

[35] Park, H. S.; Huh, S. H.; Kim, M. S.; Lee, S. H.; Choi, E. J. Nitric Oxide Negatively Regulates C-Jun N-Terminal Kinase/Stress-Activated Protein Kinase by Means of S-Nitrosylation. *Proc. Natl. Acad. Sci. USA* 2000, 97, 14382-14387.

[36] Zaman, K.; McPherson, M.; Vaughan, J.; Hunt, J.; Mendes, F.; Gaston, B.; Palmer, L. A. S-Nitrosoglutathione Increases Cystic Fibrosis Transmembrane Regulator Maturation. *Biochem. Biophys. Res. Commun*. 2001, 284, 65-70.

[37] Zaman, K.; Palmer, L. A.; Doctor, A.; Hunt, J. F.; Gaston, B. Concentration-Dependent Effects of Endogenous S-Nitrosoglutathione on Gene Regulation by Specificity Proteins Sp3 and Sp1. *Biochem. J.* 2004, 380, 67-74.
[38] Palmer, L. A.; Gaston, B.; Johns, R. A. Normoxic Stabilization of Hypoxia-Inducible Factor-1 Expression and Activity: Redox-Dependent Effect of Nitrogen Oxides. *Mol. Pharmacol.* 2000, 58, 1197-1203.
[39] Stitt, M. S.; Wasserloos, K. J.; Tang, X.; Liu, X.; Pitt, B. R.; St Croix, C. M. Nitric Oxide-Induced Nuclear Translocation of the Metal Responsive Transcription Factor, MTF-1 Is Mediated by Zinc Release From Metallothionein. *Vascul. Pharmacol.* 2006, 44, 149-155.
[40] Matthews, J. R.; Botting, C. H.; Panico, M.; Morris, H. R.; Hay, R. T. Inhibition of NF-KappaB DNA Binding by Nitric Oxide. *Nucleic Acids Res.* 1996, 24, 2236-2242.
[41] Reynaert, N. L.; Ckless, K.; Korn, S. H.; Vos, N.; Guala, A. S.; Wouters, E. F.; van, d., V; Janssen-Heininger, Y. M. Nitric Oxide Represses Inhibitory KappaB Kinase Through S-Nitrosylation. *Proc. Natl. Acad. Sci. USA* 2004, 101, 8945-8950.
[42] Schroeder, R. A.; Cai, C.; Kuo, P. C. Endotoxin-Mediated Nitric Oxide Synthesis Inhibits IL-1beta Gene Transcription in ANA-1 Murine Macrophages. *Am. J. Physiol* 1999, 277, C523-C530.
[43] Hongo, F.; Garban, H.; Huerta-Yepez, S.; Vega, M.; Jazirehi, A. R.; Mizutani, Y.; Miki, T.; Bonavida, B. Inhibition of the Transcription Factor Yin Yang 1 Activity by S-Nitrosation. *Biochem. Biophys. Res. Commun.* 2005, 336, 692-701.
[44] Fang, K.; Johns, R.; Macdonald, T.; Kinter, M.; Gaston, B. S-Nitrosoglutathione Breakdown Prevents Airway Smooth Muscle Relaxation in the Guinea Pig. *Am. J. Physiol Lung Cell Mol. Physiol.* 2000, 279, L716-L721.
[45] Grasemann, H.; Gaston, B.; Fang, K.; Paul, K.; Ratjen, F. Decreased Levels of Nitrosothiols in the Lower Airways of Patients With Cystic Fibrosis and Normal Pulmonary Function. *J. Pediatr.* 1999, 135, 770-772.
[46] Johnson, M. A.; Macdonald, T. L.; Mannick, J. B.; Conaway, M. R.; Gaston, B. Accelerated S-Nitrosothiol Breakdown by Amyotrophic Lateral Sclerosis Mutant Copper,Zinc-Superoxide Dismutase. *J. Biol. Chem.* 2001, 276, 39872-39878.
[47] Schonhoff, C. M.; Matsuoka, M.; Tummala, H.; Johnson, M. A.; Estevez, A. G.; Wu, R.; Kamaid, A.; Ricart, K. C.; Hashimoto, Y.; Gaston, B.; Macdonald, T. L.; Xu, Z.; Mannick, J. B. S-Nitrosothiol Depletion in Amyotrophic Lateral Sclerosis. *Proc. Natl. Acad. Sci. USA* 2006, 103, 2404-2409.
[48] Opazo, C.; Barria, M. I.; Ruiz, F. H.; Inestrosa, N. C. Copper Reduction by Copper Binding Proteins and Its Relation to Neurodegenerative Diseases. *Biometals* 2003, 16, 91-98.
[49] Heiss, C.; Lauer, T.; Dejam, A.; Kleinbongard, P.; Hamada, S.; Rassaf, T.; Matern, S.; Feelisch, M.; Kelm, M. Plasma Nitroso Compounds Are Decreased in Patients with Endothelial Dysfunction. *J. Am. Coll. Cardiol.* 2006, 47, 573-579.
[50] MacMicking, J. D.; North, R. J.; LaCourse, R.; Mudgett, J. S.; Shah, S. K.; Nathan, C. F. Identification of Nitric Oxide Synthase As a Protective Locus Against Tuberculosis. *Proc. Natl. Acad. Sci. USA* 1997, 94, 5243-5248.

[51] Gaston, B.; Reilly, J.; Drazen, J. M.; Fackler, J.; Ramdev, P.; Arnelle, D.; Mullins, M. E.; Sugarbaker, D. J.; Chee, C.; Singel, D. J.; . Endogenous Nitrogen Oxides and Bronchodilator S-Nitrosothiols in Human Airways. *Proc. Natl. Acad. Sci. USA* 1993, 90, 10957-10961.

[52] Jourd'heuil, D.; Gray, L.; Grisham, M. B. S-Nitrosothiol Formation in Blood of Lipopolysaccharide-Treated Rats. *Biochem. Biophys. Res. Commun.* 2000, 273, 22-26.

[53] Taylor-Robinson, A. W.; Looker, M. Sensitivity of Malaria Parasites to Nitric Oxide at Low Oxygen Tensions. *Lancet* 1998, 351, 1630.

[54] Westenberger, U.; Thanner, S.; Ruf, H. H.; Gersonde, K.; Sutter, G.; Trentz, O. Formation of Free Radicals and Nitric Oxide Derivative of Hemoglobin in Rats During Shock Syndrome. *Free Radic. Res. Commun.* 1990, 11, 167-178.

[55] Kosaka, H.; Seiyama, A. Increased Oxygen Dissociation by Nitric Oxide from RBC. *Adv. Exp. Med. Biol.* 1997, 428, 349-354.

[56] Ottesen, L. H.; Harry, D.; Frost, M.; Davies, S.; Khan, K.; Halliwell, B.; Moore, K. Increased Formation of S-Nitrothiols and Nitrotyrosine in Cirrhotic Rats During Endotoxemia. *Free Radic. Biol. Med.* 2001, 31, 790-798.

[57] Gu, Z.; Kaul, M.; Yan, B.; Kridel, S. J.; Cui, J.; Strongin, A.; Smith, J. W.; Liddington, R. C.; Lipton, S. A. S-Nitrosylation of Matrix Metalloproteinases: Signaling Pathway to Neuronal Cell Death. *Science* 2002, 297, 1186-1190.

[58] Hara, M. R.; Agrawal, N.; Kim, S. F.; Cascio, M. B.; Fujimuro, M.; Ozeki, Y.; Takahashi, M.; Cheah, J. H.; Tankou, S. K.; Hester, L. D.; Ferris, C. D.; Hayward, S. D.; Snyder, S. H.; Sawa, A. S-Nitrosylated GAPDH Initiates Apoptotic Cell Death by Nuclear Translocation Following Siah1 Binding. *Nat. Cell Biol.* 2005, 7, 665-674.

[59] Milsom, A. B.; Jones, C. J.; Goodfellow, J.; Frenneaux, M. P.; Peters, J. R.; James, P. E. Abnormal Metabolic Fate of Nitric Oxide in Type I Diabetes Mellitus. *Diabetologia* 2002, 45, 1515-1522.

[60] Tyurin, V. A.; Liu, S. X.; Tyurina, Y. Y.; Sussman, N. B.; Hubel, C. A.; Roberts, J. M.; Taylor, R. N.; Kagan, V. E. Elevated Levels of S-Nitrosoalbumin in Preeclampsia Plasma. *Circ. Res.* 2001, 88, 1210-1215.

[61] Uehara, T.; Nakamura, T.; Yao, D.; Shi, Z. Q.; Gu, Z.; Ma, Y.; Masliah, E.; Nomura, Y.; Lipton, S. A. S-Nitrosylated Protein-Disulphide Isomerase Links Protein Misfolding to Neurodegeneration. *Nature* 2006, 441, 513-517.

[62] Boullerne, A. I.; Rodriguez, J. J.; Touil, T.; Brochet, B.; Schmidt, S.; Abrous, N. D.; Le, M. M.; Pua, J. R.; Jensen, M. A.; Mayo, W.; Arnason, B. G.; Petry, K. G. Anti-S-Nitrosocysteine Antibodies Are a Predictive Marker for Demyelination in Experimental Autoimmune Encephalomyelitis: Implications for Multiple Sclerosis. *J. Neurosci.* 2002, 22, 123-132.

[63] Calabrese, V.; Scapagnini, G.; Ravagna, A.; Bella, R.; Foresti, R.; Bates, T. E.; Giuffrida Stella, A. M.; Pennisi, G. Nitric Oxide Synthase Is Present in the Cerebrospinal Fluid of Patients With Active Multiple Sclerosis and Is Associated With Increases in Cerebrospinal Fluid Protein Nitrotyrosine and S-Nitrosothiols and With Changes in Glutathione Levels. *J. Neurosci. Res.* 2002, 70, 580-587.

[64] Hilliquin, P.; Borderie, D.; Hernvann, A.; Menkes, C. J.; Ekindjian, O. G. Nitric Oxide As S-Nitrosoproteins in Rheumatoid Arthritis. *Arthritis Rheum.* 1997, 40, 1512-1517.

[65] Rocks, S. A.; Davies, C. A.; Hicks, S. L.; Webb, A. J.; Klocke, R.; Timmins, G. S.; Johnston, A.; Jawad, A. S.; Blake, D. R.; Benjamin, N.; Winyard, P. G. Measurement of S-Nitrosothiols in Extracellular Fluids From Healthy Human Volunteers and Rheumatoid Arthritis Patients, Using Electron Paramagnetic Resonance Spectrometry. *Free Radic. Biol. Med.* 2005, 39, 937-948.

[66] Corradi, M.; Montuschi, P.; Donnelly, L. E.; Pesci, A.; Kharitonov, S. A.; Barnes, P. J. Increased Nitrosothiols in Exhaled Breath Condensate in Inflammatory Airway Diseases. *Am. J. Respir. Crit Care Med.* 2001, 163, 854-858.

[67] Gaston, B.; Singel, D.; Doctor, A.; Stamler, J. S. S-Nitrosothiol Signaling in Respiratory Biology. *Am. J. Respir. Crit Care Med.* 2006, 173, 1186-1193.

[68] Moriel, P.; Pereira, I. R.; Bertolami, M. C.; Abdalla, D. S. Is Ceruloplasmin an Important Catalyst for S-Nitrosothiol Generation in Hypercholesterolemia? *Free Radic. Biol. Med.* 2001, 30, 318-326.

[69] Ferrier, M. L.; Combet, S.; van, L. M.; Stoenoiu, M. S.; Cnops, Y.; Lameire, N.; Devuyst, O. Inhibition of Nitric Oxide Synthase Reverses Changes in Peritoneal Permeability in a Rat Model of Acute Peritonitis. *Kidney Int.* 2001, 60, 2343-2350.

[70] de, F. L.; Malpeli, G.; Scarpa, A.; Janin, A.; Muchitsch, E. M.; Roncada, P.; Leboeuf, C.; Corrocher, R.; Beuzard, Y.; Brugnara, C. Protective Effects of S-Nitrosoalbumin on Lung Injury Induced by Hypoxia-Reoxygenation in Mouse Model of Sickle Cell Disease. *Am. J. Physiol Lung Cell Mol. Physiol* 2006, 291, L457-L465.

[71] Cokic, V. P.; Smith, R. D.; Beleslin-Cokic, B. B.; Njoroge, J. M.; Miller, J. L.; Gladwin, M. T.; Schechter, A. N. Hydroxyurea Induces Fetal Hemoglobin by the Nitric Oxide-Dependent Activation of Soluble Guanylyl Cyclase. *J. Clin. Invest* 2003, 111, 231-239.

[72] Ng, E. S.; Jourd'heuil, D.; McCord, J. M.; Hernandez, D.; Yasui, M.; Knight, D.; Kubes, P. Enhanced S-Nitroso-Albumin Formation From Inhaled NO During Ischemia/Reperfusion. *Circ. Res.* 2004, 94, 559-565.

[73] Mittermayr, R.; Valentini, D.; Fitzal, F.; Hallstrom, S.; Gasser, H.; Redl, H. Protective Effect of a Novel NO-Donor on Ischemia/Reperfusion Injury in a Rat Epigastric Flap Model. *Wound. Repair Regen.* 2003, 11, 3-10.

[74] Tao, L.; Gao, E.; Bryan, N. S.; Qu, Y.; Liu, H. R.; Hu, A.; Christopher, T. A.; Lopez, B. L.; Yodoi, J.; Koch, W. J.; Feelisch, M.; Ma, X. L. Cardioprotective Effects of Thioredoxin in Myocardial Ischemia and Reperfusion: Role of S-Nitrosation [Corrected]. *Proc. Natl. Acad. Sci. USA* 2004, 101, 11471-11476.

[75] Salas, E.; Langford, E. J.; Marrinan, M. T.; Martin, J. F.; Moncada, S.; de Belder, A. J. S-Nitrosoglutathione Inhibits Platelet Activation and Deposition in Coronary Artery Saphenous Vein Grafts in Vitro and in Vivo. *Heart* 1998, 80, 146-150.

[76] Gluckman, T. L.; Grossman, J. E.; Folts, J. D.; Kruse-Elliott, K. T. Modulation of Endotoxin-Induced Cardiopulmonary Dysfunction by S-Nitroso-Albumin. *J. Endotoxin. Res.* 2002, 8, 17-26.

[77] Ratjen, F.; Gartig, S.; Wiesemann, H. G.; Grasemann, H. Effect of Inhaled Nitric Oxide on Pulmonary Function in Cystic Fibrosis. *Respir. Med.* 1999, 93, 579-583.

[78] Moya, M. P.; Gow, A. J.; Califf, R. M.; Goldberg, R. N.; Stamler, J. S. Inhaled Ethyl Nitrite Gas for Persistent Pulmonary Hypertension of the Newborn. *Lancet* 2002, 360, 141-143.

In: Oxidative Stress: Clinical and Biomedical Implications ISBN: 978-1-60021-888-0
Editors: B. M. Matata, M. M. Elahi, pp. 171-182

Chapter XI

Flavonoids, Endothelial Function, and Cardiovascular Disease

Machha Ajay[1] and Mohd Rais Mustafa[2]
[1]Lakehead University, 955, Oliver Road,
Thunder Bay, P7B 5E1, Ontario, Canada
[2]Department of Pharmacology, Faculty of Medicine,
University of Malaya, Kuala Lumpur, 50603, Malaysia

Abstract

Endothelial dysfunction, mostly described as a reduction in endothelium-dependent vascular relaxation, is closely associated with the development and progression of cardiovascular diseases. Epidemiological data suggest that dietary consumption of flavonoids is associated with reduced risk of cardiovascular disease. Experimental studies investigating the relationship between flavonoids and occurrence of risk factors for cardiovascular disease are strongly supportive of the health protective benefits of these polyphenolic compounds. A multitude of biological actions are purportedly associated with the flavonoids; the most potential is the protective effects on endothelial function. In this chapter we first addressed possible mechanisms of actions of the flavonoids in modulating the various etiological factors underlying the development of endothelial dysfunction. This will be followed by our recent findings investigating the cardiovascular protective effects of the flavonoids in hypertension and diabetes mellitus.

1. Introduction

Cardiovascular disease is the principal cause of death and disability in both developed and developing countries, accounting roughly for 20% of all deaths per annum globally. The state of the cardiovascular system is dependent on the structural integrity and health of the heart and blood vessels. The tunica intima, the innermost layer of the blood vessels, consists of a single layer of endothelial cells and connective tissue. These endothelial cells are both a passive filter and a metabolically active organ that synthesizes and release several vasoactive substances into blood and adjacent smooth muscle cells, regulating vascular homeostasis. Endothelium-derived vasoactive substances include vasodilators, such as nitric oxide, prostacyclin and a yet unidentified endothelium-derived hyperpolarizing factor, and also vasoconstrictors such as free radicals, cyclooxygenase products, and endothelin-I. Emerging studies have shown that cardiovascular risk factors including hypertension and diabetes mellitus are closely associated with impairment of vascular homeostasis provoked by endothelial dysfunction. Investigations into the mechanisms of endothelial dysfunction in hypertension and diabetes mellitus have demonstrated that reduced bioactivity and/or bioavailability of endothelium-derived nitric oxide (EDNO) are closely interdependent [Endemann and Schiffrin, 2004]. Several factors, chief among which are increased production of reactive oxygen and nitrogen species or oxidants, low-density lipoprotein (LDL) cholesterol oxidation, and decreased production of nitric oxide via endothelial nitric oxide synthase (eNOS), have been identified as important causative factors in reducing the bioavailability and/or bioactivity of EDNO and the subsequent development of endothelial dysfunction [Endemann and Schiffrin, 2004; Forstermann, 2005]. These observations have led investigators to identify potential pharmacological agents capable of enhancing the bioactivity and/or bioavailability of EDNO in the vasculature. Several pharmacological agents including antihypertensive drugs, antioxidant vitamins, carotenoids and flavonoids have been suggested to provide vascular protection through various mechanisms such as reduction in oxidative stress, blood pressure, or improving hyperglycemia. Improvement in endothelial function following these interventions may result from promotion of vascular relaxation, inhibition of vasoconstriction, reduction in the production of free radicals, or other mechanisms that protect the endothelium from injury [Amudha *et al.* 2002].

2. Flavonoids

Flavonoids comprised a large group of naturally occurring, low molecular weight compounds found almost in most plants [Rice-Evans and Packer, 1998; Harborne and Williams, 2000]. More than 8000 individual flavonoids have been identified in plants and these were found to vary in type, and quantity of the favonoids due to variations in plant growing conditions [Pietta, 2000]. The broad spectrum of biological activities within the group, and the multiplicity of actions displayed by certain individual compounds have propelled the flavonoids into one of the most intriguing classes of biologically active compounds, termed as “bioflavonoids”.

2.1. Chemical Classification

The flavonoid aglycone consists of a benzene ring (A) condensed with a six membered ring (C), which carries a phenyl ring (B) as a substituent. The six-membered ring condensed with benzene ring is either an α-pyrone ring or its dihydroderivative (Figure 1) [Harborne and Williams, 2000; Raj Narayana et al. 2001]. Based on the differences in principal chemical substitutions and degree of unsaturation, flavonoids may be classified into four main groups: flavones, flavonols, flavonones, and isoflavonoids (Figure 1). The other subclasses of flavonoids, which comprise relatively smaller number of compounds, includes anthocyanidins, catechins, chalcones and dihydroflavonols.

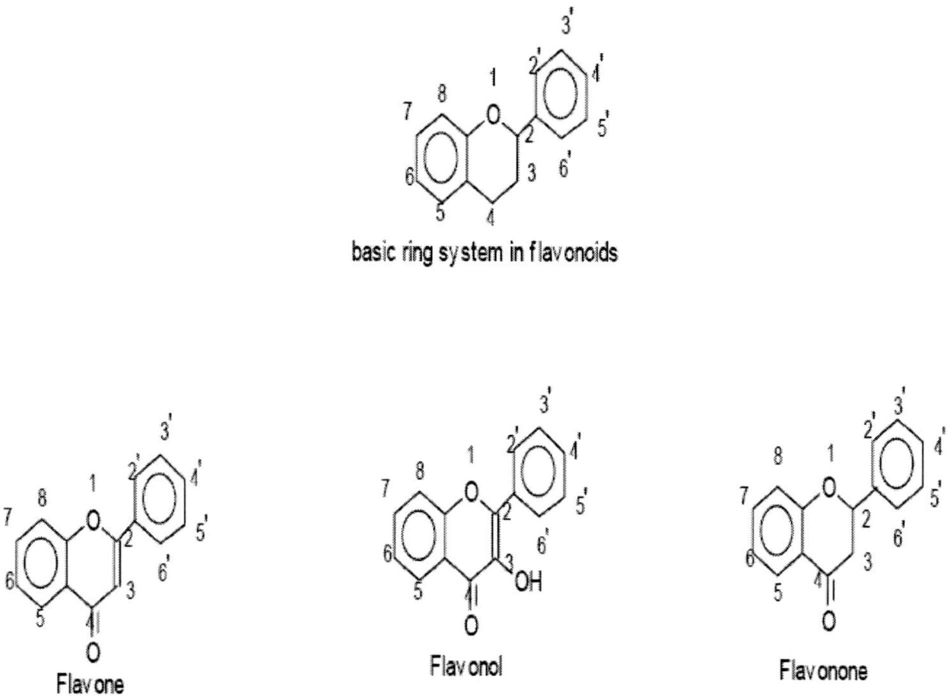

Figure 1. Structures of different sub-classes of flavonoids.

2.2. Absorption and Metabolism

Flavonoids are generally taken as a constituent of various foods and the pertinent question that always arises is whether sufficient amount is absorbed to reach therapeutic levels in the body. The major dietary sources of flavonoids include fruits and fruit products (e.g. apples, citrus fruits, cherries, grapes), vegetables (e.g. onions, broccoli, spinach, tomatoes), beverages (e.g. green tea, black tea, red wine) and herbs (e.g. *Hibiscus sabdariffa, Morinda citrifolia, Silybum marianum*) [Harborne and Williams, 2000]. The daily flavonoid intake in the diet is highly variable with estimations ranging from 23 mg day^{-1} (only flavonols plus flavones, Hertog et al. 1993a) to more than 500 mg day^{-1} (total flavonoids, Mananch et al. 1996). Rice-Evans (2001) demonstrated that dietary flavonoids are absorbed as glycosides and also confirmed that a polyphenolic plasma concentration required to attain about 50 % of maximal relaxation *in vitro* may be detected in human subjects. Studies so far have shown that there is great variability in the preferential metabolic pathways of flavonoids among individuals, which may be due to the differences in gut microflora populations, which mainly contribute to cleavage of flavonoid glycosides into free flavonoids [Hollman and Katan,

1997; Skibola and Smith, 2000; Heim et al. 2002]. Hollman and colleagues (1997) demonstrated that peak concentration of flavonoids occur approximately 2 hr after ingestion of a test food and falls to less than 25 % after 24-48 hrs of administration. Scalbert and Williamson (2000) reported that the plasma concentrations of flavonoids are usually < 10 μmol/L even after a bolus dose, and the steady-state levels are mainly < 1 μmol/L. Morand and colleagues (1998) further demonstrated that bacterial ring scission of flavonoids results in degradation products and various phenolic acids, some of which still exhibit similar activities to their parent molecule. A second important site of flavonoid biotransformation is the liver, an organ which is rich in phase I and II metabolizing enzymes. Flavonoids are gradually hydroxylated and/or *O*-demethylated by cytochrome P450 enzymes and then subjected to conjugation reactions (glucuronidation, sulphation, *O*-methylation) catalyzed by phase II enzymes [Silva et al. 1997; Rice-Evans, 2001; Heim et al. 2002]. Phase II conjugation reactions in the liver gives rise to glucuronides, sulfates, methyl conjugates and small quantities of free aglycones. Mananch et al. (1996) demonstrated that continuously high intakes of the flavonoid, quercetin in rats decreased absorption but increased the formation of the conjugated form, which has more antioxidant activity than the aglycone form..

2.3. Flavonoids and Cardiovascular Disease

The first biological activity of the flavonoids was reported in 1938 by Szent-Gyorgyi. Since then a variety of biological activities of the flavonoids have been reported and a number of reviews have appeared including pharmacological intervention studies. The antioxidant ability of flavonoids in scavenging free radicals highlights one of their major health promoting functions [Harborne and Williams, 2000; Nijveldt et al. 2001]. Research on the cardiovascular effects of the flavonoids received strong impetus with the discovery of the "French paradox" that people from the Mediterranean had less cardiovascular mortality rate in association with consumption of flavonoid-rich foods despite a high saturated fat intake. Subsequently, several population based epidemiological investigations demonstrated an inverse correlation between flavonoid-rich diets and occurrence of cardiovascular risks [Hertog et al., 1993b; Knekt et al., 1996; Yochum et al., 1999]. For example, Hertog et al. (1993b) reported a lower risk of death from coronary heart disease with higher intake of dietary flavonoids. Over the last decade, extensive research have largely focused on determining the mechanisms underlying these beneficial effects of the flavonoids in experimental animal models as well as in patients with risk factors for development of CVD. The findings from these studies strongly suggest that the vascular protective effects of the flavonoids are closely associated with modulation of endothelial function [Duarte et al., 2001; Nijveldt et al. 2001; Engler et al., 2004]. The possible mechanisms underlying the modulatory roles of the flavonoids on endothelial functions are discussed in the following sections.

2.3.1. Antioxidant Effects of Flavonoids on Free Radicals or Oxidants

An imbalance between antioxidants and production of free radicals results in oxidative stress and ultimately to the development of endothelial dysfunction by several mechanisms other than involving oxidative modification of LDL (see below). It is proposed that elevation in the production of free radicals "uncouples" the endothelial nitric oxide (eNOS) activity by oxidation of the co-factor tetrahydrobiopterin and switches the eNOS activity to produce superoxide anions. Such mechanism has been implicated in the development of the endothelial dysfunction in diabetes, hypercholesterolemia, obesity, and atherosclerosis [Amudha et al., 2002; Endemann and Schiffrin, 2004; Forstermann, 2005]. Production of superoxide anions in concert with nitric oxide inactivates the later converting it to peroxynitrite, a weak vasodilator and potent toxic radical, thereby increasing arteriolar vasoconstriction and aggravation of peripheral hemodynamic resistance [Palmer et al., 1987; Endemann and Schiffrin, 2004]. Flavonoids may directly suppress the production of free radicals or scavenge free radicals secondary to their release from endogenous enzymatic sources including "uncoupled" eNOS. For instance, flavonoids prevented superoxide-mediated attenuation of endothelium-dependent relaxations in isolated blood vessels suggesting that the *in vivo* effects of these compounds are at least in part due to the scavenging of superoxide anions and/or its reactive metabolites [Woodman and Chan, 2004]. Jiang et al. (2003) reported that supplementation with isoflavone metabolites protected eNOS function in apolipoprotein-E deficient mice through both lipid-lowering and superoxide scavenging actions. Flavonoids can prevent injury caused by reactive oxygen and nitrogen species in several ways. Flavonoids may directly couple with the free radicals, resulting in more stable and less-reactive radicals [Robak and Gryglewski, 1996]. Flavonoids may also scavenge radical species by acting as chain-breaking antioxidants [Nijveldt et al. 2001]. In addition, these polyphenols may recycle other chain-breaking antioxidants such as ∝-tocoferol by donating a hydrogen atom to the tocopheryl radical. Moreover, flavonoids can directly scavenge metabolites and reactive products of superoxide anions including peroxynitrite [Aldini et al. 2003]. The antioxidant activity of the flavonoids *in vivo* may be due to their inhibitory actions on certain endogenous prooxidant free radical generating enzymes [Robak and Gryglewski, 1996; Nijveldt et al. 2001]. Furthermore, flavonoids can inhibit protein kinase C activity through competition with binding of ATP to the nucleotide binding site on the kinases [Varga et al. 2004].

On the other hand, nitric oxide, with the presence of a free electron in its structure, may itself be considered as a free radical. Activation of the inducible form of nitric oxide synthase (iNOS) by inflammation-mediating cytokines leads to the elevated production of nitric oxide, which in turn leads to the development of oxidative stress followed by endothelial dysfunction [Briones et al. 2000; Nagareddy et al. 2005]. Flavonoids are potent scavengers of the nitric oxide generated by iNOS activation [Matsuda et al. 2003]. Kalin and colleagues (2002) demonstrated that chronic administration of flavonoids or flavonoid-rich diets (e.g. grape seeds) prevented the increased expression of pro-inflammatory markers such as ICAM-1 and E-selectin in patients with systemic sclerosis and reduced their plasma oxidative stress levels.

2.3.2. Inhibition of Low-Density Lipoprotein (LDL) Oxidation

Depletion of antioxidants and the concomitant increase in endogenous free radicals and oxidants promotes and propagate the oxidation of LDL in the vasculature. Oxidative modification of LDL plays a pivotal role in increasing vascular resistance. For instance, exogenous L-arginine restored endothelium dependent relaxation in the presence of oxidized LDL, suggesting that oxidized LDL may negate the activity of eNOS, leading to endothelial dysfunction [Kugiyama et al. 1990]. In addition, oxidized LDL induces cellular and nuclear damage promoting apoptotic cell deaths [Jeong et al. 2005]. The oxidized forms of LDL are chemotactic for vascular smooth muscle cells and cytotoxic towards aortic endothelial cells [Heinecke, 1998]. It is reasonable to suggest that flavonoids exert their protective effects against LDL-mediated endothelial dysfunction by two fundamental mechanisms. Firstly, flavonoids or foods-rich in flavonoids can modify plasma lipid levels. Results from several clinical trials have reported a serum LDL lowering effects following intakes of soy isoflavones [Zhuo et al. 2004; Zhan and Ho, 2005]. Zern et al. [2005] reported that consumption of lyophilized grape powder rich in flavonoids for four weeks reduced plasma LDL cholesterol levels in pre- and post-menopausal women through alterations in lipoprotein metabolism, oxidative stress and inflammatory markers. Kim and colleagues (2003) showed that the flavonoids, hesperetin and naringin reduced the plasma cholesterol level in high cholesterol fed animals via inhibition of hepatic HMG-CoA reductase activity. Secondly, flavonoids can prevent LDL oxidation by LDL-specific and/or cell- or tissue-specific antioxidant actions [Stein et al. 1999; Duarte et al. 2001; Kamada et al., 2005; Hirano-Ohmori et al. 2005]. Kamada and colleagues (2005) suggested that quercetin glucosides accumulate in the aorta, and these metabolites of quercetin attenuated both lipid peroxidation and hyperlipidemia. A number of mechanisms including scavenging or inhibition of reactive oxygen and nitrogen species (see above paragraph) are likely to contribute to the positive actions of the flavonoids on LDL oxidation. Other potential mechanism includes up-regulation or increasing of antioxidant defenses. In regards to the latter mechanism, it is reported that flavonoids can stimulate the expression of endogenous antioxidant enzymes, such as superoxide dismutase and catalase [Ying et al. 2004]. More recently, it is also reported that dietary flavonoids may contribute to the protection of the integrity of endothelial cells from toxic effects of oxidized LDL not only by scavenging free radicals but also by maintaining effective eNOS activity [Steffen et al. 2005].

2.3.3. Flavonoids and eNOS-Mediated Nitric Oxide Synthesis

It is now reasonably characterized that under cardiovascular pathological conditions or conditions of oxidative stress, the bioavailability of EDNO may be reduced mainly by two mechanisms. First, free radicals produced at increased rates under these conditions incapacitate nitric oxide. The second mechanism involves free radical-mediated reduction in the activity and expression of eNOS contributing to decreased synthesis and release of EDNO. In addition to and independent of their antioxidant effects on LDL oxidation and free radical production, flavonoids may also augment the production of EDNO through stimulation of the expression and activity of eNOS. For instance, in isolated endothelium-intact blood vessels, flavonoids elicited concentration-dependent relaxation which was sensitive to inhibition of eNOS activity [Ajay et al. 2003]. Using electron paramagnetic resonance, administration of red wine polyphenols was found to improve EDNO production [Stoclet et al. 1999]. Intake of cocoa flavonoids, quercetin or catechin-rich food has been shown to increase the expression or activity of eNOS [Benito et al. 2002, Heiss et al. 2003]. The exact mechanism of eNOS activation by the flavonoids is not fully understood; it has been suggested to be mediated by an extracellular calcium-dependent mechanism via activation of N-ethyl-maleimide sensitive pathway [Andriambeloson et al. 1999]. Recently, it was also reported that the up-regulation of eNOS expression by red wine is likely due to the changes in both transcriptional and post-transcriptional (mRNA-stabilizing) events [Wallerath et al. 2005].

2.3.4. Other Effects of Flavonoids

Platelets are key participants in atherogenesis and produces pro-inflammatory mediators such as platelet activating factor and serotonin. Flavonoids have been shown to modulate the process of thrombosis in the cardiovascular systems, contributing to reduction in the rate of atherosclerosis and coronary heart disease. The actions of these polyphenols may be attributed to reducing platelet activation, aggregation and adhesion by nitric oxide-dependent and -independent mechanisms including enhancing endothelial prostacyclin synthesis/release, inhibition of hydrogen peroxide-mediated phospholipase C activation, inhibition of the metabolism of arachidonic acid, and reducing the synthesis of thromboxane [Tzeng et al. 1991; Pearson, 2005]. In addition, flavonoids may modulate endothelial function by inhibiting endothelial synthesis of contractile substances including endothelin-1, a vasoactive peptide that is crucial in the development of coronary atherosclerosis [Corder et al. 2001].

2.4. Flavonoids, Endothelial Function, Hypertension and Diabetes Mellitus

Hypertension and diabetes mellitus are two common major diseases closely associated with the cardiovascular system. Hypertension and diabetes mellitus are associated with increased oxidative stress, as exhibited by the enhanced production of superoxide anions and/or its reactive metabolites [Endemann and Schiffrin, 2004]. Increased production of superoxide anions in hypertension contributes to high blood pressure and vice versa. Flavonoids exerted vasodilator activities via stimulating endogenous nitric oxide production from the endothelium in isolated blood vessels [Ajay et al., 2003]. Chronic administration of quercetin, which represents 60 % of total dietary intake of flavonoids, significantly reduced systolic blood pressure as well as enhanced endothelium-dependent relaxations in hypertensive rats [Ajay and Mustafa, 2005]. Moreover, acute exposure to quercetin at 10 μM, a maximal achievable physiological concentration [Scalbert and Williamson, 2000], markedly improved endothelium-dependent relaxations to acetylcholine in hypertensive and diabetic rat isolated aortas [Ajay et al., 2006a, b]. In contrast, quercetin was without effect on the relaxations to acetylcholine in aortas from normotensive rats where the production of superoxide anions was thought to be too little to alter bioavailability of the EDNO [Ajay et al., 2006a, b]. Collectively, these findings suggest that flavonoids can prevent the detrimental effects of the free radicals on endothelial function in hypertension. A major enzymatic source of superoxide anions in hypertensive and diabetic vessels is NAD(P)H-oxidases. In aortas of normal rats, acute exposure to 10 μM of quercetin normalized impaired relaxations to acetylcholine in the presence of superoxide anions generated by activation of NADH/NAD(P)H-oxidase with β-NADH (300 μM) (data submitted for publication). In addition, quercetin totally abolished the transient contractions induced by β-NADH in normal rat aortas. These findings further confirm that quercetin modulated endothelium-dependent relaxations in hypertensive and diabetic rat aortas either by inhibiting the production of superoxide anions via NADH/NAD(P)H-oxidase activity and/or by scavenging of superoxide anions secondary to their release via the NADH/NAD(P)H-oxidase pathway. Direct evidence of LDL oxidation and oxidative stress in hypertension and diabetes is based on the measurement of lipid peroxidation markers such as plasma malonaldehyde (MDA). We found that chronic administration of quercetin reduced significantly the plasma MDA *plus* 4-hydroxyalkenals (4-HNE) content in hypertensive and diabetic rats (data submitted for publication). These findings provide indirect evidence that chronic oral administration of quercetin reduced lipid peroxidation possibly via reduction in the superoxide anions and other free radicals. Increased oxidative stress in hypertension and diabetes mellitus may be caused not only by the enhanced production of reactive oxygen and nitrogen species but also by lowered antioxidant defense capacity. Several investigations have demonstrated that the activities of endogenous antioxidant enzymes including superoxide dismutase are reduced in experimental models of hypertension and diabetes [Giugliano et al. 1996; Inukai et al. 2002]. We found that chronic oral administration of quercetin (10 mg/kg^{-1} of body weight) enhanced significantly superoxide dismutase activity and total antioxidant capacity of plasma in hypertensive and diabetic rats, suggesting that the beneficial vascular effects of quercetin and related flavonoids may result from the enhancement in endogenous antioxidant defense capacity.

Conclusion

Over production of free radicals, mainly superoxide anions via excessive stimulation of NAD(P)H-oxidases results in oxidative stress, endothelial dysfunctions and ultimately to development of cardiovascular related diseases. Daily consumptions of the flavonoids may confer protection to the vascular endothelium in hypertension and diabetes mellitus via stimulation of EDNO production, reduction in the production and/or scavenging of free radicals and hence attenuation in oxidative stress, reduction in blood pressure and/or inhibition of hyperglycemia.

References

Ajay, M; Gilani, AU; Mustafa, MR. Relaxant effects of flavonoids on vascular smooth muscles of the isolated rat thoracic aorta. *Life Sciences* 2003, 74, 603-12.

Ajay, M; Mustafa, MR. Chronic treatment with flavonoids prevents endothelial dysfunction in spontaneously hypertensive rat aorta. *Journal of Cardiovascular Pharmacology* 2005, 46, 36-40.

Ajay, M; Achike, FI; Mustafa, AM; Mustafa, MR. Direct effects of quercetin on impaired reactivity of spontaneously hypertensive rat aortas: comparative study with ascorbic acid. *Clinical and Experimental Pharmacology and Physiology* 2006a, 33 (4), 345-50.

Ajay, M; Achike, FI; Mustafa, AM; Mustafa, MR. Effect of quercetin on altered vascular reactivity in aortas isolated from streptozotocin-induced diabetic rats. *Diabetes Research and Clinical Practice* 2006, 73(1), 1-7.

Aldini, G; Carini, M; Piccoli, A; Rossoni, G; Facino, RM. Procyanidins from grape seeds protect endothelial cells from peroxynitrite damage and enhance endothelium-dependent relaxation in human artery: new evidences for cardio-protection. *Life Sciences*, 2003, 73(22), 2883-98.

Amudha, K; Chee, KH; Lang, CC. Endothelial dysfunction in cardiovascular diseases. *Medical Progress* 2002, 29, 19-27.

Andriambeloson, E; Stoclet, JC; Andriantsitohaina, R. Mechanism of endothelial nitric oxide-dependent vasorelaxation induced by wine polyphenols in rat thoracic aorta. *Journal of Cardiovascular Pharmacology* 1999, 33(2), 248-54.

Benito, S; Lopez, D; Saiz, MP; Buxaderas, S; Sanchez, J; Puig-Parellada, P; Mitjavila, MT. A flavonoid-rich diet increases nitric oxide production in rat aorta. *British Journal of Pharmacology* 2002, 135(4), 910-16.

Briones, AM; Alonso MJ; Marin, J; Balfagon, G; Salaices, M. Influence of hypertension on nitric oxide synthase expression and vascular effects of lipopolysaccharide in rat mesenteric arteries. *British Journal of Pharmacology* 2000, 131(2), 185-94.

Corder, R; Douthwaite, JA; Less, DM; Khan, NQ; Viseu Dos Santos, AC; Wood, EG; Carrier, MJ. Endothelin-1 synthesis reduced by red wine. *Nature* 2001, 414(6866), 863-64.

Duarte, J; Perez-Palencia, R; Vargas, F; Ocete, MA; Perez-Vizcaino, F; Zarzuelo, A; Tamargo, J. Antihypertensive effects of the flavonoid quercetin in spontaneously hypertensive rats. *British Journal of Pharmacology* 2001, 133(1), 117-24.

Endemann, DH; Schiffrin, EL. Endothelial dysfunction. *Journal of American Society of Nephrology* 2002, 15(8),1983-92.

Engler, MB; Engler, MM; Chen, CY; Malloy, MJ; Browne, A; Chiu, EY; Kwak, HK; Milbury, P; Paul, SM; Blumberg, J; Mietus-Snyder, ML. Flavonoid-rich dark chocolate improves endothelial function and increases plasma epicatechin concentrations in healthy adults. *Journal of the American College of Nutrition* 2004, 23(3), 197-204.

Forstermann, U. Endothelial NO synthase as a source of NO and superoxide. *European Journal of Clinical Pharmacology* 2005, 62(113), 5-12.

Giugliano, D; Ceriello, A; Paolisso, G. Oxidative stress and diabetic vascular complications. *Diabetes Care* 1996, 19, 257-67.

Harborne, JB; Williams, CA. Advences in flavonoid research since 1992. *Phytochemistry* 2000, 55, 481-504.

Heim, KE; Tagliaferro, AR; Bobilya, DJ. Flavonoid antioxidants: chemistry, metabolism and structure-activity relationships. *Journal of Nutritional Biochemistry* 2002, 13, 572-84.

Heinecke, JW. Oxidants and antioxidants in the pathogenesis of atherosclerosis: implications for the oxidized low density lipoprotein hypothesis. *Atherosclerosis* 1998, 141(1), 1-15.

Heiss, C; Kleinbongard, P; Dejam, A; Perre, S; Schroeter, H; Sies, H; Kelm, M. Acute consumption of flavanol-rich cocoa and the reversal of endothelial dysfunction in smokers. *Journal of the American College of Cardiology* 2005, 46(7), 1276-83.

Hertog, MGL; Hollman, PCH; Katan, MB; Kromhout, D. Intake of potentially anticarcinogenic flavonoids and their determinants in adults in The Netherlands. *Nutrition and Cancer* 1993a, 20, 21-9.

Hertog, MGL; Feskens, EJM; Hollman, PCH; Katan, MB; Klomhout, D. Dietary antioxidant flavonoids and risk of coronary heart disease: The Zutphen Elderly study. *Lancet* 1993b, 342, 1007-11.

Hirano-Ohmori, R; Takahashi, R; Momiyama, Y; Taniquchi, H; Yonemura, A; Tamai, S; Umegaki, K; Nakamura, H; Kondo, K; Ohsuzu, F. Green tea consumption and serum malondialdehyde-modified LDL concentrations in healthy subjects. *Journal of the American College of Nutrition* 2005, 24(5), 342-6.

Hollman, PCH; Katan, MB. Absorption, metabolism and health effects of dietary flavonoids in man. *Biomedicine and Pharmacotherapy* 1997, 51(8), 305-10.

Inukai, T; Takanashi, K; Tayama, K; Aso, Y; Takemura, Y. High glucose concentrations abolish the superoxide dismutase response of leukocytes to ascorbic acid or troglitazone in type 2 diabetes mellitus. *Life Sciences* 2002, 70, 2391-401.

Jeong, YJ; Choi, YJ; Kwon, HM; Kang, SW; Park, HS; Lee, M; Kang, YH. Differential inhibition of oxidized LDL-induced apoptosis in human endothelial cells treated with different flavonoids. *British Journal of Nutrition* 2005, 93(5), 581-91.

Jiang, F; Jones, GT; Husband, AJ; Dusting, GJ. Cardiovascular protective effects of synthetic isoflavone derivatives in apolipoprotein e-deficient mice. *Journal of Vascular Research* 2003, 40(3), 276-84.

Kalin, R; Righi, A; Del Rosso, A; Bagchi, D; Generini, S; Cerinic MM; Das, DK. Activin, a grape seed-derived proanthocyanidin extract, reduces plasma levels of oxidative stress and adhesion molecules (ICAM-1, VCAM-1 and E-selectin) in systemic sclerosis. *Free Radical Research* 2002, 36(8), 819-25.

Kamada, C; da Silva, EL; Ohnishi-Kameyama, M; Moon, JH; Terao, J. Attenuation of lipid peroxidation and hyperlipidemia by quercetin glucoside in the aorta of high cholesterol-fed rabbit. *Free Radical Research* 2005, 39(2), 185-94.

Kim, HK; Jeong, TS; Lee, MK; Park, YB; Choi, MS. Lipid-lowering efficacy of hesperetin metabolites in high-cholesterol fed rats. *Clinica Chimica Acta* 2003, 327(1-2), 128-37.

Knekt, P; Jarvinen, R; Reunanen, A; Maatela, J. Flavonoid intake and coronary mortality in Finland; a cohort study. *British Medical Journal* 1996, 312:478-81.

Kugiyama, K; Kerns, SA; Morrisett, JD; Roberts, R; Henry, PD. Impairment of endothelium-dependent arterial relaxation by lysolecithin in modified low-density lipoproteins. *Nature* 1990, 344(6262), 160-2.

Manach, C; Regerat, F; Texier, O; Agullo, G; Demigne, C; Remesy, C. Bioavaliability, metabolism and physiological impact of 4-oxo- flavonoids. *Nutrition Research* 1996, 16, 517 - 44.

Matsuda, H; Morikawa, T; Ando, S; Toguchida, Y; Yoshikawa, M. Structural requirements of flavonoids for nitric oxide production inhibitory activity and mechanism of action. *Bioorganic and Medicinal Chemistry* 2003; 11(9), 1995-2000.

Morand, C; Crespy, V; Manach, C; Besson, C; Demigne, C; Remesy, C. Plasma metabolites of quercetin and their antioxidant properties. *American Journal of Physiology* 1998, 275(1 pt 2), R212-9.

Nagareddy, PR; Xia, Z; McNeill, JH; MacLeod, KM. Increased expression of iNOS is associated with endothelial dysfunction and impaired pressor responsiveness in streptozotocin-induced diabetes. *American Journal of Physiology Heart Circulation Physiology* 2005, 289(5), H2144-52.

Nijveldt, RJ; van Nood, E; van Hoorn, DE; Boelens, PG; van Norren, K; van Leeuwen, PA. Flavonoids: a review of probable mechanisms of action and potential applications. *American Journal of Clinical Nutrition* 2001, 74(4), 418-25.

Palmer, RMJ; Ferrige, AG; Moncoda, S. Nitric oxide release accounts for the biological activity of endothelium-derived relaxing factor. *Nature* 1987, 327, 524-6.

Pearson, DA; Holt, RR; Rein, D; Paglieroni, T; Schmitz, HH; Keen, CL. Flavanols and platelet reactivity. *Clinical and Developmental Immunology* 2005, 12(1), 1-9.

Pietta, PG. Flavonoids as antioxidants. *Journal of Natural Products* 2000, 63(7), 1035-42.

Raj Narayana, K; Sripal Reddy, M; Chaluvadi, MR; Krishna, DR. "Bioflavonoids Classification, Pharmacological, Biochemical Effects and Therapeutic Potential. *Indian Journal of Pharmacology* 2001, 33: 2-16.

Rice-Evans, CA; Packer, L. *Flavonoids in health and disease*. New York: Marcel Dekker Inc.;1998.

Rice-Evans, CA. Flavonoid antioxidants. *Current Medicinal Chemistry* 2001, 8(7), 797-807.

Robak, J; Gryglewski, RJ. Flavonoids are scavengers of superoxide anions. *Biochemical Pharmacology* 1988, 37(5), 837-41.

Scalbert, A; Williamson, G. Dietary intake and bioavailability of flavonoids. *Journal of Nutrition* 2000, 130(8S Suppl), 2073S-85S.

Silva, ID; Rodrigues, AS; Gaspar, J; Laires, A; Rueff, J. Metabolism of galangin by rat cytochromes P450: relevance to the genotoxicity of galangin. *Mutation Research* 1997, 393(3), 247-57.

Skibola, CF; Smith, MT. Potential health impacts of excessive flavonoid intake. *Free Radical Biology and Medicine* 2000, 29, 375-83.

Steffen, Y; Schewe, T; Sies, H. Epicatechin protects endothelial cells against oxidized LDL and maintains NO synthase. *Biochemical and Biophysical Research Communications* 2005, 331(4), 1277-83.

Stein, JH; Keevil, JG; Wiebe, DA; Aeschlimann, S; Folts, JD. Purple grape juice improves endothelial function and reduces the susceptibility of LDL cholesterol to oxidation in patients with coronary artery disease. *Circulation* 1999, 100(10), 1050-55.

Stoclet, JC; Kleschyov, A ; Andriambeloson, E ; Diebolt, M ; Andriantsitohaina, R. Endothelial NO release caused by red wine polyphenols. *Journal of Physiology and Pharmacology* 1999, 50(4), 535-40.

Tzeng, SH; Ko, WC; Ko, FN; Teng, CM. Inhibition of platelet aggregation by some flavonoids. *Thrombosis Research* 1991, 64(1), 91-100.

Varga, Z; Ujhelyi, L; Kiss, A; Balla, J; Czompa, A; Antus, S. Effect of silybin on phorbol myristate actetate-induced protein kinase C translocation, NADPH oxidase activity and apoptosis in human neutrophils. *Phytomedicine* 2004, 11(2-3), 206-12.

Wallerath, T; Li, H; Godlet-Ambrust, U; Schwarz, PM; Forstermann, U. A blend of polyphenolic compounds explains the stimulatory effect of red wine on human endothelial NO synthase. *Nitric Oxide* 2005, 12(2), 97-104.

Woodman, OL; Chan, ECH. Vascular and anti-oxidant actions of flavonols and flavones. *Clinical and Experimental Pharmacology and Physiology* 2004, 31(11), 786-90.

Ying, CJ; Sun, FX; Zhang, SL; Zhang, XP; Mao, LM; Zuo, XZ; Yao, P. ROS-related enzyme expressions in endothelial cells regulated by tea polyphenols. *Biomedical and Environmental Sciences* 2004, 17(1), 33-9.

Yochum, L; Kushi, LH; Meyer, K; Folsom, AR. Dietary flavonoid intake and risk of cardiovascular disease in postmenopausal women. *American Journal of Epidemiology* 1999,149(10), 943-9.

Zern, TL; Wood, RJ; Greene, C; West, KL; Liu, Y; Aggarwal, D; Shachter, NS; Fernandez, ML. Grape polyphenols exert a cardioprotective effect in pre- and postmenopausal women by lowering plasma lipids and reducing oxidative stress. *Journal of Nutrition* 2005, 135(8), 1911-7.

Zhan, S; Ho, SC. Meta-analysis of the effects of soy protein containing isoflavones on the lipid profile. *American Journal of Clinical Nutrition* 2005, 81(2), 397-408.

Zhuo, XG; Melby, MK; Watanabe, S. Soy isoflavone intake lowers serum LDL cholesterol: a meta-analysis of 8 randomized controlled trials in humans. *Journal of Nutrition* 2004, 134(9), 2395-2400.

In: Oxidative Stress: Clinical and Biomedical Implications ISBN: 978-1-60021-888-0
Editors: B. M. Matata, M. M. Elahi, pp. 183-201

Chapter XII

NO Regulatory Networks in Bacteria Focusing on Pathogenic Bacteria

Jorge Membrillo Hernández**, David S. Zamorano-Sánchez, Gerardo Croda-Garcia and Elizabeth Hernández-Urzúa
Laboratorio de Microbiología y Genética Molecular,
Departamento de Biología Molecular y Biotecnología
Instituto de Investigaciones Biomédicas.
Universidad Nacional Autónoma de México. PO Box 70-228,
Ciudad Universitaria, Mexico City, 04510, Mexico

Abstract

NO has an important role as part of the innate host response against bacterial infections. Nitric oxide can interact and damage different cellular structures and kill bacteria quickly. In this chapter we summarize the current knowledge on the effects of NO on the bacterial cellular structures, chemical reactions and transcriptional responses focusing on the most studied bacterial gene response to NO$^{\cdot}$, the Flavohemoglobins.

1. Introduction

NO$^{\cdot}$ is a free radical and a component of the innate host response against bacterial infections. During this process, toxic amounts of NO$^{\cdot}$ are produced by host cells through the enzyme iNOS (inducible Nitric Oxide Synthase) to battle invading bacteria [1]. The antimicrobial effects of NO$^{\cdot}$ and reactive nitrogen intermediates (known as RNS, reactive nitrogen species) are known to involve cellular targets such as DNA, proteins, and lipids [1]. On the other hand, NO$^{\cdot}$ has been shown to be important due to its well-established role as

* E-mail: jmh@biomedicas.unam.mx

immunomodulator. NO$^{\cdot}$ in fact is indispensable for normal clearance of diverse pathogens including viral, fungal, bacterial and parasitic microorganisms [2]. Activated leukocytes can generate local amounts of NO$^{\cdot}$ in the micromolar range through the action of the iNOS [3]. NO$^{\cdot}$ can interact directly with invading organisms and surrounding host tissues, or can be further oxidized into highly reactive RNS. NO$^{\cdot}$ and RNS in general can then interact with proteins and be responsible for the nitrosation of protein thiols and the nitrosylation of metal centres, particularly iron [4]. RNS can mediate nitration of protein tyrosine residues, trigger the lipid peroxidation reactions and cause de-amination of DNA bases [5-7]. Bacterial cells exposed to NO$^{\cdot}$ in fact have to deal with several environmental stresses. NO$^{\cdot}$ stress responses include the maintenance and replenishment of thiol and metal pools, DNA repair mechanisms, increased enzymatic NO$^{\cdot}$ detoxification activities, and probably a quicker protein turnover of damaged proteins [8-12]. In this chapter we focused on the different mechanisms that bacteria have evolved to cope with a sudden increase in RNS concentrations with especial emphasis on the role of flavohemoglobins.

2. NO Chemistry. Reactive Nitrogen Species Bacteria Deal with

The natural living style of bacteria is attached to a surface in a structure called biofilm rather than living in planktonic state [13]. Most of the recent knowledge on the effect of NO$^{\cdot}$ on pathogenic (and non pathogenic) bacteria has been mainly obtained from cultures of free-living bacteria or by experimental animal models using a short-time exposition to pathogenic bacteria [14, 15].

The main chemical reactions of NO$^{\cdot}$ in biology predominantly involve transition metal ions, dioxygen and its redox related species such as superoxide. Nitric oxide reacts with transition metals, (which differ among them in their oxidation state by one electron) to form nitrosyl complexes. Whilst it does not react with transition metals such as zinc and its reactions with copper and cobalt to form nitrosyl complexes are too slow to have a significant *in vivo* effect, the most biologically relevant reaction of NO$^{\cdot}$ is with iron [16].

NO reactions with iron. NO$^{\cdot}$ binds reversibly to ferrous ion. A comparison for the association rates between NO and O_2, both having similar size, charge and hydrophobicity, in haem-containing proteins shows that NO$^{\cdot}$ and O_2 have similar rates (in the order of 10^7-10^8 $M^{-1}s^{-1}$), however they have very different disassociation rates (10^{-5} s^{-1} for NO and 15 s^{-1} for O_2). As result of the tight bonding of NO to ferrous haem, it has been suggested that is easier to remove the haem group from the protein rather than dissociate the NO$^{\cdot}$ from it.

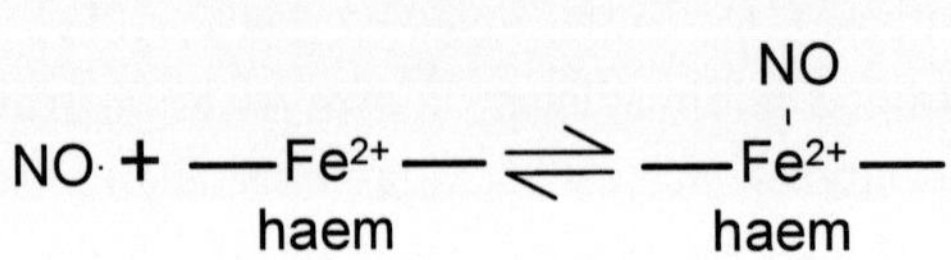

An unusual property of $NO^{\cdot}$ is that it can also bind reversibly to ferric ion, although the association rates are strongly influenced by the protein structure at the haem pocket [17].

$$NO + Fe^{3+}(haem) \rightleftharpoons Fe^{3+}\text{-}NO(haem)$$

NO reactions with dioxygen. NO reacts with oxygen because dioxygen contains two unpaired electrons with parallel spins in separate orbitals, one in each oxygen atom, therefore the nitric oxide unpaired electron forms a weak bond with one of the two unpaired electrons of dioxygen to form the nitrosyldioxyl radical.

$$NO^{\cdot} + O_2 \longrightarrow ONOO^{\cdot}$$

The nitrosyldioxyl radical can also react with another molecule of NO to form nitrogen dioxide. This is a strong one-electron oxidant and at low concentrations initiates free radical oxidation of unsaturated lipids, thiols and proteins.

$$ONOO^{\cdot} + NO^{\cdot} \rightleftharpoons ONOONO \longrightarrow 2\,{}^{\cdot}NO_2$$

NO reactions with superoxide. Nitric oxide also reacts with superoxide to form peroxynitrite ($ONOO^-$) which is a very reactive oxidising species with a rate constant of 1.9 X 10^{10} $M^{-1}s^{-1}$[18].

$$NO^{\cdot} + O_2^{\cdot -} \longrightarrow ONOO^-$$

NO^- (nitroxyl anion) reactions. The nitroxyl anion can undergo several reactions with a variety of cellular targets among these targets are: dioxygen, thiols and transition metals. The reaction of nitroxyl anion and dioxygen also produced peroxynitrite ($OONO^-$) with a rate constant of 5.7 X10^7 $M^{-1}s^{-1}$, although this reaction is slower than the reaction of nitric oxide with superoxide, the *in vivo* concentration of oxygen is several orders of magnitude higher than those of superoxide (between 5-200 μM for dioxygen and < 66 pM for superoxide), therefore this suggest that the reaction of dioxygen with the nitrosyl anion is the principal route of formation of peroxynitrite *in vivo*.

$$NO^- + O_2 \longrightarrow ONOO^-$$

NO^+ Reactions. The reaction of nitrosonium cation with thiols producing proteins *S*-nitrosated occurs significantly more frequently than *N*-nitrosation or *C*-nitrosation of proteins [19], besides *in vivo* only have been found reservoirs of S-NO, but not of N-NO or C-NO [20]. The mechanism that produced proteins *S*-nitrosated appears to be a transnitrosation, i.e. the transfer of the nitrosomium cation from S-nitrosothiols such as S-nitrosoglutathione to

other thiol, preferentially over nitrogen or carbon, via a nucleophilic attack of the thiolate anion upon the nitrogen group:

$$\text{Protein-SH} + \text{RS-NO} \rightarrow \text{Protein-SNO} + \text{R-SH}$$

The reactivity of the thiols towards *S*-nitrosation depends on the microenvironment that is around the cysteines, because the most susceptible cysteines are located within of acid-base and overall hydrophobic structural motifs [21, 22].

Peroxynitrite. Peroxynitrite is a remarkably stable RNS, this stability is due to its fold into the *cis*-conformation, which does not allow it to isomerizes in a much stable form such as nitrate. However when protonated, peroxynitrite is able to isomerise to *trans*-conformation to form the *trans*-peroxynitrous acid that can form a excited state that react like hydroxyl radical plus nitronium cation (NO_2^+; [16]).

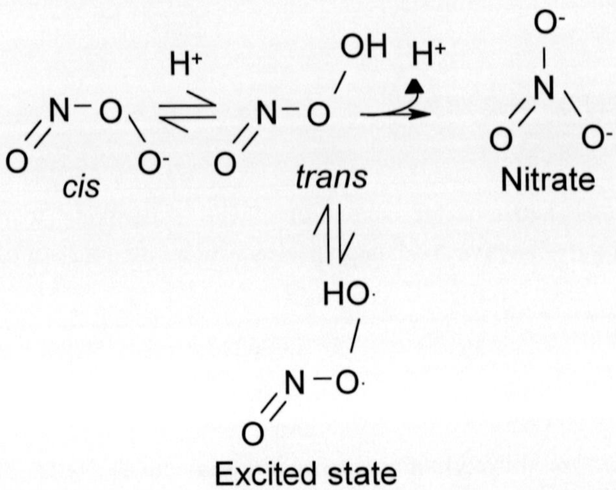

The generation of NO_2^+ and $\cdot OH$ is thought to be catalysed by transition metal such as iron; the NO_2^+ formed can attack phenolic compounds including tyrosine to produce nitrophenols.

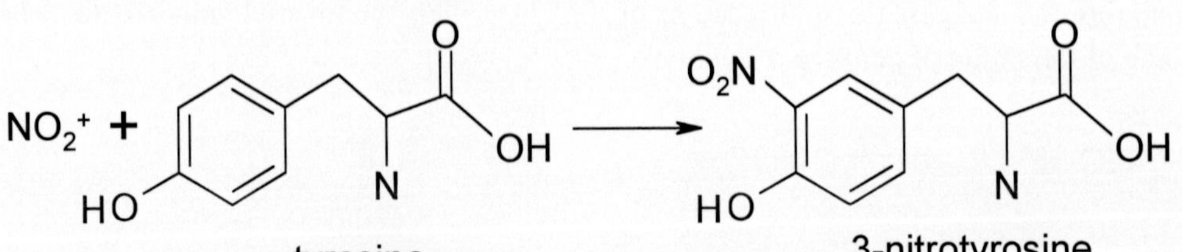

These reactions are more likely to occur in low-molecular weight iron complexes than in metalloproteins because of the accessibility of the phenolic compound to the iron atom [16, 17, 23]. It is important to mention that peroxynitrite mediates one electron transfer reactions that produce free radical species as the case mentioned above or lipid peroxidation [23].

3. RNS Damage to Bacterial Cellular Structures

Pathogens are organisms that use the host resources for their own benefit causing damage within the host. There are free-living pathogens like bacteria and pathogens like viruses that depend completely on the host resources. To assure the continuity of life different mechanisms of defence and defence evasion have been developed within the living organisms. There are specialized cells known as phagocytes that play important roles in the antimicrobial response, these cells are activated by the presence of bacterial signals, they phagocyte the strange cells with the concomitant production of reactive oxygen species (ROS) and reactive nitrogen species (RNS) a process usually known as the respiratory burst [24, 25]. ROS and RNS can react directly with any kind of biomolecule, causing reversible or irreversible modifications that could originate loss of biological function. The targets of damage may be specific for the source or type and concentration of the reactive species generated. The main source for generation of nitric oxide in phagocytic cells is the NOS, a membrane protein that uses L-arginine and oxygen as substrates to generate nitric oxide using NADPH as a cofactor. There are constitutive and inducible isoenzymes that can be dependent or independent on calcium. There is not a clear direct relationship between NO chemistry and its biological activity, as NO generates a diverse array of new species with modified reactivity and functions. These NO derived species may serve for NO storage or exert toxic effects because of their particular chemistry properties. The lipophilicity of NO allows it to diffuse across the membrane and through the cytoplasm. NO is soluble in water but does not react with it [26]. The reactivity and toxicity of nitric oxide per se is modest in comparison to oxidants derived from it. Nitric oxide could accomplish its toxic effects through reactions with superoxide. Nitric oxide and superoxide react together near the diffusion limit to yield peroxynitrite [27-29]. It was reported that exposure to 1 mM NO for 1 h did not decreased cell viability in *E.coli* in contrast to a severe effect of peroxynitrite (1mM) exposure in cell viability during the first 5 seconds of the treatment [27]. This argues for an indirect role of nitric oxide in the toxic effects developed in cell systems where superoxide could be also being produced. It is worth noting that superoxide reacts with nitric oxide three times faster than with superoxide dismutase and thus nitric oxide is the only known biological molecule that can outcompete endogenous superoxide dismutases for superoxide [27]. The damage by RNS to macromolecules in the bacterial cells is exerted on DNA, proteins and lipids (Figure 1).

N_2O_3 can be generated by NO autooxidation, and can react with amines, thiols, and available nucleophiles and directly damage DNA by deamination of DNA bases [30, 31]. Both spontaneous and nitrosative deamination may occur via nucleophilic aromatic substitution, but the nitrosative pathway would be favoured by the better leaving group, i.e. – $N_2OH_2^+$ or N_2^+ versus NH_3^+. Unrepaired DNA modification may lead to mutagenesis. The G:C to A:T transition is the predominant mutation resulting from NO exposure and this could be originated by either cytosine or guanine deamination [32]. Peroxynitrite induces the formation of 8-hidroxydeoxyguanosine or 8-nitroguanine with a mutagenic effect. Proteins have a variety of vital functions in the cell, serving as structural blocks and functioning as catalysers of unfavourable energetic metabolic processes. Inactivation or inhibition of protein activity could lead to a catastrophic end for any cell. Nitric oxide and its derivatives could generate reversible and irreversible modifications on protein structure that may alter its functionality, some of these alterations have been analysed in essential enzymes, and loss of function have been associated with a decrease in bacterial cell viability. Nitric oxide and its derivatives can directly interact with reactive thiols, they could also disassemble Fe-S clusters by releasing iron from metalloproteins driving them inactive, recent data suggest that NO inactivation of proteins with iron sulphur clusters like aconitase is indirect probably mediated by the reduction of iron or affecting sulphide availability [33].

Other targets for oxidation within the proteins are tyrosyl radicals, phenolic and aromatic amino acid residues and amines. For strict aerobic bacteria the maintenance of an accurate electron flux through its respiratory chain is a matter of life or death. Many of the enzymes forming the respiratory complexes are haem proteins like some cytochrome systems, NO-haem interactions can inactivate these enzymes and have a lethal effect for the cell [15]. Other haem proteins like catalase that are directly involve in oxidative stress defence could also be inactivated leading to an inaccurate response and susceptibility to the antibacterial response mounted by the host immune system [34].

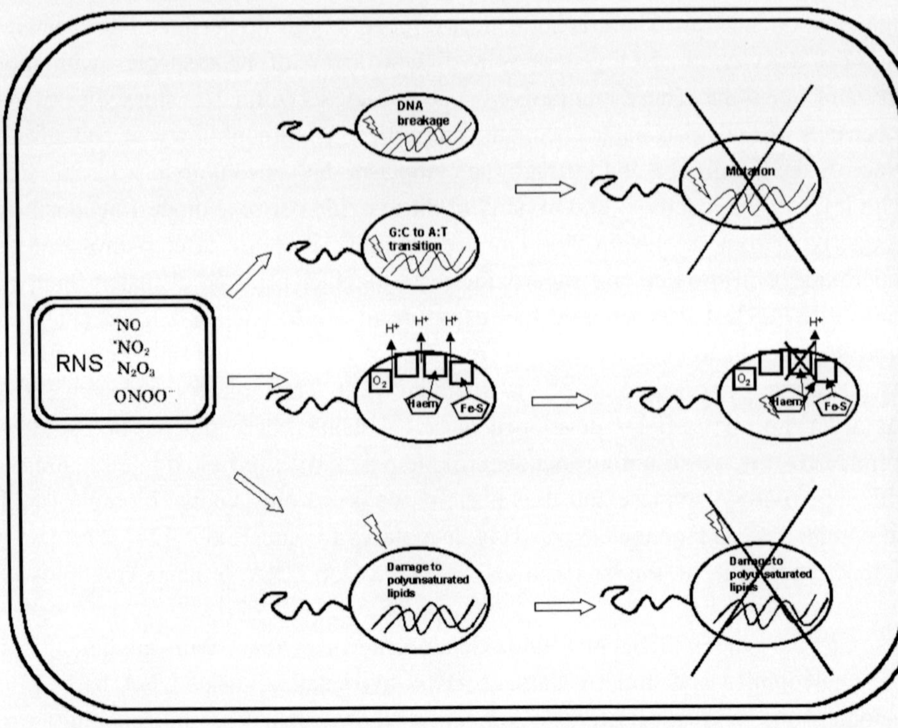

Figure 1. Cell targets for RNS that alter cell viability. An important NO target for bactericidal activity is DNA (top right) Proteins (middle) and lipids (bottom right). Insertion of mutations or breaking of the DNA can cause dangerous alterations in the cell physiology. In vitro NO is able to deaminate deoxynucleosides, deoxynucleotides, and intact DNA at physiological pH. A similar effect was proved to occur in vivo in Salmonella typhimurium by treating the strain with NO releasing compounds [6]. NO can also alter lipids with fatal consequences or proteins, the latter can be, in few cases, reversible.

In *Salmonella enterica* DNA replication is inhibited by nitric oxide and S-nitrosothiols by means of zinc mobilisation from metal-proteins of the replication machinery [12]. Proteins like glyceraldehyde-3-phosphate dehydrogenase and the gamma-glutamylcysteinyl synthetase are inactivated *in vitro* by S-nitrosylation. However critical thiol targets in NO mediated antibacterial activity have been not fully described *in vivo* [35]. Tyrosine nytration also occurs *in vivo* produced by peroxynitrite [36] and tyrosine nitration may lead to loss of function of nitrated proteins [37]. Peroxynitrite is an effective oxidant of aromatic molecules and organosulfur compounds like free amino acids and peptide residues. Methionine is converted to methionine sulphoxide or to ethylene and dimethyldisulphide. Formaldehyde could be formed by dimethly sulfoxide oxidation. Tyrosine and Tryptophan can undergo one-electron oxidation to radical cations [38-40]. Peroxynitrite could also negatively affect the antioxidant response by inhibiting important enzymatic defence proteins like superoxide dismutase [39, 41, 42] and glutaredoxin [41].

Bacterial membranes differ in their lipid composition with their eukaryotes counterparts, this may be an important trait for differential damage by reactive nitrogen or oxygen induced lipid peroxidation. Peroxynitrite dependent lipid peroxidation has been assessed, using as a model phosphatidylcholine liposomes and linolenic acid emulsions, and it shows to be dependent on NO levels, NO could serve as lipid peroxidation terminator because of its ability to react with alky- peroxides and lipid peroxides that promote propagation [43]. The monounsaturated lipids found in bacterial membranes are unreactive *in vitro*. Photosynthetic bacteria possess polyunsaturated lipids in their tylakoid membrane that are prone to oxidation [44]. Not all bacteria have phosphatidylcholine as part of their membrane lipids. The reaction of peroxynitrite not only leads to peroxide formation but also formation of nitrite-, nitro-, nitrosoperoxo-(nitritolinolenate, nitrosoperoxolinolenate, hydroxylnitrosoperoxolinolenate and hydrioerixinitrosoperoxolinolenate) and/or nitrated lipid oxidation adducts [43].

All this potentially damaging reactions of reactive nitrogen species raise the question of how microbial pathogen infections could persist even in such a hazardous environment. Now it is time to discuss some of the most recent and well known mechanisms for nitric oxide detoxification in pathogenic bacterial systems.

4. Genetic Responses to Nitric Oxide in Phatogenic Bacteria

Recent studies using microarrays to examine the response of the *E. coli* transcriptome to sources of NO or NO+ have shown that only three genes or transcription units (*hmp*, *norVW* and *nrdH*) are consistently up-regulated [11, 45, 46]. The *hmp* and *norVW* genes encode enzymes with established roles in NO metabolism. Hmp belongs to a group of globin-like oxygen binding proteins that recently received increasing attention due to their role in protecting bacterial cells from nitrosative stress [29]. The transcription of the flavohaemoglobin *hmp* gene is increased upon exposure to NO [47]. Hmp has Nitric oxide dioxygenase (NOD) activity with the ability to oxidise NO to NO_3^-, thereby, protecting the bacteria from the toxic effect of NO [48]. In this section we will focus on the current knowledge of the regulatory networks involved in the regulation of the different *hmp* genes.

Hmp from Salmonella enterica serovar Typhimurium. Salmonella enterica serovar Typhimurium causes human invasive disease, particularly in immunocompromised individuals. The invasion is the initial step in the pathogenesis of *Salmonella*, and subsequent survival within macrophages is the key feature of virulence of this bacterium. Macrophages phagocytise invading bacteria and kill them by a number of mechanisms, including production of antimicrobial peptides, lysosomal enzymes, reactive oxygen species (ROS) and reactive nitrogen species (RNS), as nitric oxide (NO). In mice, failure to generate NO leads to susceptibility to infection by *Salmonella*. Both nitrosative and oxidative stress are activated by murine macrophages, produced ROS and RNS during intracellular killing of *Salmonella* [49].

Salmonella responses to the intracellular environmental stresses confer resistance to RNS. These include DNA repair systems, small thiol molecules, and detoxifying enzymes such as superoxide dismutase and catalases [50]. The other enzymes involved in the detoxification of NO are the flavohaemoglobins.

The protein Hmp (encoded by the *hmp* gene) of *Salmonella* confers protection against nitrosative stress and contributes to intracellular survival following phagocytosis. Hmp protects *Salmonella* from killing within human macrophages and this has been attributed to the NO detoxification activity of the protein [14, 51]. In *Salmonella* the mutation of the *hmp* gene leads to hypersensitivity to nitrosative stress generated by GSNO or S-nitrosylated N-acetylcysteine but not to oxidative stress generated by Paraquat (PQ) or H_2O_2 [52]. The *hmp* gene is induced by spermine NONOate, an NO donor. The NO transcriptional regulation of *hmp* in *Salmonella* was firstly reported to involve the Fur protein, a global transcriptional regulator in charge of genes involved in the iron-uptake processes [53]. However, these observations were recently retracted due to the discovery that the published studies erroneously employed an Φ(*iroC-lacZ*) rather than an Φ(*hmp-lacZ*) reporter fusion [54]. On the other hand our lab recently reported that the regulation of *hmp* in S*almonella* is independent of the superoxide-responding two-component system SoxRS and of the peroxide-responding OxyR system [55, 56].

Analyses of the Φ(*hmp-lacZ*) fusion of *Salmonella* demonstrated that *hmp* is induced by sodium nitroprusside (SNP), an NO donor, and methyl viologen (Paraquat PQ), a superoxide-generating agent even in the absence of Fur [56]. Analysis of the *hmp* expression demonstrated that in absence of the *ramA* gene, the hmp_{St} induction by PQ decreases but not in presence of SNP (Figure 2; [56]). The *ramA* gene appears to be specific for *Salmonella serovars* and is absent in many other gram-negative microorganisms; a notable exception is *Klebsiella pneumoniae* and *Enterobacter aerogenes* [57-61]. RamA belongs to the AraC family of transcriptional regulators, in fact, RamA can recognise the well defined DNA Sox boxes [61]. The fact that the regulatory region of the *hmp* gene lacks a clear Sox box suggests that RamA may be acting indirectly on *hmp* [56].

Recently the product of the *yjeB* gene (*nsrR*) was identified as the main regulator of *hmp* in response to NO [51].

Hmp from Escherichia coli. The flavohaemoglobin Hmp of *Escherichia coli* was the first haemoglobin described in microorganisms to a molecular level [62-64]. Only few years ago, Ilari and co-workers determined the molecular structure of Hmp by X-ray crystallography [65]. While the haem domain adopts the classical globin-fold, the flavin reductase domain presents binding sites for NAD(P)H and FAD [65].

The *hmp* genomic context (Figure 3) includes *glyA* (encoding serine hydroxymetyl transferase), which is divergently transcribed from *hmp* [66, 67] and the gene downstream *hmp* is *glnB* (encoding protein PII involved in the glutamine biosynthesis pathway) [68]. The first evidence of the involvement of flavohaemoglobins in the nitrosative stress response was given by the observation of an abrupt increase in the transcriptional activity of the *hmp* gene upon exposure to NO [47]. This was further supported by the increase sensitivity to NO-releasing agents of defined *hmp* mutant strain of *E. coli* [69, 70]. Biochemical evidence showing that the purity Hmp protein has NO dioxygenase or denitrosylase activity, forming nitrates under aerobic conditions [48, 71-73] and also, with much reduced activity, reducing NO to N_2O under anaerobic conditions [74]. Together these observations strongly suggested that one of the main functions of flavohaemoglobin is to provide protection from the toxicity of NO.

The analysis of the transcriptional regulation of an Φ(*hmp*$_{Ec}$-*lacZ*) fusion determined that the induction of *hmp* by NO or nitrosating agents, SNP or GSNO, is independent of the SoxRS system [69]. This gene is also is induced by PQ in a SoxRS independent manner [69, 70]. Paradoxically, the overexpression of the Hmp generates superoxide and cause oxidative stress [75].

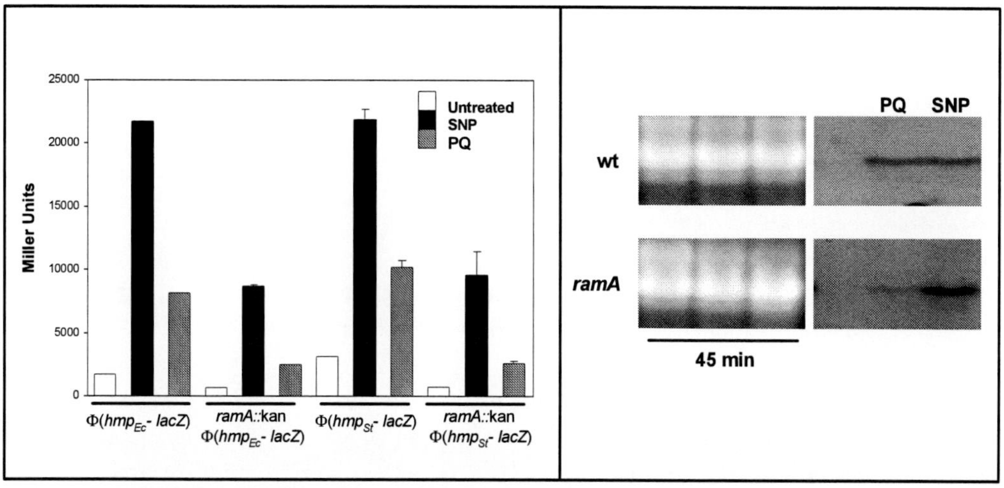

Figure 2. Expression of the *hmp* gene in *Salmonella* in response to SNP (an NO donor) and PQ (methyl viologen, a superoxide-generating agent). Left panel: Plasmids bearing the Φ(*hmp*$_{Ec}$-*lacZ*) (regulatory region of the *Escherichia coli hmp* gene) and Φ(*hmp*$_{St}$-*lacZ*) (regulatory region of the *Salmonella hmp* gene) fusions were transformed into wild-type and its *ramA*::*kan* derivative *Salmonella* strains. β–Galactosidase activity is shown. Right panel: Northern blot analyses using total RNA from untreated (first lanes), treated with PQ (200 μM; middle lanes) or SNP (200 μM; right lanes) *Salmonella* cultures. The *hmp* ORF was used as a probe. Northern blot analyses were carried out in wild-type (wt; top) and *ramA*::*kan* (*ramA*; bottom) strains. Left: Ethidium bromide stained agarose gel as a control of RNA loading; Right: autoradiography.

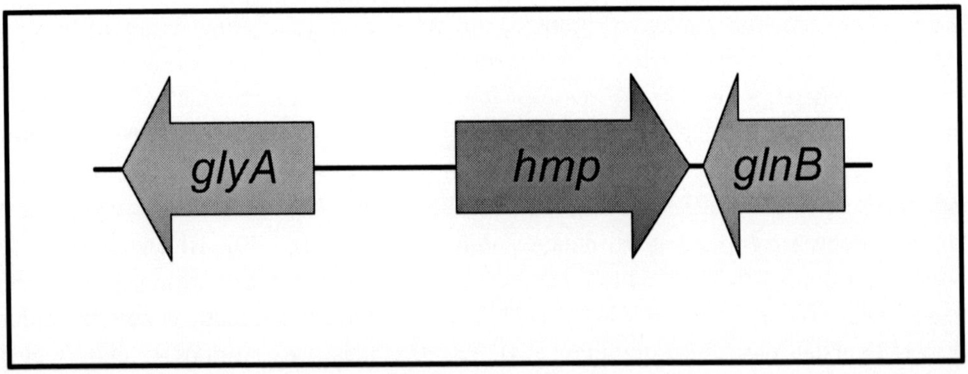

Figure 3. Genomic context of *hmp* in the *Escherichia coli* chromosome.

Several transcriptional factors have been experimentally involved in the *hmp* regulation: Fnr (transcriptional regulator of genes necessary under anaerobic conditions) is a negative regulator of *hmp* in cultures grown in the absence of oxygen. Using gel retardation and DNase I protection assays an Fnr binding site was found between -2 to +11 on the promoter region of *hmp*. Inactivation by nitrosylation of the Fe-S cluster of Fnr with the formation of a dinitrosyl-iron-cysteine complex lowers the affinity of binding of Fnr to the "Fnr box" near the translational start codon ATG [76]. In the presence of oxygen, different regulators are involved in the expression of *hmp* [47, 76], for example, expression of *hmp* is induced in cells entering the stationary phase. This induction is mediated by the stationary phase-specific sigma subunit (σ^s) of RNA polymerase [77].

In earlier reports, the induction of *hmp* expression by NO was shown to be dependent on MetR, which is the global regulator of the methionine biosynthetic pathway in *E. coli* [78]. But later it was shown that this regulation was in fact indirectly due to the fact that MetR is a transcriptional factor of *glyA* and indirectly affects *hmp* expression (Figure 3, [78]). Recently it was unequivocally demonstrated that MetR is not necessary for the *hmp* induction by NO [11].

A 10- to 15-fold induction in *hmp* expression was observed in aerobically or anaerobically grown in presence of the iron-chelating agent 2,2'-dipyridyl suggesting the involvement of an iron regulatory protein [47, 79]. At the same time, D'Autreaux and co-workers [80] demonstrated that Fur (a global regulator for the iron up-take in *E. coli*), is directly inactivated by iron nitrosylation [80]. A connexion between these two finding was recently reported by studying an Φ(*hmp*$_{Ec}$-*lacZ*) fusion in a *fur* strain. Indeed Fur is a mild regulator of *hmp* expression [56].

This year, several reports identified the product of *yjeB* (*nsrR*), NsrR, as the master regulator of the *hmp* expression in *E. coli;* interestingly orthologous genes in *Salmonella* and *Bacillus subtilis* have been also identified [51, 81, 82]. NsrR, is a transcriptional repressor of the Rrf2 family that binds to the *hmp* promoter region on the transcriptional start site. The NsrR binding sequence is present in other promoter regions of genes that are induced by NO, like *ytfE* and *ygbA*. Interestingly, the β-galactosidase activity of an Φ(*hmp*$_{Ec}$-*lacZ*) fusion is elevated up to 10-fold in the basal levels. It has been proposed that NsrR is inactivated by NO by disrupting its Fe-S cluster (Figure 4). The slight induction of *hmp* by SNP and PQ in an *yjeB* (*nsrR*) mutant strain may suggest the involvement of additional transcriptional factors, but clearly, NsrR is the main transcriptional factor responsible for NO induction (Figure 4).

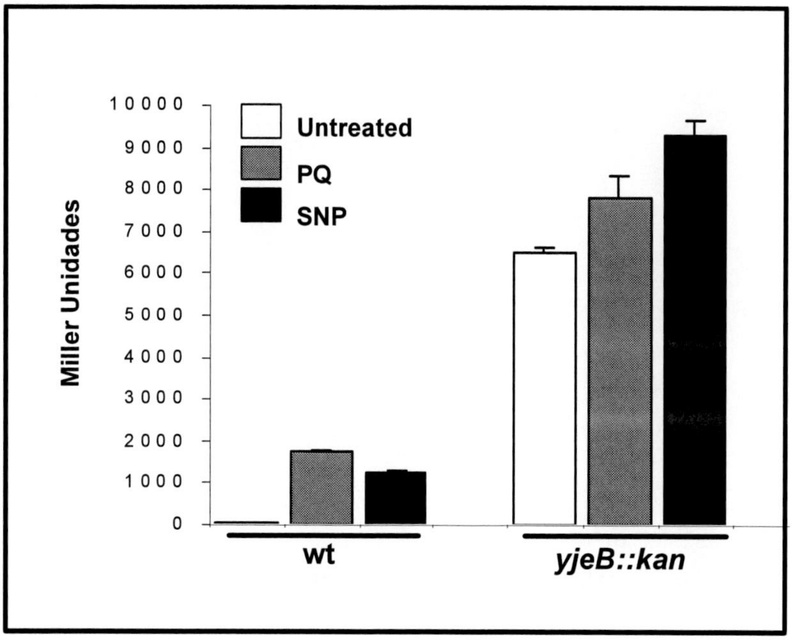

Figure 4. β-galactosidase activity of an Φ(hmpEc-lacZ) in the presence of 200μM PQ or SNP in a wild-type (wt) strain or an yjeB (nsrR) mutant in Escherichia coli.

HMP from Mycobacterium tuberculosis. Mycobacterium tuberculosis is the single most important infectious pathogen in the world and causes over 3 million deaths annually. *M. tuberculosis* persists in a dormant form in the body by resisting the effects of the human immune response and chemotherapy. It has been reported that one-third of the human population may be latently infected with dormant tubercle bacilli which can reactivate, causing tuberculosis later in life. An understanding of the molecular basis by which *M. tuberculosis* becomes dormant is a prerequisite for the identification of molecules in dormant bacteria which can be targeted by new drugs [83].

Genomic studies with the *M. tuberculosis* genome identified an ORF that is 30.8% identical to Hmp of *E. coli*. Transcriptional regulation studies during the growth cycle showed that hum_{Mtb} expression is rather weak during exponential phase, but as the culture enters late exponential phase, when oxygen tension begins to decrease as a result of bacterial consumption, hum_{Mtb} transcription increases and reaches its maximum level at early stationary phase [83].

As in other cases, hmp_{Mt} expression is abruptly increased upon exposure to NO [83]. This is particularly interesting because it is known that after infection, *M. tuberculosis* initially multiplies within the host macrophages until circulation of infected macrophages is restricted to form calcified granuloma [84]. Within the activated macrophages, there are inducible nitric oxide synthases (NOS) that rapidly generate copious amounts of NO to poison pathogenic microorganisms. NO, in this case, exerts its cytotoxic and genotoxic effects through its ability to damage DNA and iron-sulphur clusters of key enzymes such as aconitase and inhibits terminal respiratory oxidases [85]. The regulatory networks involved in hmp_{Mt} transcriptional regulation are still unknown. Moreover, some reports have determined that in several mycobacterial species there are proteins called truncated haemoglobins. These proteins have also been involved in the defence of toxicity by NO but little is known on the regulation of their genes [85-87].

FHP from Pseudomonas aeruginosa. P.aeruginosa is an opportunistic pathogen with the ability to reduce nitrate to N_2 by nitrification under anaerobic conditions. This bacterium causes persistent infections in the lungs of cystic fibrosis patients that are frequently associated with the emergence of antibiotic-resistant subpopulations of bacteria. A genomic analysis of *P. aeruginosa* strain PAO1 identified the flavohaemoglobin (*fhp*) gene. Like in other bacteria, the *fhp* promoter is up-regulated by NO but in this case, this induction is dependent on the transcriptional regulator FhpR. The gene *fhp*R is located upstream of flavohaemoglobin in a divergent direction. *fhp*R is similar to the Nor genes, which regulate the NO-responsive expression of flavorubredoxin in *E. coli*. In fact, recently it was shown that *fhp* has an important role in the *P. aeruginosa* inducible resistance to NO under aerobic conditions [88].

Flavorubredoxin NorVW. E. coli bears an additional NO detoxifying mechanism under anaerobic conditions that involves the flavorubredoxin protein (FlavoRb). This protein has an NO reductase activity [89, 90]. FlavoRb is a member of the family of A-type flavoproteins, identified in a number of anaerobes and facultative anaerobes. The members of this family have in common a two-domain core, built by a β-lactamase-like domain at the N-terminal region and harbouring a di-iron catalytic site, and a flavodoxin-like domain, containing one FMN moiety. The FlavoRb of *E. coli* has an extra domain at the C-terminal, containing a rubredoxin-like centre [91].

The FlavoRb is encoded by the *norV* gene, downstream of this gene there is the *norW* gene that codes for an NADH:(flavo)rubredoxin oxidoreductase with a FAD domain. RT-PCR assays showed that *norV* and *norW* are part of the same transcriptional unit [91]. Upstream of this operon and divergently transcribed is the *norR* gene, encoding the transcriptional regulator of the *norVW* operon. NorR is a σ^{54}-dependent transcriptional organised into three domains, an N-terminal GAF domain that is a site for an interaction with a small molecule ligand, a central domain that is predicted to interact with σ^{54} factor and to hydrolase ATP, and a C-terminal DNA binding domain [92].

Binding of NO stimulates the ATPase activity of NorR, allowing the activation of transcription by RNA polymerase. Transcriptional activation by σ^{54}-dependent activators is usually controlled by sensory modules, and in the case of NorR the N-terminal GAF domain is the probable site of signal sensing. In a mutant strain expressing a mutated NorR in the GAF domain, the transcription of Φ(*norV-lacZ*) fusion is constitutive independently of the signal, strongly suggesting that the function of the GAF domain is to sense the signal for NO [93].

Conclusion

Enteric bacteria have multiple regulatory systems that respond to NO and RNS, and there is growing evidence that more NO-responsive regulators remain to be discovered. The challenge for the future will be to elucidate the mechanisms for NO sensing and also to get closer to the natural habitats of bacteria such as the biofilm communities. It is evident that bacteria has evolved defence mechanisms to cope with NO, so certainly, NO has been around for long time and flavohaemoglobins and flavorubredoxins are not the only proteins involved in the resistance to RNS, we are still in the dawn of the discovery of new resistance mechanisms, especially of those bacteria that have chosen to live in communities or within higher organisms.

References

[1] Fang FC (2004). Antimicrobial reactive oxygen and nitrogen species: concepts and controversies. *Nat. Rev. Microbiol.*2:820-832.

[2] De Groote MA, Ochsner UA, Shiloh MU, Nathan C, McCord JM, Dinauer MC, Libby SJ, Vazquez-Torres A, Xu Y, Fang FC (1997). Periplasmic superoxide dismutase protects *Salmonella* from products of phagocyte NADPH-oxidase and nitric oxide-synthase. *Proc. Nat. Acad. Sci. USA* 94:13997-14001.

[3] Lewis RS, Tamir S, Tannenbaum SR, Deen WM (1995). Kinetic Analysis of the Fate of Nitric Oxide Synthesized by Macrophages *in Vitro*. *J. Biol. Chem.* 270:29350-29355.

[4] Wink DA, Mitchel JB (1998). Chemical biology of nitric oxide: Insights into regulatory, cytotoxic, and cytoprotective mechanisms of nitric oxide. *Free Radic. Biol. Med.* 25:434-456.

[5] Radi R, Beckman JS, Bush KM, Freeman BA (1991). Peroxynitrite-induced membrane lipid peroxidation: The cytotoxic potential of superoxide and nitric oxide. *Arch. Biochem. Biophys.* 288:481-487.

[6] Wink DA, Kasprzak KS, Maragos CM, Elespuru RK, Misra M, Dunams TM, Cebula TA, Koch WH, Andrews AW, Allen JS, *et al.* (1991). DNA deaminating ability and genotoxicity of nitric oxide and its progenitors. *Science* 254:1001-1003.

[7] Schopfer FJ, Baker PRS, Freeman BA (2003). NO-dependent protein nitration: a cell signaling event or an oxidative inflammatory response? *TIBS* 28:646-654.

[8] Ohno H, Zhu G, Mohan VP, Chu D, Kohno S, Jacobs WR, Chan J (2003). The effects of reactive nitrogen intermediates on gene expression in *Mycobacterium tuberculosis*. *Cell. Microbiol.* 5:637-648.

[9] Firoved AM, Wood SR, Ornatowski W, Deretic V, Timmins GS (2004). Microarray Analysis and Functional Characterization of the Nitrosative Stress Response in Nonmucoid and Mucoid *Pseudomonas aeruginosa*. *J. Bacteriol.*186:4046-4050.

[10] Moore CM, Nakano MM, Wang T, Ye RW, Helmann JD (2004). Response of Bacillus subtilis to Nitric Oxide and the Nitrosating Agent Sodium Nitroprusside. *J. Bacteriol.* 186:4655-4664.

[11] Mukhopadhyay P, Zheng M, Bedzyk LA, LaRossa RA, Storz G (2004). Prominent roles of the NorR and Fur regulators in the *Escherichia coli* transcriptional response to reactive nitrogen species. *Proc. Nat. Acad. Sci. USA* 101:745-750.

[12] Schapiro JM, Libby SJ, Fang FC (2003). Inhibition of bacterial DNA replication by zinc mobilization during nitrosative stress. Proc. Nat. Acad. Sci. USA 100:8496-8501.

[13] Costerton JW, Lewandowski Z, Caldwell DE, Korber DR, Lappin-Scott HM (1995). Microbial biofilms. *Annu. Rev. Microbiol.* 49:711-745.

[14] Stevanin TM, Poole RK, Demoncheaux EAG, Read RC (2002). Flavohemoglobin Hmp protects *Salmonella enterica* serovar Typhimurium from nitric oxide-related killing by human macrophages. *Infect.Immun.*70:4399-4405.

[15] Stevanin TM, Ioannidis N, Mills CE, Kim SO, Hughes MN, Poole RK (2000). Flavohemoglobin Hmp affords inducible protection for *Escherichia coli* respiration, catalyzed by cytochromes *bo'* or *bd,* from nitric oxide. *J. Biol. Chem.* 275:35868-35875.

[16] Beckman JS, Koppenol WH (1996). Nitric oxide, superoxide, and peroxynitrite: the good, the bad, and ugly. *AJP - Cell Physiol.* 271:1424-1437.

[17] Cooper CE (1999). Nitric oxide and iron proteins. Biochim. Biophys. *Acta Bioenergetics* 1411:290-309.

[18] Kissner R, Nauser T, Bugnon P, Lye PG, Koppenol WH (1997). Formation and Properties of Peroxynitrite as Studied by Laser Flash Photolysis, High-Pressure Stopped-Flow Technique, and Pulse Radiolysis. *Chem. Res. Toxicol.* 10:1285-1292.

[19] Simon DI, Mullins ME, Jia L, Gaston B, Singel DJ, Stamler JS (1996). Polynitrosylated proteins: Characterization, bioactivity, and functional consequences. *Proc. Nat. Acad. Sci. USA* 93:4736-4741.

[20] Di Simplicio P, Franconi F, Frosali S, Di Giuseppe D. (2003). Thiolation and nitrosation of cysteines in biological fluids and cells. *Amino Acids* V25:323-339.

[21] Miersch S, Mutus B (2005) Protein S-nitrosation: Biochemistry and characterization of protein thiol-NO interactions as cellular signals. *Clin. Biochem*. 38:777-791.

[22] Romero JM, Bizzozero OA (2006). Extracellular S-nitrosoglutathione, but not S-nitrosocysteine or N2O3, mediates protein S-nitrosation in rat spinal cord slices. *J. Neurochem*. 99:1299-1310.

[23] Patel RP, McAndrew J, Sellak H, White CR, Jo H, Freeman BA, Darley-Usmar VM (1999). Biological aspects of reactive nitrogen species. *Biochim. Biophys. Acta Bioenergetics* 1411:385-400.

[24] Fang FC, Vazquez-Torres A (2002). Nitric oxide production by human macrophages: there's NO doubt about it. Am. J. Physiol. Lung Cell. *Mol. Physiol.* 282:941-943.

[25] Vazquez-Torres A, Fang FC (2001). *Salmonella* evasion of the NADPH phagocyte oxidase. *Microb. Infect*. 3:1313-1320.

[26] Poole RK (2005). Nitric oxide and nitrosative stress tolerance in bacteria. *Biochem. Soc. Trans.* 33:176-180.

[27] Brunelli L, Crow JP, Beckman JS (1995). The Comparative Toxicity of Nitric Oxide and Peroxynitrite to *Escherichia coli. Arch. Biochem. Biophys.* 316:327-334.

[28] Hughes MN (1999). Relationships between nitric oxide, nitroxyl ion, nitrosonium cation and peroxynitrite. Biochim. Biophys. *Acta Bioenergetics* 1411:263-272.

[29] Poole RK, Hughes MN (2000). New functions for the ancient globin family: bacterial responses to nitric oxide and nitrosative stress. *Mol. Microbiol.* 36:775-783.

[30] Rayat S, Wu Z, Glaser R (2004). Nitrosative Guanine Deamination: Ab Initio Study of Deglycation of *N*-Protonated 5-Cyanoimino-4-oxomethylene-4,5-dihydroimidazoles. *Chem. Res. Toxicol.* 17:1157-1169.

[31] Rayat S, Qian M, Glaser R (2005). Nitrosative Cytosine Deamination. An Exploration of the Chemistry Emanating from Deamination with Pyrimidine Ring-Opening. *Chem. Res. Toxicol.* 18:1211-1218.

[32] Caulfield JL, Wishnok JS, Tannenbaum SR (1998). Nitric Oxide-induced Deamination of Cytosine and Guanine in Deoxynucleosides and Oligonucleotides. *J. Biol. Chem.* 273:12689-12695.

[33] Hentze MW, Kuhn LC (1996). Molecular control of vertebrate iron metabolism: mRNA-based regulatory circuits operated by iron, nitric oxide, and oxidative†stress. *Proc. Nat. Acad. Sci. USA* 93:8175-8182.

[34] Kim YM, Bergonia HA, Muller C, Pitt BR, Watkins WD, Lancaster JRJ (1995). Nitric oxide and intracellular heme. *Adv. Pharmacol.* 34:277-291.

[35] Fang FC (1997).: Mechanisms of Nitric Oxide-related Antimicrobial Activity. *J. Clin. Invest.* 99:2818-2825.

[36] Crow JP, Beckman JS (1995). Reactions between nitric oxide, superoxide, and peroxynitrite: footprints of peroxynitrite *in vivo*. *Adv. Pharmacol.* 34:17-43.

[37] Greenacre S, Ischiropoulos H (2001). Tyrosine nitration: Localisation, quantification, consequences for protein function and signal transduction. *Free Radical Res.* 34:541-581.

[38] Szabo C (2003). Multiple pathways of peroxynitrite cytotoxicity. *Toxicol. Lett.* 140-141:105-112.

[39] Ischiropoulos H, Zhu L, Chen J, Tsai M, Martin JC, Smith CD, Beckman JS (1992). Peroxynitrite-mediated tyrosine nitration catalyzed by superoxide dismutase. *Arch. Biochem. Biophys.* 298:431-437.

[40] Ramezanian MS, Padmaja S, Koppenol WH, Lester P (1996). Nitration and hydroxylation of phenolic compounds by peroxynitrite: Meth. *Enzymol.* 269: 195-201.

[41] Yamakura F, Taka H, Fujimura T, Murayama K (1998). Inactivation of Human Manganese-superoxide Dismutase by Peroxynitrite Is Caused by Exclusive Nitration of Tyrosine 34 to 3-Nitrotyrosine. *J. Biol. Chem.* 273:14085-14089.

[42] MacMillan-Crow LA, Crow JP, Thompson JA (1998). Peroxynitrite-Mediated Inactivation of Manganese Superoxide Dismutase Involves Nitration and Oxidation of Critical Tyrosine Residues. *Biochemistry* 37:1613-1622.

[43] Rubbo H, Radi R, Trujillo M, Telleri R, Kalyanaraman B, Barnes S, Kirk M, Freeman B (1994). A: Nitric oxide regulation of superoxide and peroxynitrite-dependent lipid peroxidation. Formation of novel nitrogen-containing oxidized lipid derivatives. *J. Biol. Chem.* 269:26066-26075.

[44] Imlay JA (2003). Pathways of oxidative damage. *Ann. Rev. Microbiol.* 57:395-418.

[45] Flatley J, Barrett J, Pullan ST, Hughes MN, Green J, Poole RK (2005). Transcriptional responses of *Escherichia coli* to S-nitrosoglutathione under defined chemostat

conditions reveal major changes in methionine biosynthesis. *J. Biol. Chem.* 280:10065-10072.

[46] Justino MC, Vicente JB, Teixeira M, Saraiva LM (2005). New genes implicated in the protection of anaerobically grown *Escherichia coli* against nitric oxide. *J. Biol. Chem.* 280:2636-2643.

[47] Poole RK, Anjum MF, Membrillo-Hernandez J, Kim SO, Hughes MN, Stewart V (1996). Nitric oxide, nitrite, and Fnr regulation of *hmp* (flavohemoglobin) gene expression in *Escherichia coli K-12*. *J. Bacteriol.* 178:5487-5492.

[48] Gardner PR, Gardner AM, Martin LA, Salzman AL (1998). Nitric oxide dioxygenase: an enzymic function for flavohemoglobin. *Proc. Natl. Acad. Sci. USA* 95:10378-10383.

[49] Vazquez-Torres A, Jones-Carson J, Mastroeni P, Ischiropoulos H, Fang FC (2000). Antimicrobial Actions of the NADPH Phagocyte Oxidase and Inducible Nitric Oxide Synthase in Experimental Salmonellosis. I. Effects on Microbial Killing by Activated Peritoneal Macrophages In Vitro. *J. Exp. Med.*192:227-236.

[50] Groote MAD, Granger D, Xu Y, Campbell G, Prince R, Fang FC (1995). Genetic and Redox Determinants of Nitric Oxide Cytotoxicity in a *Salmonella typhimurium* Model. *Proc. Nat. Acad. Sci. USA* 92:6399-6403.

[51] Bang IS, Liu L, Vazquez-Torres A, Crouch ML, Stamler JS, Fang FC (2006). Maintenance of nitric oxide and redox homeostasis by the *salmonella* flavohemoglobin Hmp. *J. Biol. Chem.* 281:28039-28047.

[52] Crawford MJ, Goldberg DE (1998). Role for the *Salmonella* flavohemoglobin in protection from nitric oxide. *J. Biol. Chem.* 273:12543-12547.

[53] Crawford MJ, Goldberg DE (1998). Regulation of *Salmonella typhimurium* flavohemoglobin gene. A new pathway for bacterial gene expression in response to nitric oxide. *J. Biol. Chem.* 273:34028-34032.

[54] Crawford MJ, Goldberg DE (2006). Regulation of the *Salmonella typhimurium* flavohemoglobin gene. *J. Biol. Chem.* 281:3752.

[55] Rodionov DA, Dubchak IL, Arkin AP, Alm EJ, Gelfand MS (2005). Dissimilatory metabolism of nitrogen oxides in bacteria: comparative reconstruction of transcriptional networks. *PLoS Comput. Biol.* 1:e55.

[56] Hernández-Urzúa E, Zamorano-Sánchez, DS, Ponce-Coria J, Morett E, Grogan, S, Poole, RK, Membrillo-Hernández, J (2006). Multiple regulators of the Flavohaemoglobin (hmp) gene of Salmonella enterica serovar Typhimurium include RamA, a transcriptional regulator conferring the multidrug resistance phenotype. *Arch. Microbiol. In press.*

[57] van der Straaten T, Janssen R, Mevius DJ, van Dissel JT (2004). *Salmonella* Gene *rma* (*ramA*) and Multiple-Drug-Resistant *Salmonella enterica* Serovar Typhimurium. Antimicrob. *Agents Chemother*. 48:2292-2294.

[58] van der Straaten T, Zulianello L, van Diepen A, Granger DL, Janssen R, van Dissel JT (2004). *Salmonella enterica* Serovar Typhimurium RamA, Intracellular Oxidative Stress Response, and Bacterial Virulence. *Infect. Immun.*72:996-1003.

[59] George AM, Hall RM, Stokes HW (1995). Multidrug resistance in *Klebsiella pneumoniae*: a novel gene, *ramA*, confers a multidrug resistance phenotype in *Escherichia coli*. *Microbiology* UK141:1909-1920.

[60] Chollet R, Chevalier J, Bollet C, Pages JM, Davin-Regli A (2004). RamA is an alternate activator of the multidrug resistance cascade in *Enterobacter aerogenes*. *Antimicrob. Agent. Chemother.* 48:2518-2523.

[61] Yassien MA, Ewis HE, Lu CD, Abdelal AT (2002). Molecular cloning and characterization of the *Salmonella enterica* serovar paratyphi B *rma* gene, which confers multiple drug resistance in *Escherichia coli. Antimicrob. Agent. Chemother.* 46:360-366.

[62] Membrillo-Hernández J, Poole RK (1997). Bacterial flavohaemoglobins: a consensus sequence and identification of a discrete enterobacterial group and of further bacterial globins. *FEMS Microbiol. Lett.* 155:179-184.

[63] Vasudevan SG, Armarego WL, Shaw DC, Lilley PE, Dixon NE, Poole RK (1991). Isolation and nucleotide sequence of the *hmp* gene that encodes a haemoglobin-like protein in *Escherichia coli K-12. Mol. Gen. Genet.* 226:49-58.

[64] Andrews SC, Shipley D, Keen JN, Findlay JB, Harrison PM, Guest JR (1992). The haemoglobin-like protein (HMP) of *Escherichia coli* has ferrisiderophore reductase activity and its C-terminal domain shares homology with ferredoxin NADP+ reductases. *FEBS Lett.* 302:247-252.

[65] Ilari A, Bonamore A, Farina A, Johnson KA, Boffi A (2002). The X-ray structure of ferric *Escherichia coli* flavohemoglobin reveals an unexpected geometry of the distal heme pocket. *J. Biol. Chem.*277:23725-23732.

[66] Plamann MD, Stauffer GV (1983). Characterization of the *Escherichia coli* gene for serine hydroxymethyltransferase. *Gene* 22:9-18.

[67] Plamann MD, Stauffer GV (1985). Characterization of a *cis*-acting regulatory mutation that maps at the distal end of the *Escherichia coli glyA* gene. *J. Bacteriol.* 161:650-654.

[68] Bueno R, Pahel G, Magasanik B (1985). Role of *glnB* and *glnD* gene products in regulation of the *glnALG* operon of *Escherichia coli. J. Bacteriol.* 164:816-822.

[69] Membrillo-Hernández J, Coopamah MD, Anjum MF, Stevanin TM, Kelly A, Hughes MN, Poole RK (1999). The flavohemoglobin of *Escherichia coli* confers resistance to a nitrosating agent, a "Nitric oxide Releaser," and paraquat and is essential for transcriptional responses to oxidative stress. *J. Biol. Chem.* 274:748-754.

[70] Membrillo-Hernández J, Kim SO, Cook GM, Poole RK (1997). Paraquat regulation of hmp (flavohemoglobin) gene expression in *Escherichia coli K-12* is SoxRS independent but modulated by sigma S. *J. Bacteriol.* 179:3164-3170.

[71] Gardner AM, Martin LA, Gardner PR, Dou Y, Olson JS (2000). Steady-state and transient kinetics of *Escherichia coli* nitric-oxide dioxygenase (flavohemoglobin). The B10 tyrosine hydroxyl is essential for dioxygen binding and catalysis. *J. Biol. Chem.* 275:12581-12589.

[72] Hausladen A, Gow A, Stamler JS (2001). Flavohemoglobin denitrosylase catalyzes the reaction of a nitroxyl equivalent with molecular oxygen. *Proc. Natl. Acad. Sci. USA* 98:10108-10112.

[73] Hausladen A, Gow AJ, Stamler JS (1998). Nitrosative stress: metabolic pathway involving the flavohemoglobin. *Proc. Natl. Acad. Sci. USA* 95:14100-14105.

[74] Kim SO, Orii Y, Lloyd D, Hughes MN, Poole RK (1999). Anoxic function for the *Escherichia coli* flavohaemoglobin (Hmp): reversible binding of nitric oxide and reduction to nitrous oxide. *FEBS Lett.* 445:389-394.

[75] Membrillo-Hernández J, Ioannidis N, Poole RK: The flavohaemoglobin (HMP) of *Escherichia coli* generates superoxide in vitro and causes oxidative stress in vivo. *FEBS Lett* 1996;382:141-144.

[76] Cruz-Ramos H, Crack J, Wu G, Hughes MN, Scott C, Thomson AJ, Green J, Poole RK (2002). NO sensing by FNR: regulation of the *Escherichia coli* NO-detoxifying flavohaemoglobin, Hmp. EMBO J 21:3235-3244.

[77] Membrillo-Hernández J, Cook GM, Poole RK (1997). Roles of RpoS (sigmaS), IHF and ppGpp in the expression of the *hmp* gene encoding the flavohemoglobin (Hmp) of *Escherichia coli K-12*. Mol. Gen. Genet. 254:599-603.

[78] Membrillo-Hernández J, Coopamah MD, Channa A, Hughes MN, Poole RK (1998). A novel mechanism for upregulation of the *Escherichia coli K-12 hmp* (flavohaemoglobin) gene by the 'NO releaser', S-nitrosoglutathione: nitrosation of homocysteine and modulation of MetR binding to the *glyA-hmp* intergenic region. Mol. Microbiol. 29:1101-1112.

[79] Anjum MF, Ioannidis N, Poole RK (1998). Response of the NAD(P)H-oxidising flavohaemoglobin (Hmp) to prolonged oxidative stress and implications for its physiological role in *Escherichia coli*. *FEMS Microbiol. Lett.* 166:219-223.

[80] D'Autreaux B, Touati D, Bersch B, Latour JM, Michaud-Soret I (2002). Direct inhibition by nitric oxide of the transcriptional ferric uptake regulation protein via nitrosylation of the iron. *Proc. Natl. Acad. Sci. USA* 99:16619-16624.

[81] Nakano MM (2006). Essential role of flavohemoglobin in long-term anaerobic survival of *Bacillus subtilis*. *J. Bacteriol.* 188:6415-6418.

[82] Bodenmiller DM, Spiro S (2006). The *yjeB* (*nsrR*) gene of *Escherichia coli* encodes a nitric oxide-sensitive transcriptional regulator. *J. Bacteriol.* 188:874-881.

[83] Hu Y, Butcher PD, Mangan JA, Rajandream MA, Coates AR (1999). Regulation of hmp gene transcription in *Mycobacterium tuberculosis*: effects of oxygen limitation and nitrosative and oxidative stress. *J. Bacteriol.* 181:3486-3493.

[84] Fenton MJ (1998). Macrophages and tuberculosis. *Curr. Opin. Hematol.* 5:72-78.

[85] Pathania R, Navani NK, Gardner AM, Gardner PR, Dikshit KL (2002). Nitric oxide scavenging and detoxification by the *Mycobacterium tuberculosis* haemoglobin, HbN in *Escherichia coli*. *Mol. Microbiol.*45:1303-1314.

[86] Fabozzi G, Ascenzi P, Renzi SD, Visca P (2006). Truncated hemoglobin GlbO from *Mycobacterium leprae* alleviates nitric oxide toxicity. *Microb. Pathog.* 40:211-220.

[87] Ouellet H, Ouellet Y, Richard C, Labarre M, Wittenberg B, Wittenberg J, Guertin M (2002). Truncated hemoglobin HbN protects *Mycobacterium bovis* from nitric oxide. *Proc. Natl. Acad. Sci. USA*;99:5902-5907.

[88] Arai H, Hayashi M, Kuroi A, Ishii M, Igarashi Y (2005). Transcriptional regulation of the flavohemoglobin gene for aerobic nitric oxide detoxification by the second nitric oxide-responsive regulator of *Pseudomonas aeruginosa*. *J. Bacteriol.* 187:3960-3968.

[89] Gardner AM, Helmick RA, Gardner PR (2002). Flavorubredoxin, an inducible catalyst for nitric oxide reduction and detoxification in *Escherichia coli. J. Biol. Chem.* 277:8172-8177.

[90] Gomes CM, Giuffre A, Forte E, Vicente JB, Saraiva LM, Brunori M, Teixeira M (2002). A novel type of nitric-oxide reductase. *Escherichia coli* flavorubredoxin. *J. Biol. Chem.* 277:25273-25276.

[91] da Costa PN, Teixeira M, Saraiva LM (2003). Regulation of the flavorubredoxin nitric oxide reductase gene in *Escherichia coli:* nitrate repression, nitrite induction, and possible post-transcription control. *FEMS Microbiol. Lett.* 218:385-393.

[92] Hutchings MI, Mandhana N, Spiro S (2002). The NorR protein of *Escherichia coli* activates expression of the flavorubredoxin gene *norV* in response to reactive nitrogen species. *J. Bacteriol.*184:4640-4643.

[93] D'Autreaux B, Tucker NP, Dixon R, Spiro S (2005). A non-haem iron centre in the transcription factor NorR senses nitric oxide. *Nature* 437:769-772.

In: Oxidative Stress: Clinical and Biomedical Implications ISBN: 978-1-60021-888-0
Editors: B. M. Matata, M. M. Elahi, pp. 203-226

Chapter XIII

Endogenous Regulation of Nitric Oxide Synthase Activity and Free Radical Generation

Scott P. Forbes[1], Yong Xia[2] and Arturo J. Cardounel[1*]
[1]The Davis Heart and Lung Research Institute,
[2]Departments of Pharmacology and Internal Medicine,
The Ohio State University College of Medicine,
Columbus, OH 43210

Abstract

Discoveries over the past ten years indicate that crucial features of neuronal communication, cardiovascular homeostasis and immune response are mediated by a simple chemical, nitric oxide (NO). Endogenous NO is generated from L-arginine by a family of three distinct calmodulin- dependent nitric oxide synthase (NOS) enzymes. NOS from endothelial cells (eNOS) and neurons (nNOS) are both constitutively expressed enzymes, whose activities are regulated by changes in intracellular calcium. Immune functions for NO are mediated by a calcium-independent inducible form of NOS (iNOS). As a potent cell-signaling, effector, and vasodilatory molecule, NO plays important roles in diverse biological processes in the brain, kidney, vasculature, and many other tissues. In addition, NOS and its catalytic products NO and superoxide appear to play a major role in the pathophysiology involved in numerous cardiovascular, immune and neurological disorders. Because of its high biological reactivity and diffusibility, multiple tiers of regulation, ranging from transcriptional to posttranslational controls, tightly control NO biosynthesis. Control of NOS activity is known to be regulated by allosteric enzyme mechanisms, by posttranslational modification, subcellular targeting and cofactor/ substrate bioavailability. Therefore, in this chapter we will discuss molecular intricacies and the enzymology of NOS regulation. This

* Address for correspondence: Arturo J. Cardounel, Ph.D, University of Florida, 1600 SW Archer Rd. M552, Gainesville, FL 32610352-273-7877Arturo Cardounel, Ph.D, University of Florida, 1600 SW Archer Rd. M552, Gainesville, FL 32610, 352-273-7877 Email: cardounel.1@ufl.edu

information will provide a global picture of the molecular mechanisms involved in the control of NOS enzymatic activity in the context of both normal physiologic function and pathophysiologic states.

Introduction

Nitric oxide synthase (NOS) plays a central role in the regulation of various biological processes through the formation of the gaseous free radical nitric oxide [66]. Endogenous nitric oxide (NO) acts as an essential signaling molecule and effector in cardiovascular, neuronal, and immune systems [12, 36, 40, 65]. In cells, NO is derived from the guanidino group of L-arginine in a reaction catalyzed by a family of NO synthases (NOSs) including neuronal NOS (nNOS), inducible NOS (iNOS), and endothelial NOS (eNOS)[67]. Whereas nNOS and eNOS are constitutively present in cells, iNOS expression requires the stimulation of microbial endotoxins or cytokines. Activation of nNOS and eNOS requires Ca^{2+}/calmodulin, hence NO production from these two isoforms is initiated and modulated by elevated intracellular free Ca^{2+}[11, 13, 97]. In contrast, iNOS is tightly bound to calmodulin (CaM) and is fully active under basal cytosolic Ca^{2+} concentrations, NO formation from iNOS appears to depend primarily on the level of enzyme [99].

All three NOS isoforms have considerable similarity in their structure and catalytic function. They share 50% homology in their amino acid sequences and structurally resemble NADPH cytochrome P-450 reductase [10, 57]. All NOSs use L-arginine, oxygen, and NADPH as substrates to synthesize NO. Tetrahydrobiopterin (BH_4), CaM, FAD, and FMN are the requisite cofactors for this catalytic process [12, 73]. As implied by the definition, the prime function of NOS is to catalyze the reaction in which L-arginine and oxygen are converted to NO and L-citrulline. However, NO synthesis is certainly not the only function of NOS. It has become clear that all three NOSs also synthesize superoxide ($O_2^{-\bullet}$) under certain circumstances (Diagram 1.). At low concentrations of L-arginine, nNOS and eNOS produce $O_2^{-\bullet}$ in a Ca^{2+}/CaM dependent manner [85, 93-95). In L-arginine-depleted cells, nNOS-derived $O_2^{-\bullet}$ has been shown to react with NO to form another potent oxidant peroxynitrite leading to cellular injury. iNOS also produces $O_2^{-\bullet}$ under L-arginine depletion[93]. BH_4 bioavailability also appears to be a central regulatory mechanism in the control of NOS derived $O_2^{-\bullet}$, however, the enzymatic catalysis of $O_2^{-\bullet}$ differs among the three NOS isoforms. Studies from purified isozymes of NOS have revealed that both eNOS and nNOS generate $O_2^{-\bullet}$ from the oxygenase domain while iNOS derived $O_2^{-\bullet}$ is largely catalyzed from the reductase domain[94, 95]. These data demonstrate that all three NOS isoforms are capable of reducing molecular oxygen to $O_2^{-\bullet}$ and that this reduction occurs from the oxygenase site of eNOS and nNOS and from the flavin-binding sites of the reductase domain with iNOS.

While the availability of enzyme substrates or cofactors can influence NO production, NO synthesis from NOS is also regulated by means of protein-protein interaction in cells. Among these interactions, CaM is the chief modulator. Following a rise in intracellular free (Ca^{2+}), CaM binds with nNOS and eNOS to initiate enzymatic activity. iNOS on the other hand is fully active under basal Ca^{2+} concentrations in resting cells due to its tight association with CaM. Beyond CaM, recent studies have identified several other proteins such as caveolin, PIN, NOSIP and Dynamin which have been demonstrated to be critical effector proteins in the regulation of NOS catalytic activity. As such, the remainder of this chapter will focus on the role of protein-protein interactions, post translational modifications and substrate/cofactor regulation as regulatory mechanisms in the control of NOS derived NO and $O_2^{-\bullet}$ production.

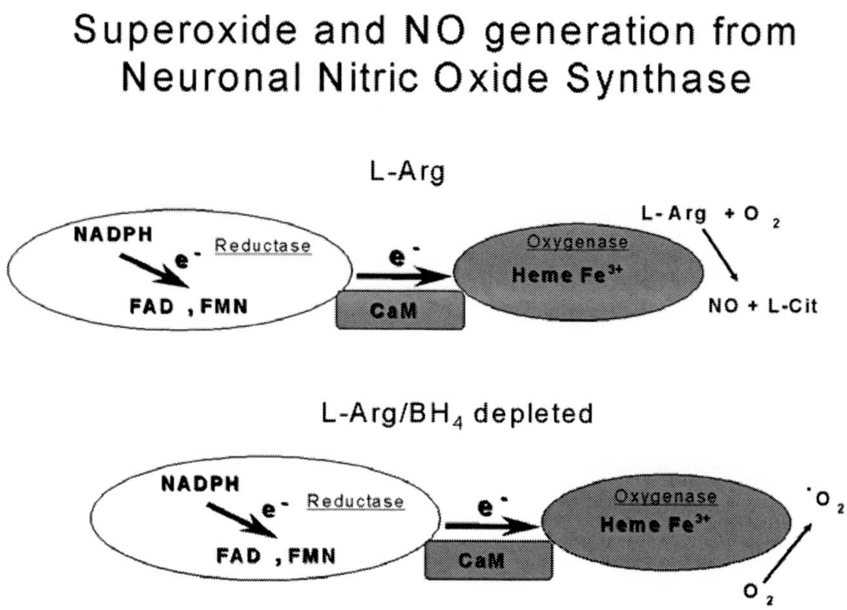

Diagram 1.

NOS Regulation by Protein-Protein Interactions

NO biosynthesis is under the regulatory influence of a variety of mechanisms ranging from transcriptional to post-transcriptional controls. Among these mechanisms, protein-protein interactions have been demonstrated to play a critical role in NOS regulation in both health and disease. Among the most widely studied of these interactions, is the role of calmodulin in the regulation of NOS activity and NO biosynthesis. Calmodulin has been demonstrated to act as an allosteric activator of the NOS isozymes with important isoform specific differences observed in regard to the calcium dependence of this protein-protein interaction. The constitutive NOS isoforms, eNOS and nNOS, are calcium dependent enzymes and are inactive at resting calcium concentrations [11, 53]. Elevations in intracellular calcium concetrations induce CAM binding resulting in activation of these NOS isoforms [86]. In contrast, the inducible NOS isozyme, iNOS, tightly binds CAM at resting calcium concentrations and is thus its activity is largely Ca-CAM independent [86].

The Ca^{2+}-CAM dependence of NO production from the NOS isoforms is well established. Its role in modulating NOS derived $O_2^{-\bullet}$ has also been examined and several studies have been published describing these interactions. Xia et.al have demonstrated that eNOS catalyzed $O_2^{-\bullet}$ formation occurs primarily at the heme center of its oxygenase domain and that $O_2^{-\bullet}$ generatinon from eNOS requires Ca2+/calmodulin and is primarily regulated by BH_4 rather than L-arginine (figure 1.)[95]. Similarly, studies published by Pou et.al demonstrated that nNOS generates $O_2^{-\bullet}$ in a Ca2+/calmodulin dependent manner and the formation of this free radical also occurs at the heme domain[68]. In contrast, iNOS derived $O_2^{-\bullet}$ is not Ca2+/calmodulin dependent and occurs primarily from the reductase site (figure 2.][94].

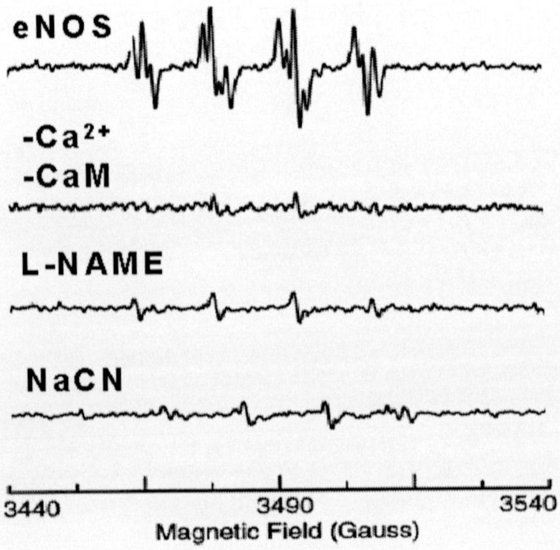

Figure 1. Ca^{2+}/calmodulin-dependent $O_2^{\cdot -}$ generation from eNOS. The composition of the reaction mixture was the same as described in the legend to Figure 2. eNOS, control spectrum showing $O_2^{\cdot -}$ generation by eNOS; L-NAME, in the presence of 1 mM L-NAME; Ca^{2+}/CaM free, in the absence of Ca^{2+}/calmodulin; NaCN, in the presence of 100 μM NaCN. EPR spectra were recorded in presence of 50 mM DMPO.

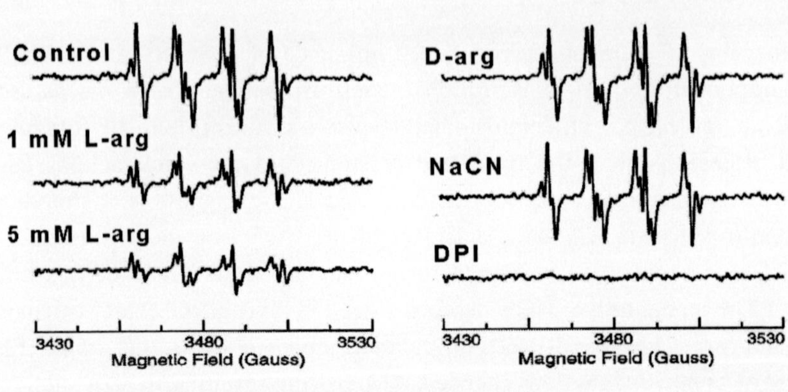

Figure 2. Effect of L-arginine, DPI, and NaCN on the $O_2^{\bullet -}$ formation from iNOS. The composition of the reaction mixture was the same as described in the legend to Figure 2. Control spectrum showing $O_2^{\bullet -}$ generation by iNOS (control); with 1 mM L-arginine (1 mM L-arg); with 5 mM L-arginine (5 mM L-arg); with 5 mM D-arginine (D-arg); with 100 μM NaCN (NaCN); with 20 μM DPI (DPI). EPR spectra were recorded in the presence of 50 mM DMPO.

In addition to these direct activation mechanisms induced upon CaM binding, it has been demonstrated by several groups that caveolin binding by NOS is regulated by CaM-NOS interactions [32, 60]. Caveolin-1, the predominant scaffolding protein of caveolae, has been shown to directly associate with and inhibit eNOS in endothelial cells. Under resting conditions, eNOS associates with caveolin-1 and this interaction maintains NOS in an inactive state. Following the intracellular rise in calcium that accompanies agonist stimulation, CaM binds NOS and caveolin dissociates form the enzyme resulting in an activated NOS-CaM complex. As the agonist invoked stimulation subsides, intracellular calcium levels drops resulting in eNOS dissociation from CaM with subsequent binding to caveolin resulting in inactivation [32, 60]. This CaM-caveolin control of eNOS does not appear to be exclusive to eNOS as similar findings have been published for the interaction of nNOS with caveolin-3[87]. Furthermore, similar to its effects on eNOS and nNOS derived NO, caveolin-1 has been demonstrated to inhibit eNOS and nNOS derived $O_2^{-\bullet}$ although the exact mechanism of these interactions is presently unclear.

The family of heat shock proteins, specifically hsp90, represents another NOS associated protein which influences the catalytic activity of all three NOS isoforms. Hsp90 is a highly abundant cytosolic protein known to serve as a molecular chaperone in protein folding and maturation events [24]. Formation of a nNOS-Hsp90 complex has been demonstrated to enhance NO formation. Hsp90 has been found to serve as an allosteric activator of eNOS[30, 34]. Garcia-Cardena et.al have demonstrated that vascular endothelial growth factor, histamine, and fluid shear stress induce association of Hsp90 with eNOS and enhance eNOS activity. Blockade of Hsp90-mediated signaling limits both agonist-induced NO production and vasorelaxation [30]. In addition, hsp90 and eNOS appear to form a multiprotein complex that includes the kinase Akt. In support of this, stimulation of endothelial cells with vascular endothelial growth factor has been shown to trigger the recruitment of eNOS and Akt to the same domain of Hsp90, facilitating Akt-driven phosphorylation of eNOS and promoting NO release [3]. Biochemical studies have shown that Hsp90 increases the activity of nNOS, increases the binding of calmodulin to nNOS and shifts the calmodulin-nNOS dose-response curve markedly to the left, thereby augmenting the maximal activity of the constitutive NOS isoforms [76].

Similar observations have been published with regards to iNOS as Yoshida and Xia have demonstrated that Hsp90 acts as an allosteric enhancer of iNOS [98]. Results from there studies demonstrated that at physiological concentrations (10-500 nm), Hsp90 dose-dependently increases iNOS activity (Figure 3.). This was a specific effect because neither denatured Hsp90 nor irrelevant bovine serum albumin affected iNOS function. Furthermore, overexpression of hsp90 enhanced NO production in iNOS-transfected cells while Hsp90 inhibition dramatically decreased iNOS derived NO formation in macrophages [98].

The effects of Hsp90 on the $O_2^{-\bullet}$ generation from NOS have also been studied wherein Hsp90 dose-dependently inhibited $O_2^{-\bullet}$ generation from nNOS with an IC_{50} of 658 nM [75]. This inhibition was not due to $O_2^{-\bullet}$ scavenging because Hsp90 did not affect the $O_2^{-\bullet}$ generated by Xanthine Oxidase. In the presence of low levels of L-arginine, where marked $O_2^{-\bullet}$ generation occurs, Hsp90 caused marked enhancement of NO synthesis from nNOS. Measurement of NOS derived superoxide demonstrated that adding Hsp90 prevented the $O_2^{-\bullet}$ production which led to the enhanced nNOS activity observed (figure 4.)[75].

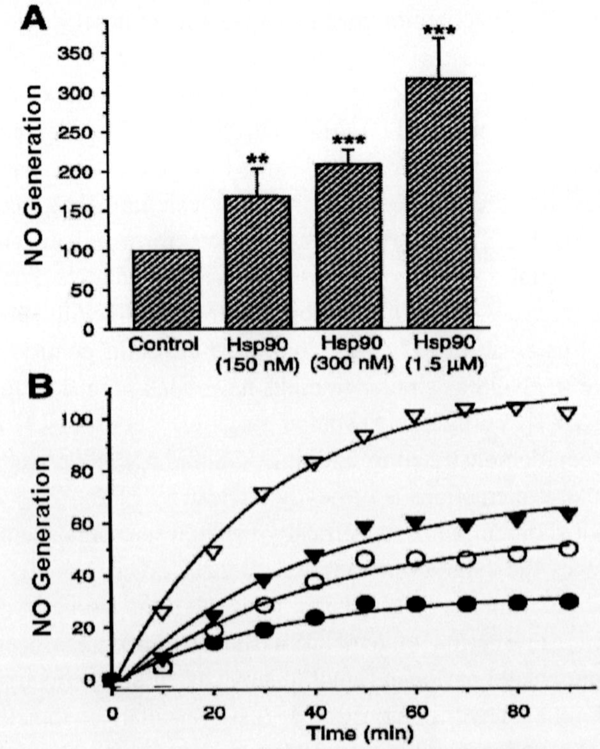

Figure 3. Effect of HSP90 on NO generation from purified nNOS. A: HSP90 increased NO generation from nNOS in a dose-dependent manner. Conversely, L-NAME totally blocked NO production. Data were shown as means ± SE (**P < 0.01; ***P < 0.001 compared with control, n = 3). B: effect of HSP90 on the time course of NO generation from nNOS. ●, control time course of NO production from nNOS; ■, L-NAME (1 mM); ○, HSP90 (150 nM); ▲, HSP90 (300 nM); △, HSP90 (1.5 μM). EPR spectra were continuously recorded at 10 1-min acquisitions from the beginning of the reaction until 90 min.

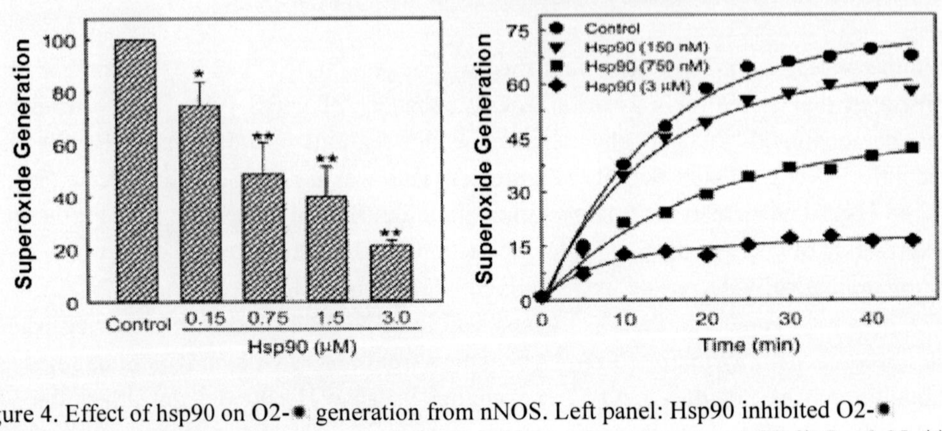

Figure 4. Effect of hsp90 on O2-• generation from nNOS. Left panel: Hsp90 inhibited O2-• generation from nNOS in a dose-dependent manner. Data are shown as mean ± SE (*, P < 0.05; **, P < 0.01 vs control; n = 3). Right panel: Effect of hsp90 on the time course of O2-• generation from nNOS. Spectra were continuously recorded at five 1-min acquisitions from the beginning of the reaction until 45 min.

In contrast, studies using cultured endothelial cells treated with the Hsp90 inhibitor, Radicicol, did not demonstrate any effect of Hsp90 on eNOS derived superoxide [90]. The effects of Hsp90 on iNOS derived O_2^- has not been examined and will be of interest in future studies.

The protein inhibitor of NOS (PIN) or dynein light chain, an 89-amino acid polypeptide isolated from a rat hippocampal cDNA library using the yeast two-hybrid system, was first identified by Jaffrey et.al and published in 1996[45]. PIN was originally reported to inhibit only the nNOS isoforms by dissociating the active nNOS homodimer. However, controversy exists as to the specificity of these results. Hemmens et al., using recombinant proteins, demonstrated that all NOS isoenzymes form contacts with and are inhibited by PIN [39]. Interestingly, this study also demonstrated that PIN inhibited uncoupled NOS NADPH oxidation suggesting PIN may also play a role in modulating NOS derived superoxide generation. In support of this hypothesis, Xia et.al performed direct measurements of the effects of PIN on NO and superoxide generation from purified nNOS using electron paramagnetic resonance spin trapping techniques. In these studies, PIN caused a dose-dependent inhibition of both NO and L-citrulline production [92]. In the absence of L-arginine, the strong $O_2^{-\bullet}$ signal that was observed from nNOS, was subsequently blocked by PIN in a dose-dependent manner [92]. Contrary to these findings, studies by Rodriguez-Crespo et al. suggest that PIN neither inhibits nNOS activity nor dissociates the nNOS dimer as originally proposed [71]. These authors suggested that PIN might function as a dynein light chain involved in nNOS axonal transport rather than as an nNOS inhibitor.

Dynamin-2 is a large GTPase that has been shown to participate in internalization events in caveolae [15, 55]. As such, Dynamin-2 resides within similar membrane compartments as eNOS and has been reported by Cao et.al to directly bind with eNOS thereby augmenting eNOS activity [15]. Dynamin and eNOS have been shown to colocalize in bovine aortic endothelial cells (BAEC) and this colocalization is predominantly distributed in the golgi membrane. Additionally, activation of endothelial cells with Ca^{2+} ionophore enhances coprecipitation of dynamin with eNOS in BAEC, suggesting the interaction between the proteins can be regulated by intracellular signals [15, 55]. These events may be involved eNOS regulation and trafficking within the vascular endothelium. However, the effects of Dyanamin on NOS catalytic activity have not yet been investigated.

Yeast two-hybrid screens have identified a novel eNOS-interacting protein (NOSIP) which associates with and has regulatory control of eNOS [21]. NOSIP is a 34-kDa protein that avidly binds to the COOH-terminal region (amino acids 366–486) of the eNOS*ox* domain [21]. The formation of the NOSIP-eNOS complex is inhibited by the scaffolding protein caveolin-1, suggesting competition between NOSIP and caveolin-1 for eNOS binding. Although NOSIP does not alter eNOS activity in vitro, it promotes translocation of eNOS from the plasma membrane to intracellular sites when coexpressed in cells, resulting in diminished calcium ionophore stimulated NO release. Thus NOSIP appears to uncouple eNOS from plasma membrane caveolae, thereby suppressing NO synthesis [21].

NOS Regulation by Phosphorylation

Phosphorylation is a major regulatory mechanism of protein function. Kinases are responsible for phosphorylation of proteins, most commonly at serine, threonine or tyrosine amino acid residues. Kinases are a large and diverse family of enzymes, and as would be expected, their actions are also highly varied. Upon phosphorylation, the activity of a protein can be increased or inhibited. In addition, cellular compartmentalization of proteins can be regulated by phosphorylation.

eNOS is phosphorylated in response to a variety of factors, including hormones, cell stress, growth factors and other cellular signaling molecules. Depending on the site of phosphorylation, eNOS activity can be increased or decreased with subsequent regulation of NOS derived NO production (figure 5). The most extensively studied site of eNOS phosphorylation is Ser^{1177} human sequence/Ser^{1179} bovine sequence.

Ser^{1177} phosphorylation rapidly increases the production of NO primarily through increasing electron flux through the reductase domain and by reducing calmodulin dissociation at low calcium levels [59]. Typically found unphosphorylated in endothelial cells, exposure to stimuli causes rapid phosphorylation of Ser^{1177}.

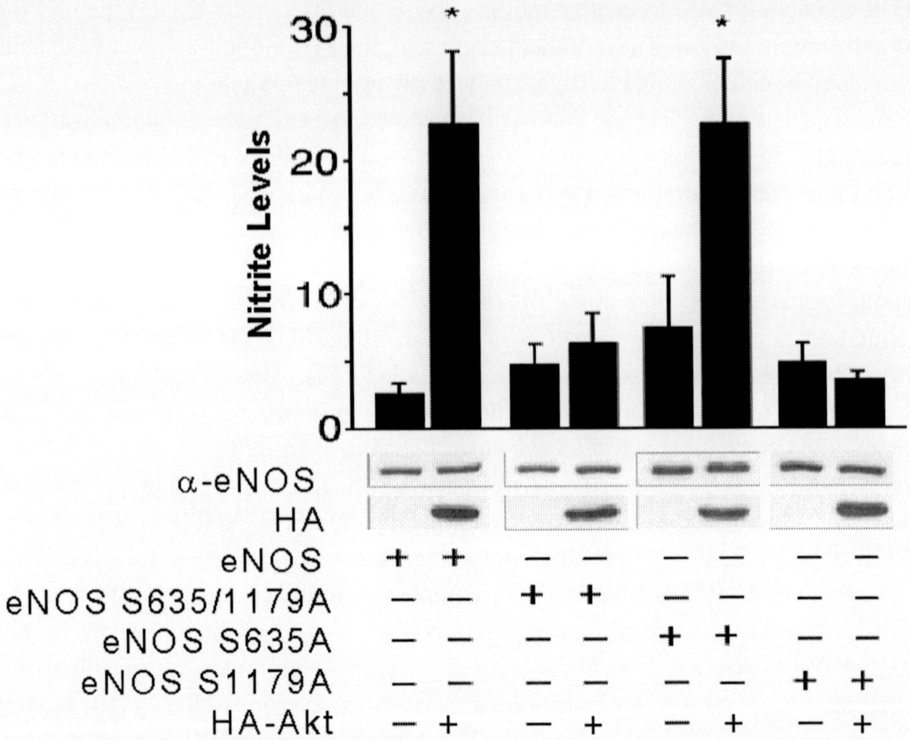

Figure 5. COS cells were transfected with plasmids for wild-type eNOS or eNOS mutants, in the absence or presence of Akt, and the expression of the proteins and production of NO (assayed as NO-2) were determined. Constructs with theS1179A mutation were not activated by Akt; the S1179D mutation resulted in a gain of function. Data are mean ± s.e.m. of 4–7 experiments; asterisks represent significant differences ($P < 0.05$).

Fluid shear stress phosphorylates Ser^{1177} [20] through both Akt/PKB [22] and PKA pathways [9]. Akt phosphorylation of Ser^{1177} is mediated through activation of phosphatidylinositol 3-kinase (PI(3)K), an upstream regulator of Akt activity. Exposure of VEGF [28], estrogen [42], sphingosine-1-phosphate [43], and hydrogen peroxide also induce Ser^{1177} phosphorylation through the PI(3)K/Akt pathway. Insulin has been found to induce Ser^{1177} phosphorylation through both the PI(3)K/Akt pathway [46] and through the AMP-activated protein kinase (AMPK) [26]. PKG has also been shown to phosphorylate Ser^{1177} [14]. Bradykinin and histamine are G protein-coupled receptor agonists, and are responsible for CaM-dependent kinase II (CaMKII) phosphorylation of Ser^{1177} [25]. CamKII Ser^{1177} phosphorylation is associated with increased intracellular Ca^{2+} levels, but PI(3)K/Akt Ser^{1177} phosphorylation derived increases in NO production are believed to be independent of intracellular Ca^{2+} levels [59].

To date, protein phosphatase 2A (PP2A) is the only phosphatase that has been shown to dephosphorylate Ser^{1177} [35] through two known mechanisms. Exposure of phorbol 12-myristate 13-acetate (PMA) has been shown to activate PP2A [62]. In addition, proteasome inhibition has been found to dephosphorylate Ser^{1177} through enhanced PP2A association with eNOS [89]. Ubiquitinated PP2A dephosphorylates Akt then is translocated to the cell membrane where it dephosphorylates eNOS and therefore NO production is decreased. A negative feedback mechanism between NO and Ser^{1177} phosphorylation exists. NO mediates cytosolic depletion of Akt through caspases resulting in decreased Ser^{1177} phosphorylation [48]. When uncoupled by depletion of BH_4, eNOS produces superoxide. Production of $O_2^{-\bullet}$ by eNOS has not been found to be regulated by Ser^{1177} phosphorylation [90] or phosphorylation at other sites on eNOS.

Thr^{495} human sequence/Thr^{497} bovine sequence is located within the calmodulin (CaM) binding domain of eNOS. Thr^{495} is constitutively phosphorylated and phosphorylation of this residue leads to a decrease in eNOS activity and decreased NO production by prevention of CaM binding [25]. PKC has been shown to be responsible for Thr^{495} phosphorylation [25, 58] through PMA [62]. Dephosphorylation of Thr^{495} is necessary for eNOS activation and CaM binding. The exact mechanism for Thr^{495} dephosphorylation is unclear. There is evidence to suggest that protein phosphatase 1 (PP1) [37, 62] PP2A [35] and calcineurin [37] are responsible for dephosphorylation of Thr^{495}. Stimuli that increase intracellular Ca^{2+} levels, such as bradykinin and histamine, are associated with alteration of Thr^{495} phosphorylation, but increases in NO production are not a result of only increased association of Ca^{2+} with CaM [24]. In addition, there is evidence that endothelial nitric-oxide synthase (eNOS) is regulated by reciprocal dephosphorylation of Thr^{497} and phosphorylation of Ser^{1179}. Using Thr^{497A} and Thr^{497A}/Ser^{1179D} mutants, Lin et.al demonstrated that Thr^{497} phosphorylation status plays a critical role in the regulation of eNOS, with both mutants generating 2-3 times more superoxide anion than WT eNOS, and both basal and stimulated interactions of Thr^{497A}/S^{1179D} eNOS with Hsp90 were reduced in co-immunoprecipitation experiments. Thus, the phosphorylation/ dephosphorylation of Thr^{497} may be an intrinsic switch mechanism that determines whether eNOS generates NO versus superoxide in cells. These results demonstrate an interrelationship between eNOS phosphorylation sites and association with Akt and heat shock protein 90 (hsp90) [3] and suggests that multiple phosphorylation sites of eNOS must act in concert to maximize production of NO or to fine tune production of NO.

Additional sites of eNOS phosphorylation have been identified but are poorly understood. Ser^{114} human sequence/Ser^{116} bovine sequence is located in the oxygenase domain of eNOS. Ser^{114} is basally phosphorylated and causes inhibition of eNOS similar to that observed with Thr^{495}. Fluid shear stress [29] and lysophosphatidic acid [24] have been shown to phosphorylate Ser^{114}, but the mechanism of phosphorylation is unclear. *Gallis et al* speculate that a proline-directed protein kinase is responsible for phosphorylation of Ser^{114} due to a neighboring proline residue. VEGF activates Ser^{114} dephosphorylation that is inhibited by cyclosporin A, suggesting that calcineurin may be responsible for dephosphorylation of Ser^{114}; dephosphorylation of Ser^{114} results in increased production of NO by eNOS [47]. This action of VEGF is in contrast to its Akt mediated phosphorylation of Ser^{1177} [28], supporting a complex relationship between eNOS phosphorylation sites. The proximity of Ser^{114} to the Zn^{2+} and BH_4 binding sites of eNOS suggest a possible role in eNOS dimerization and $O_2^{-\bullet}$ production due to eNOS uncoupling, but this is yet to be shown.

Ser^{615} human sequence/Ser^{617} bovine sequence and Ser^{633} human sequence/Ser^{635} bovine sequence have also been shown to be phosphorylated. Both residues are located in the CaM auto-inhibitory loop within the FMN binding domain of eNOS. Ser^{615} is rapidly phosphorylated upon exposure to stimulus resulting in increased eNOS activity by increased Ca^{2+}/CaM association, Ser^{633} is then phosphorylated and eNOS activity is increased to a level similar to that caused by Ser^{1177} phosphorylation [63]. Bradykinin, ATP, and VEGF are responsible for phosphorylation of both sites [63] and fluid shear stress causes Ser^{633} phosphorylation via PKA [31]. PKG has also been associated with Ser^{633} phosphorylation [14].

Tyrosine residues have been reported to be phosphorylated in primary cultured endothelial cells [23], but in passaged endothelial cell cultures tyrosine phosphorylation has not been detected [20, 61]. Tyrosine phosphorylation may be involved in the association of eNOS with scaffolding and regulatory proteins [23, 38].

There is evidence to suggest that phosphorylation of eNOS can result in cellular compartmentalization of eNOS. Bradykinin induced phosphorylation has been found to translocate eNOS from the plasma membrane to the cytosol [61], but this is not a universal finding [70]. Bradykinin stimulated tyrosine phosphorylation of eNOS appears to target eNOS to the cytoskeleton [88]. Other research has found that hydrogen peroxide stimulated tyrosine phosphorylation of eNOS causes association with caveolin-1 [31]. Ser^{1177} phosphorylated eNOS has been found to be active in two distinct subcellular locations, caveolin-1-enriched plasma membranes and in the membrane of the Golgi apparatus [27]. The effect of eNOS phosphorylation on translocation is poorly understood, but it is clear that P-eNOS is found and is active in multiple locations within the cell.

NOS Regulation by Endogenous Methylarginines

In addition to the molecular mechanisms previously described for NOS regulation including susbtrate/cofactor bioavailability, protein-protein interaction and post-translational modifications, NOS activity is also influenced by the methylated arginine derivatives,

ADMA and L-NMMA. The biological significance of these guanidino-methylated arginine derivitives has been known since the inhibitory actions of N^G-methyl-L-arginine (L-NMMA) on macrophage induced cytotoxicity were first demonstrated [41]. It was subsequently realized that these effects were mediated through inhibition of nitric oxide (NO) release. In this regard, intracellular accumulation of ADMA has been implicated as causative factor in the development of a variety of pathological conditions through the impairment of the normal physiological functions of NO [8, 56, 81, 82].

ADMA and L-NMMA are endogenous inhibitors of NOS and are derived from the proteolysis of methylated arginine residues on various proteins. The methylation is carried out by a group of enzymes referred to as protein-arginine methyl transferase's (PRMT's) [50]. Subsequent proteolysis of proteins containing methylarginine groups leads to the release of free methylarginine into the cytoplasm and if sufficient levels are reached NO production from NOS would be inhibited [54, 82]. To date, six different isoforms of the enzyme have been identified with each subtype exhibiting various levels of activity, substrate specificity and tissue distribution [50, 78]. These methylarginines are subsequently degraded by the enzyme dimethylarginine dimethylamino-hydrolase (DDAH) which catalyzes the hydrolysis of ADMA or L-NMMA to L-citrulline and dimethylamine or monomethyl amine [1, 44, 79].

In support of the role of ADMA and L-NMMA as endogenous inhibitors of NOS function, several studies have demonstrated that L-arg supplementation enhances endothelial function through increased NO bioavailability [2, 5, 33, 91]. This effect could be due to increased substrate availability. However, when one takes into consideration that the intracellular levels of L-arg are typically more than 50 times higher than the K_m of the enzyme, increased NO generation would not be expected with L-arg supplementation [18]. This observation of increased NO bioavailability with L-arg supplementation has been termed the "arginine paradox". One hypothesis put forth to explain this paradox is that L-arg supplementation overcomes the endogenous inhibitory actions of cellular methylarginines [4, 19, 50, 80, 81]. In support, by directly measured NO production from endothelial cells and plotting these values as a function of the experimentally measured intracellular ADMA/L-arg or L-NMMA/L- arg ratio, the cellular kinetics of ADMA and L-NMMA inhibition were generated. When these values were plotted against the curve for competitive inhibition, a strong correlation was observed demonstrating that cellular inhibition closely follows what would be predicted from the K_i [16].

Based on the K_i's for ADMA and L-NMMA, it was then predicted that under normal physiological conditions endogenous methylarginine levels would inhibit NO production by ~ 10-20 %. These results demonstrate the importance of ADMA and L-NMMA in NOS regulation and support a role for methylarginines in the "arginine paradox." In addition to these basal effects, methylarginine levels are known to increase in a variety of pathological settings. There are numerous studies demonstrating that under pathological conditions methylarginine levels are increased by 3 to 9 fold [6, 7, 51, 52, 56, 74, 77, 81], which based on enzyme kinetic data, would be expected to inhibit NO generation from NOS by 30 - 70 % [16]. The equation for inhibition can thus be extrapolated to studies aimed at measuring cellular methylarginine levels in disease and allow for the prediction of the extent of NOS inhibition.

In vitro studies have demonstrated that the methylarginines, ADMA and L-NMMA, dose dependently inhibit eNOS-derived NO production from both isolated enzyme and cellular systems (figures 6-7) [16, 18]. The results obtained from these studies translate to the whole organ as it has been demonstrated that methylarginines modulate vascular relaxation [16, 49]. Measurements of intracellular methylarginine levels suggest that normal physiological concentrations would only have a modest effect on NOS activity. However, under pathological conditions such as vascular injury and restenosis, where endogenous methylarginines are elevated, methylarginine-mediated NOS inhibition with secondary endothelial dysfunction has been observed [16, 49]. These studies demonstrated that cellular methylarginines are elevated under pathological conditions and reach concentrations sufficient to significanty inhibit NO production. Together, these results represent a major step forward in our understanding of the regulation, impact, and role of methylarginines, in disease.

In addition to inhibition of NO generation, methylarginines may have other important effects on NOS function. Several groups have reported that when cells are depleted of the NOS substrate L-arginine (L-arg) or the cofactor tetrahydrobiopterin (BH_4), NOS switches from production of NO to the $O_2^{-\bullet}$ anion radical, $O_2^{-\bullet}$ [68, 69, 72, 84, 85, 93, 95, 96].

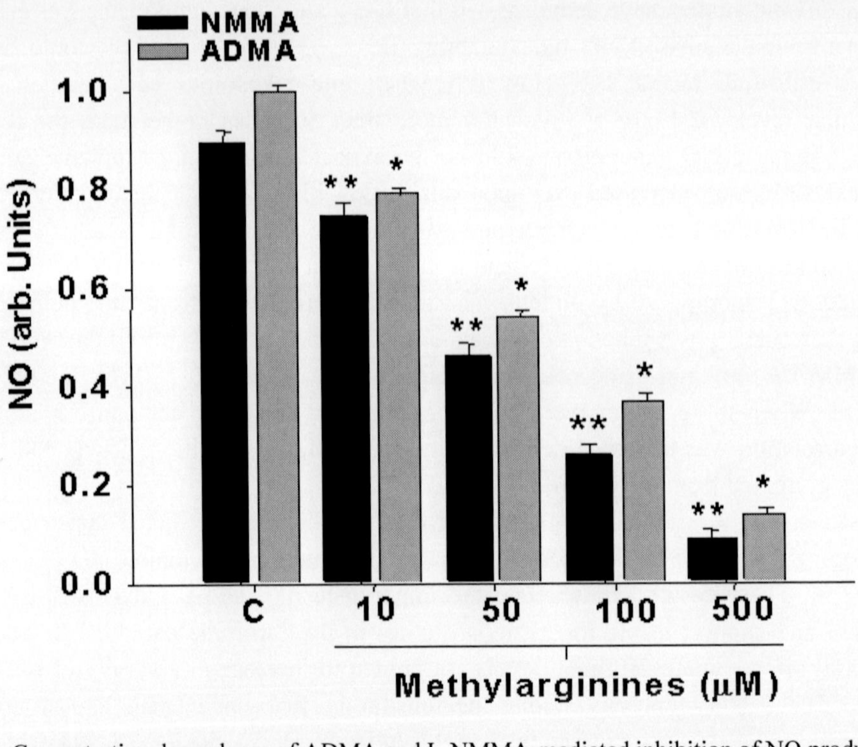

Figure 6. Concentration dependence of ADMA and L-NMMA-mediated inhibition of NO production from purified eNOS. NO generation from eNOS (10 µg/mL) was measured by EPR spin trapping with the Fe-MGD complex. effects of ADMA on NO generation. B, effects of L-NMMA on NO generation. Dose dependent inhibition of NO production was seen with concentrations of ADMA and L-NMMA ranging from 10-500 µM. Results are the mean ± S.D. of 4-5 measurements. * Significantly different at $p<0.05$ as compared to control.

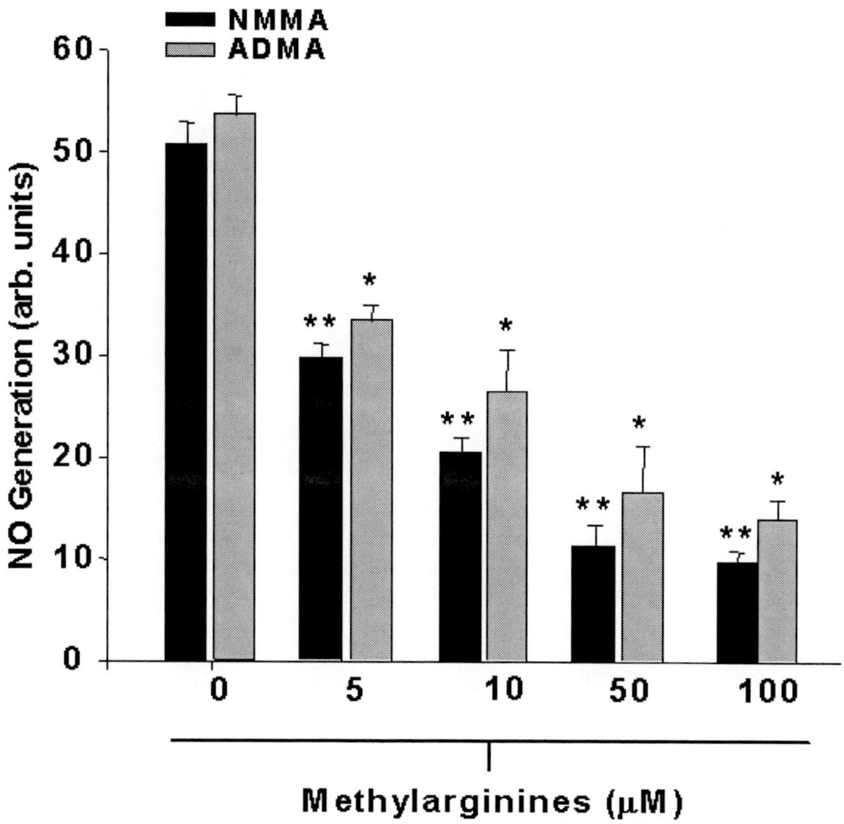

Figure 7. Effects of ADMA and L-NMMA on endothelial cell NO production. NO generation from calcium ionophore A23187 (1 μM) stimulated bovine aortic endothelial cells (1 X 106) was measured by EPR spin trapping with the Fe-MGD complex. The effects of ADMA (5-100 μM) and L-NMMA (5-100 μM) on NO generation from BAECs. Dose dependent inhibition of NO production was seen with concentrations of ADMA and L-NMMA ranging from 5-100 μM. Results are the mean ± S.D. of 4 measurements. *Significantly different at $p<0.05$ as compared to control.

In the absence of either of these requisite substrates or co-factors, NOS mediated NADPH oxidation is uncoupled from NO synthesis and results in the reduction of O_2 to form $O_2^{-\bullet}$ [64, 68, 69, 93]. Superoxide exerts cellular effects on signaling and function that are quite different and often opposite to those of NO. Thus, $O_2^{-\bullet}$ is another very important NOS product and it appears that its production is also regulated by methylarginines.

In this regard, recent findings have demonstrated that the endogenous L-arg derivatives, ADMA and L-NMMA, play an important role in the regulation of both NO and $O_2^{-\bullet}$ release from NOS. Results have shown that ADMA and L-NMMA differentially modulate NOS derived $O_2^{-\bullet}$ and these effects were BH_4 dependent [17].

In the presence of BH_4, ADMA selectively and dose dependently inhibited $O_2^{-\bullet}$ generation from the nNOS, while L-NMMA had no effect despite their structural similarities. However, when NOS was depleted of BH_4, ADMA no longer had an effect on $O_2^{-\bullet}$, while L-NMMA dose dependently enhanced $O_2^{-\bullet}$ production from the enzyme [17]. These results support the role of methylarginines as important regulatory molecules in the regulation of nNOS-mediated $O_2^{-\bullet}$ and NO generation.Similar observations have been observed in regard to the effects of methylarginines on eNOS derived $O_2^{-\bullet}$. When eNOS is uncoupled through depletion of the cofactor BH_4, the NOS derived superoxide that is observed is significantly enhanced in the presence of both ADMA and L-NMMA (figures 8-9).

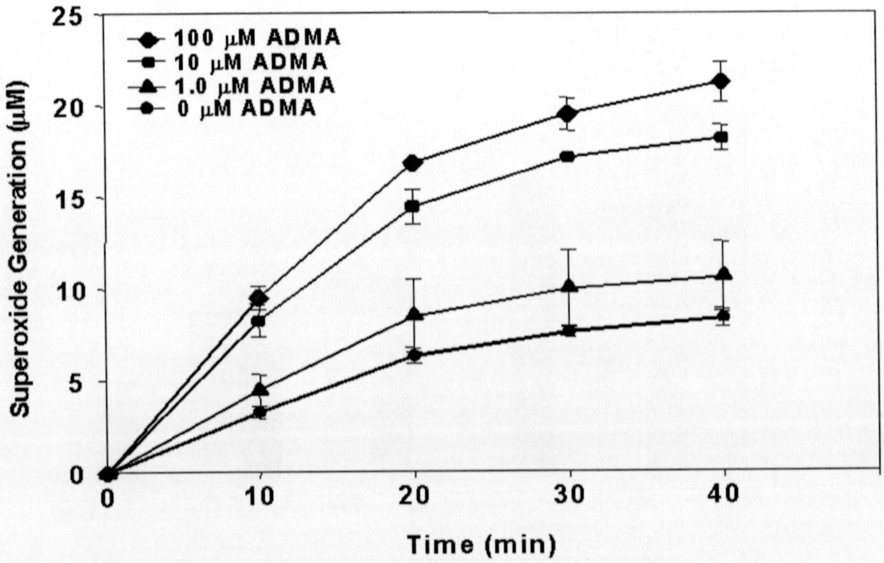

Figure 8. Effects of ADMA on eNOS derived superoxide production. Superoxide generation from BH4-free eNOS (10 µg/mL) was measured by EPR spin trapping with the DEPMPO. ADMA (1-100 µM) dose dependently increased superoxide production with a greater than two fold increase observed at 100 µM L-NMMA.

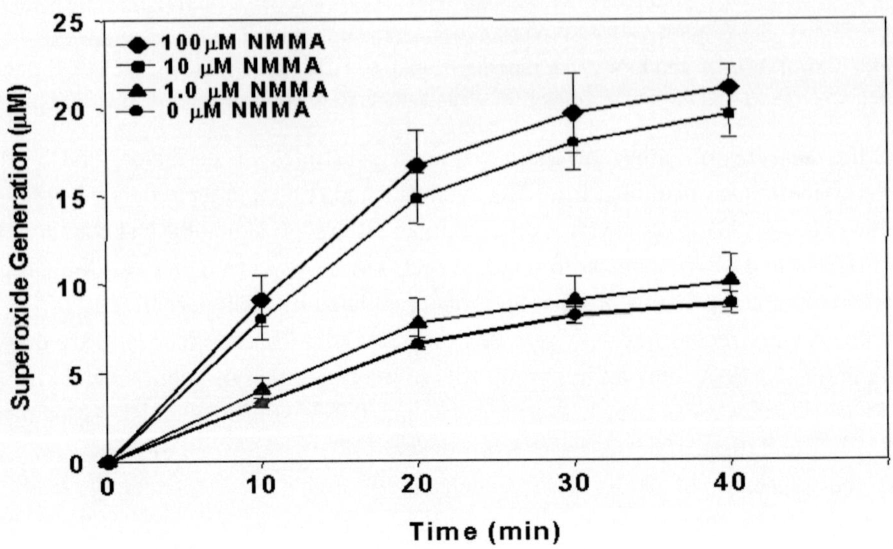

Figure 9. Effects of L-NMMA on eNOS derived superoxide production. Superoxide generation from BH4-free eNOS (10 µg/mL) was measured by EPR spin trapping with the DEPMPO. L-NMMA (1-100

μM) dose dependently increased superoxide production with a greater than two fold increase observed at 100 μM L-NMMA.

These results differ form those observed with nNOS, wherein, only L-NMMA had a potentiating effect on NOS derived $O_2^{-\bullet}$. The mechanism for this methylargininie mediated enhancement of eNOS derived superoxide involves the heme center of the oxygenase domain. NADPH consumption studies demonstrate that binding of L-NMMA or ADMA to BH_4-free eNOS increases NADPH consumption rate and studies of midpoint potential reveal that methylarginine binding to NOS increases the heme reduction potential and thereby facilitates electron transfer through the heme, resulting in increased $O_2^{-\bullet}$ generation from the enzyme. These results would suggest that under conditions of decreased BH_4 bioavailabiltiy, as occurs under a variety of pathological conditions, methylarginines would shift the balance of NO and $O_2^{-\bullet}$ production from NOS to a largely oxidant producing enzyme.

In support of this, prior cellular studies provide evidence suggesting that L-NMMA shifts the balance of NO and $O_2^{-\bullet}$ generation and enhances $O_2^{-\bullet}$ in cells. It has previously been observed that while L-NMMA is as effective as ADMA in inhibiting NO generation, it is less effective in preventing excitotoxic neuronal death [18]. Another interesting study has previously reported that L-NMMA induced enhanced $O_2^{-\bullet}$ generation in cardiac myocytes subjected to ischemia-reperfusion injury [83]. These results were previously unexplained however, ischemia-reperfusion has been shown to result in BH_4 oxidation, so the increased $O_2^{-\bullet}$ production induced by L-NMMA is likely NOS-derived and BH_4 dependent. In the future, further studies will be needed to characterize the effects of endogenous levels of methyl-arginines on $O_2^{-\bullet}$production in cells and tissues, as well as the modulatory action of BH_4 on methylarginine-NOS interactions in *in vivo* models of physiology and disease.

Because of the importance of NOS and the involvement of NO and $^{\cdot}O_2^-$ in a variety of neurological and cardiovascular disorders, methylarginines may play a critical role in modulating the balance of NO and oxygen radical formation, as well as oxidant injury in these diseases.

Conclusion

The regulation of the NOS isozymes is a multifactorial process that involves mechanisms acting through diverse pathways and include but are not limited to 1.) the formation of NOS signaling complexes, 2.) cellular localization and trafficking, 3.) regulation of susbstrate/cofactor bioavailability and 4.) phosphorylation. Although the data presented in this chapter have elucidated a number of the molecular mechanisms involved in NOS regulation through the use of in-vitro models and cell systems, many question remain as to how these pathways interact and how are these regulatory steps affected in disease processes involving NOS dysfunction.

References

[1] Achan V, Tran CT, Arrigoni F, Whitley GS, Leiper JM, and Vallance P. all-trans-Retinoic acid increases nitric oxide synthesis by endothelial cells: a role for the induction of dimethylarginine dimethylaminohydrolase. *Circ. Res.* 90: 764-769, 2002.

[2] Aji W, Ravalli S, Szabolcs M, Jiang XC, Sciacca RR, Michler RE, and Cannon PJ. L-arginine prevents xanthoma development and inhibits atherosclerosis in LDL receptor knockout mice. *Circulation* 95: 430-437., 1997.

[3] Bauer PM, Fulton D, Boo YC, Sorescu GP, Kemp BE, Jo H, and Sessa WC. Compensatory phosphorylation and protein-protein interactions revealed by loss of function and gain of function mutants of multiple serine phosphorylation sites in endothelial nitric-oxide synthase. *J. Biol. Chem.* 278: 14841-14849, 2003.

[4] Boger RH. Asymmetric dimethylarginine, an endogenous inhibitor of nitric oxide synthase, explains the andquot;L-arginine paradoxandquot; and acts as a novel cardiovascular risk factor. *J. Nutr.* 134: 2842S-2847S; discussion 2853S, 2004.

[5] Boger RH, Bode-Boger SM, Brandes RP, Phivthong-ngam L, Bohme M, Nafe R, Mugge A, and Frolich JC. Dietary L-arginine reduces the progression of atherosclerosis in cholesterol-fed rabbits: comparison with lovastatin. *Circulation* 96: 1282-1290., 1997.

[6] Boger RH, Bode-Boger SM, Sydow K, Heistad DD, and Lentz SR. Plasma concentration of asymmetric dimethylarginine, an endogenous inhibitor of nitric oxide synthase, is elevated in monkeys with hyperhomocyst(e)inemia or hypercholesterolemia. *Arterioscler. Thromb. Vasc. Biol.* 20: 1557-1564., 2000.

[7] Boger RH, Bode-Boger SM, Szuba A, Tsao PS, Chan JR, Tangphao O, Blaschke TF, and Cooke JP. Asymmetric dimethylarginine (ADMA): a novel risk factor for endothelial dysfunction: its role in hypercholesterolemia. *Circulation* 98: 1842-1847., 1998.

[8] Boger RH, Maas R, Schulze F, and Schwedhelm E. Elevated levels of asymmetric dimethylarginine (ADMA) as a marker of cardiovascular disease and mortality. *Clin. Chem. Lab. Med.* 43: 1124-1129, 2005.

[9] Boo YC, Sorescu G, Boyd N, Shiojima I, Walsh K, Du J, and Jo H. Shear stress stimulates phosphorylation of endothelial nitric-oxide synthase at Ser1179 by Akt-independent mechanisms: role of protein kinase A. *J. Biol. Chem.* 277: 3388-3396, 2002.

[10] Bredt DS, Hwang PM, Glatt CE, Lowenstein C, Reed RR, and Snyder SH. Cloned and expressed nitric oxide synthase structurally resembles cytochrome P-450 reductase. *Nature* 351: 714-718, 1991.

[11] Bredt DS and Snyder SH. Isolation of nitric oxide synthetase, a calmodulin-requiring enzyme. *Proc. Natl. Acad. Sci. USA* 87: 682-685, 1990.

[12] Bredt DS and Snyder SH. Nitric oxide mediates glutamate-linked enhancement of cGMP levels in the cerebellum. *Proc. Natl. Acad. Sci USA* 86: 9030-9033, 1989.

[13] Busse R and Mulsch A. Calcium-dependent nitric oxide synthesis in endothelial cytosol is mediated by calmodulin. *FEBS Lett.* 265: 133-136, 1990.

[14] Butt E, Bernhardt M, Smolenski A, Kotsonis P, Frohlich LG, Sickmann A, Meyer HE, Lohmann SM, and Schmidt HH. Endothelial nitric-oxide synthase (type III) is activated and becomes calcium independent upon phosphorylation by cyclic nucleotide-dependent protein kinases. *J. Biol. Chem.* 275: 5179-5187, 2000.

[15] Cao S, Yao J, McCabe TJ, Yao Q, Katusic ZS, Sessa WC, and Shah V. Direct interaction between endothelial nitric-oxide synthase and dynamin-2. Implications for nitric-oxide synthase function. *J. Biol. Chem.* 276: 14249-14256, 2001.

[16] Cardounel AJ, Cui H, Samouilov A, Johnson W, Kearns P, Tsai AL, Berka V, and Zweier JL. Evidence for the Pathophysiological Role of Endogenous Methylarginines in Regulation of Endothelial NO Production and Vascular Function. *J. Biol. Chem.* 282: 879-887, 2007.

[17] Cardounel AJ, Xia Y, and Zweier JL. Endogenous methylarginines modulate superoxide as well as nitric oxide generation from neuronal nitric-oxide synthase: differences in the effects of monomethyl- and dimethylarginines in the presence and absence of tetrahydrobiopterin. *J. Biol. Chem.* 280: 7540-7549, 2005.

[18] Cardounel AJ and Zweier JL. Endogenous methylarginines regulate neuronal nitric-oxide synthase and prevent excitotoxic injury. *J. Biol. Chem.* 277: 33995-34002. Epub 32002 Jun 33928., 2002.

[19] Cooke JP and Oka RK. Atherogenesis and the arginine hypothesis. *Curr. Atheroscler. Rep.* 3: 252-259., 2001.

[20] Corson MA, James NL, Latta SE, Nerem RM, Berk BC, and Harrison DG. Phosphorylation of endothelial nitric oxide synthase in response to fluid shear stress. *Circ. Res.* 79: 984-991, 1996.

[21] Dedio J, Konig P, Wohlfart P, Schroeder C, Kummer W, and Muller-Esterl W. NOSIP, a novel modulator of endothelial nitric oxide synthase activity. *Faseb J.* 15: 79-89, 2001.

[22] Dimmeler S, Fleming I, Fisslthaler B, Hermann C, Busse R, and Zeiher AM. Activation of nitric oxide synthase in endothelial cells by Akt-dependent phosphorylation. *Nature* 399: 601-605, 1999.

[23] Fleming I, Bauersachs J, Fisslthaler B, and Busse R. Ca2+-independent activation of the endothelial nitric oxide synthase in response to tyrosine phosphatase inhibitors and fluid shear stress. *Circ. Res.* 82: 686-695, 1998.

[24] Fleming I and Busse R. Molecular mechanisms involved in the regulation of the endothelial nitric oxide synthase. *Am. J. Physiol. Regul. Integr. Comp. Physiol.* 284: R1-12, 2003.

[25] Fleming I, Fisslthaler B, Dimmeler S, Kemp BE, and Busse R. Phosphorylation of Thr(495) regulates Ca(2+)/calmodulin-dependent endothelial nitric oxide synthase activity. *Circ. Res.* 88: E68-75, 2001.

[26] Fleming I, Schulz C, Fichtlscherer B, Kemp BE, Fisslthaler B, and Busse R. AMP-activated protein kinase (AMPK) regulates the insulin-induced activation of the nitric oxide synthase in human platelets. *Thromb. Haemost.* 90: 863-871, 2003.

[27] Fulton D, Fontana J, Sowa G, Gratton JP, Lin M, Li KX, Michell B, Kemp BE, Rodman D, and Sessa WC. Localization of endothelial nitric-oxide synthase

phosphorylated on serine 1179 and nitric oxide in Golgi and plasma membrane defines the existence of two pools of active enzyme. *J. Biol. Chem.* 277: 4277-4284, 2002.

[28] Fulton D, Gratton JP, McCabe TJ, Fontana J, Fujio Y, Walsh K, Franke TF, Papapetropoulos A, and Sessa WC. Regulation of endothelium-derived nitric oxide production by the protein kinase Akt. *Nature* 399: 597-601, 1999.

[29] Gallis B, Corthals GL, Goodlett DR, Ueba H, Kim F, Presnell SR, Figeys D, Harrison DG, Berk BC, Aebersold R, and Corson MA. Identification of flow-dependent endothelial nitric-oxide synthase phosphorylation sites by mass spectrometry and regulation of phosphorylation and nitric oxide production by the phosphatidylinositol 3-kinase inhibitor LY294002. *J. Biol. Chem.* 274: 30101-30108, 1999.

[30] Garcia-Cardena G, Fan R, Shah V, Sorrentino R, Cirino G, Papapetropoulos A, and Sessa WC. Dynamic activation of endothelial nitric oxide synthase by Hsp90. *Nature* 392: 821-824, 1998.

[31] Garcia-Cardena G, Fan R, Stern DF, Liu J, and Sessa WC. Endothelial nitric oxide synthase is regulated by tyrosine phosphorylation and interacts with caveolin-1. *J. Biol. Chem.* 271: 27237-27240, 1996.

[32] Garcia-Cardena G, Martasek P, Masters BS, Skidd PM, Couet J, Li S, Lisanti MP, and Sessa WC. Dissecting the interaction between nitric oxide synthase (NOS) and caveolin. Functional significance of the nos caveolin binding domain in vivo. *J. Biol. Chem.* 272: 25437-25440, 1997.

[33] Girerd XJ, Hirsch AT, Cooke JP, Dzau VJ, and Creager MA. L-arginine augments endothelium-dependent vasodilation in cholesterol-fed rabbits. *Circ. Res.* 67: 1301-1308, 1990.

[34] Gratton JP, Fontana J, O'Connor DS, Garcia-Cardena G, McCabe TJ, and Sessa WC. Reconstitution of an endothelial nitric-oxide synthase (eNOS), hsp90, and caveolin-1 complex in vitro. Evidence that hsp90 facilitates calmodulin stimulated displacement of eNOS from caveolin-1. *J. Biol. Chem.* 275: 22268-22272, 2000.

[35] Greif DM, Kou R, and Michel T. Site-specific dephosphorylation of endothelial nitric oxide synthase by protein phosphatase 2A: evidence for crosstalk between phosphorylation sites. *Biochemistry* 41: 15845-15853, 2002.

[36] Gruetter CA, Barry BK, McNamara DB, Gruetter DY, Kadowitz PJ, and Ignarro L. Relaxation of bovine coronary artery and activation of coronary arterial guanylate cyclase by nitric oxide, nitroprusside and a carcinogenic nitrosoamine. *J Cyclic Nucleotide Res.* 5: 211-224, 1979.

[37] Harris MB, Ju H, Venema VJ, Liang H, Zou R, Michell BJ, Chen ZP, Kemp BE, and Venema RC. Reciprocal phosphorylation and regulation of endothelial nitric-oxide synthase in response to bradykinin stimulation. *J. Biol. Chem.* 276: 16587-16591, 2001.

[38] Hellermann GR, Flam BR, Eichler DC, and Solomonson LP. Stimulation of receptor-mediated nitric oxide production by vanadate. *Arterioscler. Thromb. Vasc. Biol.* 20: 2045-2050, 2000.

[39] Hemmens B, Woschitz S, Pitters E, Klosch B, Volker C, Schmidt K, and Mayer B. The protein inhibitor of neuronal nitric oxide synthase (PIN): characterization of its action on pure nitric oxide synthases. *FEBS Lett.* 430: 397-400, 1998.

[40] Hibbs JB, Jr., Taintor RR, Vavrin Z, and Rachlin EM. Nitric oxide: a cytotoxic activated macrophage effector molecule. *Biochem. Biophys. Res. Commun.* 157: 87-94, 1988.

[41] Hibbs JB, Jr., Vavrin Z, and Taintor RR. L-arginine is required for expression of the activated macrophage effector mechanism causing selective metabolic inhibition in target cells. *J. Immunol.* 138: 550-565, 1987.

[42] Hisamoto K, Ohmichi M, Kurachi H, Hayakawa J, Kanda Y, Nishio Y, Adachi K, Tasaka K, Miyoshi E, Fujiwara N, Taniguchi N, and Murata Y. Estrogen induces the Akt-dependent activation of endothelial nitric-oxide synthase in vascular endothelial cells. *J. Biol. Chem.* 276: 3459-3467, 2001.

[43] Igarashi J and Michel T. Sphingosine 1-phosphate and isoform-specific activation of phosphoinositide 3-kinase beta. Evidence for divergence and convergence of receptor-regulated endothelial nitric-oxide synthase signaling pathways. *J. Biol. Chem.* 276: 36281-36288, 2001.

[44] Ito A, Tsao PS, Adimoolam S, Kimoto M, Ogawa T, and Cooke JP. Novel mechanism for endothelial dysfunction: dysregulation of dimethylarginine dimethylaminohydrolase. *Circulation* 99: 3092-3095, 1999.

[45] Jaffrey SR and Snyder SH. PIN: an associated protein inhibitor of neuronal nitric oxide synthase. *Science* 274: 774-777, 1996.

[46] Kim F, Gallis B, and Corson MA. TNF-alpha inhibits flow and insulin signaling leading to NO production in aortic endothelial cells. *Am. J. Physiol. Cell Physiol.* 280: C1057-1065, 2001.

[47] Kou R, Greif D, and Michel T. Dephosphorylation of endothelial nitric-oxide synthase by vascular endothelial growth factor. Implications for the vascular responses to cyclosporin A. *J. Biol. Chem.* 277: 29669-29673, 2002.

[48] Kou R PP, and Michel T. Phosphorylation of the endothelial isoform of nitric oxide synthase at serine 116: Identification of a novel path for eNOS regulation by lysophosphatidic acid. *Circulation* 104: 509, 2001.

[49] Leiper J, Nandi M, Torondel B, Murray-Rust J, Malaki M, O'Hara B, Rossiter S, Anthony S, Madhani M, Selwood D, Smith C, Wojciak-Stothard B, Rudiger A, Stidwill R, McDonald NQ, and Vallance P. Disruption of methylarginine metabolism impairs vascular homeostasis. *Nat. Med.* 13: 198-203, 2007.

[50] Leiper J and Vallance P. Biological significance of endogenous methylarginines that inhibit nitric oxide synthases. *Cardiovasc Res.* 43: 542-548., 1999.

[51] Lenzen H, Tsikas D, and Boger RH. Asymmetric dimethylarginine (ADMA) and the risk for coronary heart disease: the multicenter CARDIAC study. *Eur. J. Clin. Pharmacol.* 62 Suppl 13: 45-49, 2006.

[52] Lin KY, Ito A, Asagami T, Tsao PS, Adimoolam S, Kimoto M, Tsuji H, Reaven GM, and Cooke JP. Impaired nitric oxide synthase pathway in diabetes mellitus: role of asymmetric dimethylarginine and dimethylarginine dimethylaminohydrolase. *Circulation* 106: 987-992., 2002.

[53] Luckhoff A, Pohl U, Mulsch A, and Busse R. Differential role of extra- and intracellular calcium in the release of EDRF and prostacyclin from cultured endothelial cells. *Br. J. Pharmacol* .95: 189-196, 1988.

[54] MacAllister RJ, Fickling SA, Whitley GS, and Vallance P. Metabolism of methylarginines by human vasculature; implications for the regulation of nitric oxide synthesis. *Br. J. Pharmacol.* 112: 43-48, 1994.

[55] Maniatis NA, Brovkovych V, Allen SE, John TA, Shajahan AN, Tiruppathi C, Vogel SM, Skidgel RA, Malik AB, and Minshall RD. Novel mechanism of endothelial nitric oxide synthase activation mediated by caveolae internalization in endothelial cells. *Circ. Res.* 99: 870-877, 2006.

[56] Marliss EB, Chevalier S, Gougeon R, Morais JA, Lamarche M, Adegoke OA, and Wu G. Elevations of plasma methylarginines in obesity and ageing are related to insulin sensitivity and rates of protein turnover. *Diabetologia* 49: 351-359, 2006.

[57] Masters BS, McMillan K, Sheta EA, Nishimura JS, Roman LJ, and Martasek P. Neuronal nitric oxide synthase, a modular enzyme formed by convergent evolution: structure studies of a cysteine thiolate-liganded heme protein that hydroxylates L-arginine to produce NO. as a cellular signal. *Faseb J.* 10: 552-558, 1996.

[58] Matsubara M, Titani K, and Taniguchi H. Interaction of calmodulin-binding domain peptides of nitric oxide synthase with membrane phospholipids: regulation by protein phosphorylation and Ca(2+)-calmodulin. *Biochemistry* 35: 14651-14658, 1996.

[59] McCabe TJ, Fulton D, Roman LJ, and Sessa WC. Enhanced electron flux and reduced calmodulin dissociation may explain andquot;calcium-independentandquot; eNOS activation by phosphorylation. *J. Biol. Chem.* 275: 6123-6128, 2000.

[60] Michel JB, Feron O, Sacks D, and Michel T. Reciprocal regulation of endothelial nitric-oxide synthase by Ca2+-calmodulin and caveolin. *J Biol Chem* 272: 15583-15586, 1997.

[61] Michel T, Li GK, and Busconi L. Phosphorylation and subcellular translocation of endothelial nitric oxide synthase. *Proc. Natl. Acad. Sci. USA* 90: 6252-6256, 1993.

[62] Michell BJ, Chen Z, Tiganis T, Stapleton D, Katsis F, Power DA, Sim AT, and Kemp BE. Coordinated control of endothelial nitric-oxide synthase phosphorylation by protein kinase C and the cAMP-dependent protein kinase. *J. Biol. Chem.* 276: 17625-17628, 2001.

[63] Michell BJ, Harris MB, Chen ZP, Ju H, Venema VJ, Blackstone MA, Huang W, Venema RC, and Kemp BE. Identification of regulatory sites of phosphorylation of the bovine endothelial nitric-oxide synthase at serine 617 and serine 635. *J. Biol. Chem.* 277: 42344-42351, 2002.

[64] Nishida CR and Ortiz de Montellano PR. Electron transfer and catalytic activity of nitric oxide synthases. Chimeric constructs of the neuronal, inducible, and endothelial isoforms. *J. Biol. Chem.* 273: 5566-5571., 1998.

[65] Palmer RM, Ferrige AG, and Moncada S. Nitric oxide release accounts for the biological activity of endothelium-derived relaxing factor. *Nature* 327: 524-526., 1987.

[66] Palmer RM, Ferrige AG, and Moncada S. Nitric oxide release accounts for the biological activity of endothelium-derived relaxing factor. *Nature* 327: 524-526, 1987.

[67] Palmer RM, Rees DD, Ashton DS, and Moncada S. L-arginine is the physiological precursor for the formation of nitric oxide in endothelium-dependent relaxation. *Biochem. Biophys. Res. Commun.* 153: 1251-1256, 1988.

[68] Pou S, Keaton L, Surichamorn W, and Rosen GM. Mechanism of superoxide generation by neuronal nitric-oxide synthase. *J. Biol. Chem.* 274: 9573-9580., 1999.

[69] Pou S, Pou WS, Bredt DS, Snyder SH, and Rosen GM. Generation of superoxide by purified brain nitric oxide synthase. *J. Biol. Chem.* 267: 24173-24176., 1992.

[70] Rizzo V, McIntosh DP, Oh P, and Schnitzer JE. In situ flow activates endothelial nitric oxide synthase in luminal caveolae of endothelium with rapid caveolin dissociation and calmodulin association. *J. Biol. Chem.* 273: 34724-34729, 1998.

[71] Rodriguez-Crespo I, Straub W, Gavilanes F, and Ortiz de Montellano PR. Binding of dynein light chain (PIN) to neuronal nitric oxide synthase in the absence of inhibition. *Arch. Biochem. Biophys*. 359: 297-304, 1998.

[72] Rosen GM, Tsai P, Weaver J, Porasuphatana S, Roman LJ, Starkov AA, Fiskum G, and Pou S. The role of tetrahydrobiopterin in the regulation of neuronal nitric-oxide synthase-generated superoxide. *J. Biol. Chem.* 277: 40275-40280. Epub 42002 Aug 40214., 2002.

[73] Schmidt HH and Murad F. Purification and characterization of a human NO synthase. *Biochem Biophys Res Commun* 181: 1372-1377, 1991.

[74] Schnabel R, Blankenberg S, Lubos E, Lackner KJ, Rupprecht HJ, Espinola-Klein C, Jachmann N, Post F, Peetz D, Bickel C, Cambien F, Tiret L, and Munzel T. Asymmetric dimethylarginine and the risk of cardiovascular events and death in patients with coronary artery disease: results from the AtheroGene Study. *Circ. Res.* 97: e53-59, 2005.

[75] Song Y, Cardounel AJ, Zweier JL, and Xia Y. Inhibition of superoxide generation from neuronal nitric oxide synthase by heat shock protein 90: implications in NOS regulation. *Biochemistry* 41: 10616-10622, 2002.

[76] Song Y, Zweier JL, and Xia Y. Heat-shock protein 90 augments neuronal nitric oxide synthase activity by enhancing Ca2+/calmodulin binding. *Biochem. J.* 355: 357-360, 2001.

[77] Tanaka M, Sydow K, Gunawan F, Jacobi J, Tsao PS, Robbins RC, and Cooke JP. Dimethylarginine dimethylaminohydrolase overexpression suppresses graft coronary artery disease. *Circulation* 112: 1549-1556, 2005.

[78] Tang J, Gary JD, Clarke S, and Herschman HR. PRMT 3, a type I protein arginine N-methyltransferase that differs from PRMT1 in its oligomerization, subcellular localization, substrate specificity, and regulation. *J. Biol. Chem.* 273: 16935-16945., 1998.

[79] Tojo A, Welch WJ, Bremer V, Kimoto M, Kimura K, Omata M, Ogawa T, Vallance P, and Wilcox CS. Colocalization of demethylating enzymes and NOS and functional effects of methylarginines in rat kidney. *Kidney Int*. 52: 1593-1601, 1997.

[80] Tsikas D, Boger RH, Sandmann J, Bode-Boger SM, and Frolich JC. Endogenous nitric oxide synthase inhibitors are responsible for the L-arginine paradox. *FEBS Lett.* 478: 1-3, 2000.

[81] Vallance P and Leiper J. Cardiovascular Biology of the Asymmetric Dimethylarginine:Dimethylarginine Dimethylaminohydrolase Pathway. *Arterioscler Thromb. Vasc. Biol.* 22: 22, 2004.

[82] Vallance P, Leone A, Calver A, Collier J, and Moncada S. Accumulation of an endogenous inhibitor of nitric oxide synthesis in chronic renal failure. *Lancet* 339: 572-575, 1992.

[83] Vanden Hoek TL, Becker LB, Shao Z, Li C, and Schumacker PT. Reactive oxygen species released from mitochondria during brief hypoxia induce preconditioning in cardiomyocytes. *J. Biol. Chem.* 273: 18092-18098., 1998.

[84] Vasquez-Vivar J, Hogg N, Martasek P, Karoui H, Pritchard KA, Jr., and Kalyanaraman B. Tetrahydrobiopterin-dependent inhibition of superoxide generation from neuronal nitric oxide synthase. *J. Biol. Chem.* 274: 26736-26742., 1999.

[85] Vasquez-Vivar J, Kalyanaraman B, Martasek P, Hogg N, Masters BS, Karoui H, Tordo P, and Pritchard KA, Jr. Superoxide generation by endothelial nitric oxide synthase: the influence of cofactors. *Proc. Natl. Acad. Sci. USA* 95: 9220-9225., 1998.

[86] Venema RC, Sayegh HS, Kent JD, and Harrison DG. Identification, characterization, and comparison of the calmodulin-binding domains of the endothelial and inducible nitric oxide synthases. *J. Biol. Chem.* 271: 6435-6440, 1996.

[87] Venema VJ, Ju H, Zou R, and Venema RC. Interaction of neuronal nitric-oxide synthase with caveolin-3 in skeletal muscle. Identification of a novel caveolin scaffolding/inhibitory domain. *J. Biol. Chem.* 272: 28187-28190, 1997.

[88] Venema VJ, Marrero MB, and Venema RC. Bradykinin-stimulated protein tyrosine phosphorylation promotes endothelial nitric oxide synthase translocation to the cytoskeleton. *Biochem. Biophys. Res. Commun.* 226: 703-710, 1996.

[89] Wei Q and Xia Y. Proteasome inhibition down-regulates endothelial nitric-oxide synthase phosphorylation and function. *J. Biol. Chem.* 281: 21652-21659, 2006.

[90] Whitsett J, Martasek P, Zhao H, Schauer DW, Hatakeyama K, Kalyanaraman B, and Vasquez-Vivar J. Endothelial cell superoxide anion radical generation is not dependent on endothelial nitric oxide synthase-serine 1179 phosphorylation and endothelial nitric oxide synthase dimer/monomer distribution. *Free Radic. Biol. Med.* 40: 2056-2068, 2006.

[91] Wu Z, Qian X, Cai D, Liu Y, Li Z, and He J. Long-term oral administration of L-arginine enhances endothelium-dependent vasorelaxation and inhibits neointimal thickening after endothelial denudation in rats. *Chin. Med. J. (Engl)* 109: 592-598., 1996.

[92] Xia Y, Berlowitz CO, and Zweier JL. PIN inhibits nitric oxide and superoxide production from purified neuronal nitric oxide synthase. *Biochim. Biophys. Acta* 1760: 1445-1449, 2006.

[93] Xia Y, Dawson VL, Dawson TM, Snyder SH, and Zweier JL. Nitric oxide synthase generates superoxide and nitric oxide in arginine-depleted cells leading to peroxynitrite-mediated cellular injury. *Proc. Natl. Acad. Sci. USA* 93: 6770-6774., 1996.

[94] Xia Y, Roman LJ, Masters BS, and Zweier JL. Inducible nitric-oxide synthase generates superoxide from the reductase domain. *J. Biol. Chem.* 273: 22635-22639., 1998.

[95] Xia Y, Tsai AL, Berka V, and Zweier JL. Superoxide generation from endothelial nitric-oxide synthase. A Ca2+/calmodulin-dependent and tetrahydrobiopterin regulatory process. *J. Biol. Chem.* 273: 25804-25808., 1998.
[96] Xia Y and Zweier JL. Superoxide and peroxynitrite generation from inducible nitric oxide synthase in macrophages. *Proc. Natl. Acad. Sci. USA* 94: 6954-6958., 1997.
[97] Xie Q and Nathan C. The high-output nitric oxide pathway: role and regulation. *J. Leukoc. Biol.* 56: 576-582, 1994.
[98] Yoshida M and Xia Y. Heat shock protein 90 as an endogenous protein enhancer of inducible nitric-oxide synthase. *J. Biol. Chem.* 278: 36953-36958, 2003.
[99] Yui Y, Hattori R, Kosuga K, Eizawa H, Hiki K, Ohkawa S, Ohnishi K, Terao S, and Kawai C. Calmodulin-independent nitric oxide synthase from rat polymorphonuclear neutrophils. *J. Biol. Chem.* 266: 3369-3371, 1991.

In: Oxidative Stress: Clinical and Biomedical Implications ISBN: 978-1-60021-888-0
Editors: B. M. Matata, M. M. Elahi, pp. 227-243

Chapter XIV

Nitric Oxide Cell Signaling in Cellular Processes

Emil Martin, Ka Bian and Ferid Murad
Brown Foundation Institute of Molecular Medicine at University of Houston Health Science Center

Abstract

Nitric Oxide emerged as a vital signaling molecule in cellular processes. At low concentrations nitric oxide exerts its functions through the recruitment of the cGMP pathways. Nitric oxide activates soluble guanylyl cyclase resulting in increased intracellular cGMP levels. This leads to recruitment of cGMP-dependent effector proteins, such as cGMP-dependent protein kinases, cGMP-dependent phosphodiesterases or cyclic nucleotide-dependent channels. The NO/cGMP pathway participates in a variety of cellular processes, including regulation of intracellular Ca^{2+} homeostasis, cytoskeletal rearrangement, transcriptional regulation, etc. When produced at high concentrations, nitric oxide alone, or after its conversion into one of the reactive nitrogen species, may directly interact with proteins and wields its direct effect. Through covalent protein modification nitric oxide or its metabolites alters protein function, stability and affects downstream processes.

1. Introduction

Nitric oxide was used in biological sciences for long time as a useful reagent to analyze the properties of hemeproteins [1]. Not until the late 1970's and early1980's did the importance of NO as messenger mediating cell communication and intracellular signals become apparent. Now NO-mediated signaling is a recognized component in various physiologic processes, e.g. smooth muscle relaxation, inhibition of platelet and leukocyte aggregation, attenuation of vascular smooth muscle cell proliferation, neurotransmission and immune defense, to name only a few. NO has also been implicated in the pathology of many inflammatory diseases, including arthritis, myocarditis, colitis, and nephritis, and a large number of pathological conditions such as amyotrophic lateral sclerosis (ALS), cancer, diabetes, and neurodegenerative diseases [2-11]. Some of these processes, e.g. smooth muscle relaxation, platelet aggregation or neurotransmission, requires only a brief production of NO at low nanomolar concentrations and are dependent on the recruitment of cGMP-dependent signaling. Others processes are associated with direct interaction of NO, or reactive nitrogen species (RNS) derived from it, with target proteins and require a more sustained production of NO at higher concentrations, but do not involve the cGMP pathway.

The goal of this chapter is to review both NO/cGMP signaling and the direct effects of NO.

Nitric Oxide Synthases – making NO. A group of specialized enzymes called nitric oxide synthase (NOS) is the main physiological source of nitric oxide. Three independent isoforms, products of three independent genes, where identified in mammals (Figure 1). NOSI, first identified in neuronal tissue, is also termed nNOS. NOSII, the second NOS identified, was initially described in activated macrophages and is also known as iNOS. NOSIII, initially detected in endothelium and also referred to as eNOS, was characterized last. These specialized enzymes catalyze the O_2-dependent oxidation of the guanidino nitrogen of L-arginine to form NO and the aminoacid L-citrulline. In addition to oxygen and L-arginine, a heme prosthetic group, reduced forms of nicotinamide adenine dinucleotide phosphate (NADPH), flavine adenine dinucleotide (FAD), mononucleotide (FMN) and tetrahydrobiopterin are required for NO synthesis. All three isoforms act as homodimers, are ubiquitously expressed and are found in soluble and/or particulate cellular fractions. The C-terminal half of NOS, also known as the "reductase domain", shares homology with the cytochrome P450 reductase and binds the nucleotides. The structure of a proteolytically stable fragment of this domain was determined [12]. The N-terminal half of the enzyme termed the "oxygenase domain" contains the heme group and binds L-arginine and the biopterin. The crystal structure of the "oxygenase domain" was also determined [13, 14]. The flow of electrons from the reductase to oxygenase domains is essential for the NO production. It appears that it is the binding of calmodulin (CaM) to a region between these domains that couples them and promotes the flow of electrons. Thus, the flux of intracellular Ca^{2+} modulates the CaM-NOS interaction and is one of the key elements regulating NOSI and NOSIII enzymes. Although calmodulin binding to NOSII is independent of Ca^{2+} fluxes, NOSII activity is still strictly dependent on the presence of bound calmodulin. The isoforms have a high degree of homology with each other and are found throughout the animal kingdom.

2. cGMP-Mediated Effects of Nitric Oxide Signaling

Soluble guanylyl cyclase – switching from NO to the second messenger cGMP. As the intracellular calcium fluxes are transient, so is, under normal conditions, the production of nitric oxide by the Ca^{2+} -dependent NOSI and III. The transient nature of NO synthesis and the relatively high rate of NO diffusion demand a sensitive and efficient downstream NO sensor. To preserve and amplify the nitric oxide signal(s), the next player in the NO/cGMP cascade, soluble guanylyl cyclase (sGC), is recruited (Figure 1).

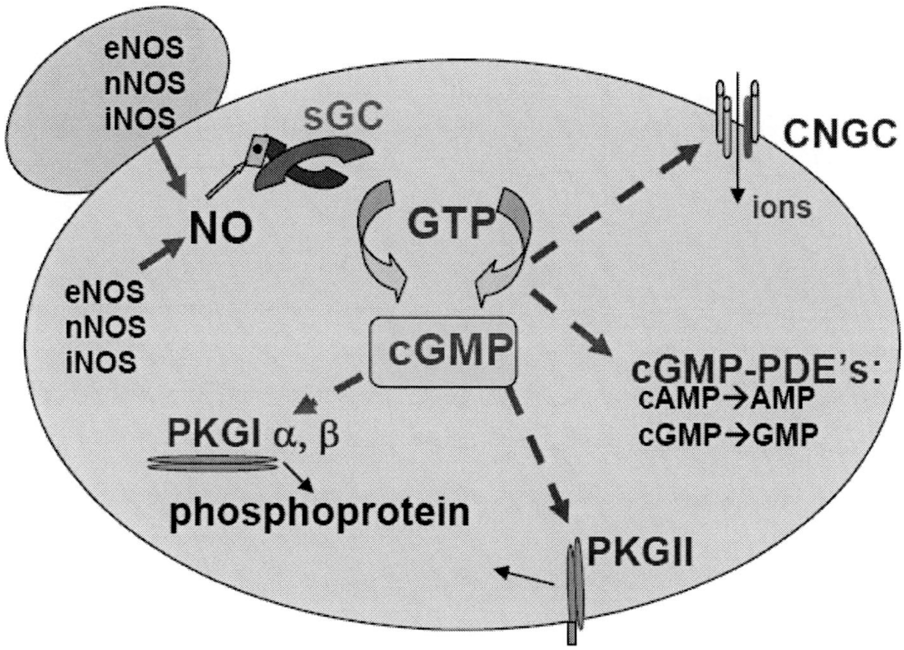

Figure 1. Schematic representation of NO/cGMP pathway. NO synthesized by NOS's from the same or adjascent cells binds and activate sGC. sGC-dependent production of cGMP results in activation of cGMP-dependent effector molecules PKG, PDE and CNGC.

sGC is a member of a large family of guanylyl cyclase receptor proteins. These guanylyl cyclases synthesize cGMP from GTP and are activated upon binding of various ligands [15]. sGC is a heterodimer composed of one α and one β subunit (16). To date, two isoforms for each sGC subunit have been identified (α1, α2, β1 and β2) in mammals [17, 18]. The most common α1β1 sGC isoform is expressed ubiquitously, although lungs, nervous system and liver exhibit the highest level of expression. The α2β1 heterodimer is more abundant in the uterus, placenta and brain [19]. Although β2 mRNA was detected in kidney and liver [20], no sGC heterodimers containing β2 subunit have been identified so far *in vivo*. Only sGC heterodimers are active [21], while α1/α1 and β1/β1 homodimers are inactive [22]. Interestingly, recombinant β2/β2 homodimer is also enzymaticaly active [23], although at non-physiologic concentrations of Mn^{2+} ion. The physiological meaning this activity remains to be determined.

Soluble cyclase is different form other members of the GC family by its cytosolic localization and its responsiveness to nitric oxide. Exposure of sGC enzyme to NO donors transforms a weak and inefficient enzyme with low cGMP output (specific activity 0.01-0.1 μmole cGMP/min/mg sGC) [24] into a robust high-output sGC with a specific activity several hundred fold higher (10-35 μmole cGMP/min/mg sGC) [24-26]. Under NO-activation and with abundant GTP, one sGC enzyme molecule is capable of generating 25-80 molecules of cGMP per second, providing an efficient amplification of NO signal.

The sGC enzyme also has a high affinity to NO. The enzyme binds NO with a high k_{ON} rate of $1.4x10^8/M^{-1}s^{-1}$, which is almost diffusion limited [27]. At the same time, in the absence of any additional factors, dissociation of NO from sGC proceeds very slowly with a k_{OFF} of only $6x10^{-4}$ s^{-1}. Based on these values, the dissociation constant ($k_D = k_{OFF}/k_{ON}$) is very small, ~$4.2x10^{-12}$ M. This high sensitivity to NO is provided by one protoporphyrin type IX heme moiety [28], which is coordinated by the histidine 105 residue in the β subunit [29]. Extensive analysis of the purified sGC by UV-Vis [24, 30], EPR (31) and resonance Raman spectroscopy [32, 33] demonstrated that in resting sGC the heme is in a ferrous five-coordinate (5C) high spin histidyl complex. Upon NO binding this complex is converted into a transient six-coordinate intermediate, which is quickly converted into a more stable 5C NO-heme complex [27] due to the loss of coordination with histidine 105. Careful measurements of NO-sGC interaction kinetics [27, 34, 35] indicate that not only the binding of NO to the heme, but also the following cleavage of the His-iron bond is accelerated by NO. These measurements suggest that a second NO binding event occurs. A recent study suggested that the NO-sGC complex formed in the absence of GTP substrate is not active, when formed under stoichiometric conditions, and binding of the second NO molecule is required to achieve full sGC activation [36]. However, in the presence of GTP substrate or even cGMP product, full activation may be achieved without additional NO [36].

A simple calculation, based on the average size of mammalian cell (~50 μm) and the low k_D for NO ($4.2x10^{-12}$ M), reveals that sGC should bind NO evn only with 0.3 molecule of NO molecule per cell is present. This exceptional sensitivity to NO coupled with the high turnover rate are the key factors assuring an efficient switch from NO signaling to cGMP signaling. Absence of any detectable affinity to oxygen is another essential factor contributing to the proper function of sGC as high affinity NO receptor. Even in pure oxygen environment sGC does not bind O_2 and maintains the heme in a ferrous form, ready to bind NO. It was postulated that the absence of hydrogen-bonding residues in the heme pocket of sGC makes this oxygen discrimination possible [37]. Although this is true for a fragment of sGC containing the heme-binding domain [38], introduction of a hydrogen-bonding residue in the intact protein did not affect the oxygen discrimination [35].

Conversion of the NO signal into an intracellular cGMP signal leads to the recruitment of several cGMP-dependent intracellular targets (Figure 1).

cGMP-dependent protein kinases – providing diversity of the NO/cGMP pathway. cGMP-dependent protein kinases (PKG or cGK), homodimeric serine/threonine kinases, are considered the main effector molecules in the NO/cGMP pathway. PKGIα, PKGIβ, and PKGII are three different isoforms of mammalian PKG, [39]. PKGIα and PKGIβ are cytosolic protein products of alternative RNA splicing and differ only in their first 100 amino acids [40, 41]. PKG II is a membrane bound [42-44] isoforms and is encoded by a separate gene.

The carboxyl-terminal region of PKG kinases contains the catalytic domain. Two tandem allosteric cGMP-binding sites change their conformation upon binding of cGMP. These binding sites are located in the amino-terminal portion of PKG in close proximity to the catalytic domain and affect its activity [45, 46]. An autoinhibitory/autophosphorylation domain, with autophosphorylation sites within the domain [47-49], is located immediately amino-terminal to the cGMP-binding sites. The autoinhibitory domain of each of the PKGs contains a pseudosubstrate sequence that interacts with the catalytic site and blocks the access of the substrate, thus maintaining the kinase in an inactive state [50]. This autoinhibition is relieved by cGMP binding and/or autophosphorylation in PKGI [48, 50, 51], or cGMP binding alone in PKG II [45, 52]. Both PKGIα and PKGIβ undergo autophosphorylation, which increases the kinase and the affinity to cyclic nucleotides [48].

Activated PKG phosphorylates a variety of intracellular targets and affects multiple intracellular processes [15]. For example, intracellular Ca^{2+} -homeostasis is affected by phosphorylation of phospholamban and IP3 receptor, cytoskeletal rearrangements are possible due to phosphorylation of myosin light chain phosphatase or vasodilator-stimulated phosphoprotein, transcriptional rate changes due to phosphorylation of transcription factors, such as NFAT. These few examples demonstrate that PKG recruitment by the NO/cGMP pathway provides a wide diversity in protein targets and cellular processes affected by nitric oxide.

Phosphodiesterases and NO/cGMP – keeping the cyclic nucleotides in check. Cyclic nucleotide phosphodiesterases (PDE's) directly regulated by cGMP play a critical role in NO/cGMP signal transduction. PDEs catalyze the breakdown of cGMP molecules, including those produced by soluble guanylyl cyclase in response to NO. By balancing and modulating the cGMP production-decomposition equilibria PDE determining the actual tissue cGMP levels.

PDEs were first discovered by Sutherland and co-workers more than five decades ago [53, 54]. Presently, more then 50 different mammalian PDEs are known. They are divided into 11 enzyme families based on coding sequence, domain structure and regulation [55]. Some of these families contain several isoforms coded by independent genes, many of which have several alternative splicing forms resulting in proteins with different amino acid sequences. The function of some PDE's is affected by changes in intracellular cGMP and, therefore, these PDE's are important elements of the cGMP-dependent NO-signaling. Modulation of PDE activity significantly changes the tissue cGMP levels, altgering the extent of physiological response initiated by NO/cGMP signaling.

Cyclic GMP may regulate PDE activity through binding to specific allosteric sites. For example, PDE2, 5, 6 and 10 contain two tandem sites that bind cGMP [56], also known as GAF domains. Binding of cGMP to these domains affects the catalytic properties of these PDE's. PDE5 is not only activated by cGMP, but is also cGMP-specific. In some systems this PDE isoform is perhaps the main mechanism that keeps the intracellular cGMP concentration in the proper range. For example, PDE5 is an important regulator of NO/cGMP signaling in platelets. Measurements of cGMP accumulation in platelets showed that under a continuous flux of NO generated by NO-donors a sharp increase of cGMP level is followed by a rapid decline of intracellular cGMP. This decline is not due to decreased catalytic activity of sGC, but is the result of increased PDE5 activity [57, 58]. NO-induced activation and phosphorylation of PDE5 is, most probably, the major mechanism that contributes to the long-lasting negative feedback loop controling the extent cGMP response in human platelets.

The complex PDE6 enzyme is comprised of two large catalytically active subunits, which are associated with smaller regulatory or inhibitory subunits. cGMP binding to the GAF domain regulates the interaction between these subunits and controls the function of PDE6 (59). PDE2 and PDE 10 degrade both cAMP and cGMP. Binding of cGMP to the GAF domains of PDE2 and PDE10 stimulates their activity and, thus, regulates intracellular concentration of both cyclic nucleotides.

Since some PDE enzymes have dual specificity, increased cGMP levels may influence intracellular cAMP as well. For example, cGMP may compete for the catalytic site of PDE1, 2 and 3 and decrease cAMP degradation [56].

CNG Channels – regulating the ion influx. Cyclic nucleotide-gated (CNG) channels constitute a special class of downstream cGMP effectors. Six CNG channel members have been identified in mammals and can be divided into two subtypes, the α subunits (CNGA1-4) and the β subunits (CNGB1 and 3) [60]. Although CNG channels are best known for their function in sensory systems, they are found not only in sensory system, such as rod (61) and cone photoreceptors [62] or olfactory sensory neurons [63], but also in the pineal gland (64), kidney, testis and heart [65], hippocampus, adrenal gland and colon [66-72]. Native CNG channels form a heterotetrameric structure composed of three α subunits and one β subunit [73-75]. The four subunits of each heterotetrameric CNG channel form a central pore through which the ions pass [73-75]. The intracellular carboxy-terminal region of the CNG channel contains a cyclic nucleotide binding domain (CNBD) and a region termed the C-linker. The CNBD shares sequence similarity with cGMP- and cAMP-binding domains of cGMP- and cAMP-dependent protein kinases and the *E. coli* catabolite activator protein (CAP) [76].

All CNG channels respond to both cAMP and cGMP, although some CNG channels display preferential affinity for one of the cyclic nucleotides. Following the binding of cGMP, CNG channels allow non-selective passage of divalent cations through the plasma membrane. The mechanism of CNG channel function is best characterized in the outer segment of rod photoreceptors. In the absence of light, cGMP levels are relatively high in cells containing rod photoreceptors and cGMP binds to CNBD. This opens the CNG channels and result in an inward flow of Ca^{+2} and Na^{+} ions and membrane depolarization.

cGMP-dependent regulation of CNG channesl is mainly associated with the function of membrane-associated (particulate) guanylyl cyclases. However, an increasing amount of evidence suggests that NO-stimulated cGMP production mediated by soluble guanylyl cyclase (sGC) may also be important for modulation of CNG channel. Retinal cells express NOSI and produce NO, which is released into the cone terminals and activate soluble guanylyl cyclase [77-80]. NO may also play a role in the regulation of the olfactory CNG channel. Exposure to NO donors stimulates olfactory sensory neurons, which coincides with increased intracellular cGMP [81]. It is not clear, however, whether the stimulation is dependent on sGC-dependent production of cGMP. Serine substitution of a specific cysteine residue within the channel abolishes this NO-induced activity. Thus, nitrosylation of cysteine thiol by NO was proposed as the meachnism of channel activation (81, 82). Interestingly, nitrosylation-dependent activation of CNG channels may be specific only to olfactory sensory neurons, since rod and cone photoreceptor channels in a *C. elegans* model were not activated in a similar manner [80, 83, 84].

Examples of cGMP-mediated NO-dependent activation of channels are also reported. For example, inhibition of NOS and scavenging of NO blocked odorant-stimulated cGMP production in rat olfactory neurons [85]. In *Xenopus* olfactory neurons, exogenously added NO or cGMP induced similar currents, suggesting that NO-dependent activation occurs via cGMP pathway [86, 87]. CNG channels in rat retinal ganglion cells were also activated by

NO [88]. It is important to note, however, that only developing and regenerating olfactory sensory neurons and not the mature olfactory sensory neurons (OSN) exhibited NOS activity [89-92]. These, often contradictory data, clearly indicate that the role of NO in modulation of CNG channels requires further investigation.

3. Direct Effects of Nitric Oxide

In contrast to physiological NO concentrations, larger amounts of NO may have toxic and pro-inflammatory effect without cGMP synthesis. These processes are often reffered to as direct effects of NO. The direct NO bioactivity is primarily attributed to the properties of reactive nitrogen species (RNS), which often affect cellular signaling by protein modification. Tyrosine nitration (covalent modification of tyrosine residues) and S-nitrosylation (covalent modification of cysteine thiol moieties) represent such direct NO signaling processes in biological systems [93].

Nitrotyrosine (NO_2-Tyr) formation – interfering with cellular signaling. Constant generation of reactive oxygen species (ROS), for example superoxide anion ($O_2^{\bullet-}$), hydroxyl radicals ($^{\bullet}OH$) and nonradical hydrogen peroxide (H_2O_2), is a consequence of aerobic metabolism. Since the discovery of nitric oxide as biologically active molecule, formation of reactive nitrogen species (RNS) and analysis of their function are key components of the NO-related research. Induction of NO synthesis by NOS2 enzyme during inflammatory processes constitutes an excellent defense mechanism against a variety of pathogens. However, excessive formation of NO has been implicated in the injury of host tissues. For example, NO molecule reacts with O_2^- siperoxide anion to form peroxynitrite ($ONOO^-$), which in its acid form, peroxynitrous acid (ONOOH), is a very unstable and highly reactive oxidizing species. Involvement of $ONOO^-$ in inflammatory diseases has been documented by multiple examples of nitrotyrosine (NO_2-Tyr) detection in various inflamed tissues (Figure 2). Another major oxidation product derived from NO is nitrite ion (NO_2^-). Nitrite can be oxidized by myeloperoxidase (MPO), lactoperoxidase (LPO), horseradish peroxidase to form reactive nitrogen intermediate(s), such as nitrogen dioxide, capable of tyrosine nitrating (Figure 2). MPO and other peroxidase may also use halides and pseudohalides as co-substrates to generate reactive intermediate hypochlorous acid, which further forms nitryl chloride, that, in turn, mediate the formation of NO2-Tyr. Thus, it is likely that tyrosine nitration occur through multiple pathways [94].

Protein nitration was initially regarded only as evidence of peroxynitrite generation in biological systems. It was later demonstrated that nitration alter protein conformation, catalytic activity, and/or susceptibility to protease digestion. It was also postulated that tyrosine nitration diminishes the efficiency of protein phosphorylation by tyrosine kinases [95].

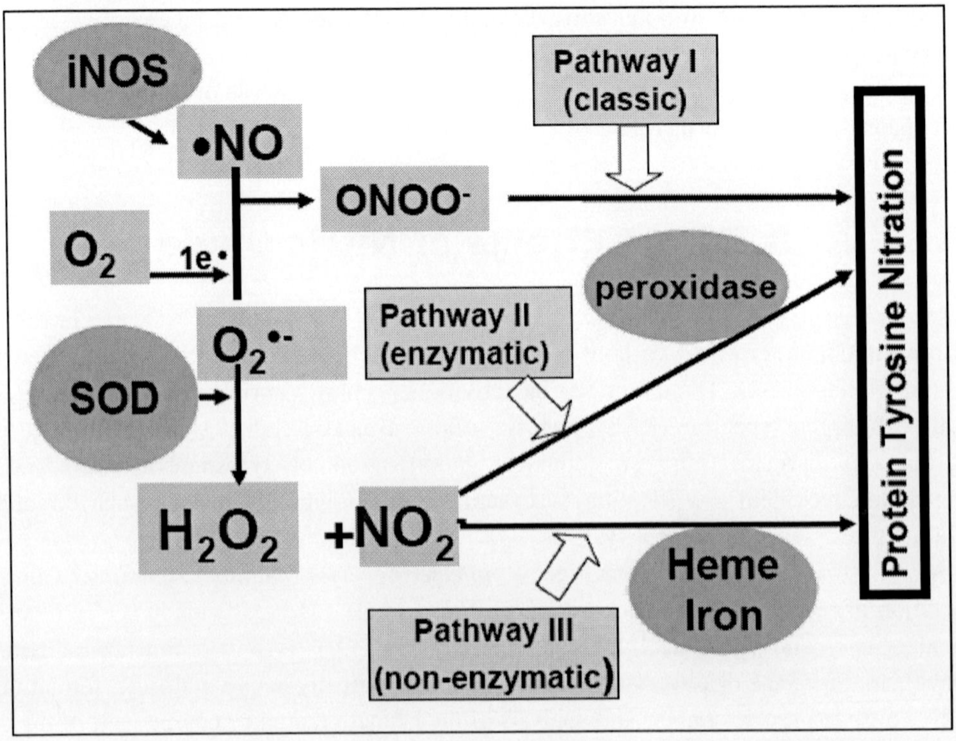

Figure 2. Pathways of nitrotyrosine formation.

Heme is a crucial prosthetic groups in a wide variety of processes. For example, oxygen transport and mitochondrial respiration, drug metabolism, steroid biosynthesis, cellular antioxidant defenses, many signal transduction events are dependent on heme[96, 97]. The capacity of iron ion to readily accept and donate electrons by interconverting between ferric (Fe^{3+}) and ferrous (Fe^{2+}) forms is fundamental for the proper function of hemo- and metalloproteins. The transition metal properties facilitate the catalytic reduction of molecular oxygen and/or the conversion of superoxide and hydrogen peroxide to free radical species. Free iron ion, free heme and hemoproteins all contribute to the formation of free radical species and subsequent protein modification.

A strong interaction between O_2 and ferrous hemoprotein often catalyses the reduction of O_2 and formation of $O_2^{\bullet-}$ [99]. Superoxide dismutase (SOD) enzymes convert $O_2^{\bullet-}$ to H_2O_2 and, thus, decrease the accumulation of reactive oxygen and reactive nitrogen species. Although SOD activity reduces the formation of $ONOO^-$, it increases the concentration of H_2O_2. Transition metals, such as iron, catalyse the formation of a strong oxidant, hydroxyl radical ($^{\bullet}OH$), from the hydrogen peroxide via the Fenton reaction. Catalase, another heme-containing enzyme, degrades H_2O_2 to water and oxygen and prevents the formation of $^{\bullet}OH$.

Recently, we tested the role of free heme and iron in NO_2-Tyr formation [98]. We found that isolated heme and free iron and other metal ions promote tyrosine nitration in the presence of hydrogen peroxide and nitrite. To evaluate the effect of SOD and catalase on protein nitration, we used both heme- and MPO-dependent pathways of NO_2-Tyr formation (Figure 2). SOD had no significant effect on protein nitration. In contrast, catalase abolished both heme and MPO-induced tyrosine nitration. This further confirms the requirement for

H_2O_2 in protein nitration by the heme/$O_2^{\bullet-}$/H_2O_2 pathway. The inhibitory effect of catalase on heme- or iron-catalyzed nitration suggests the involvement of the Fenton reaction [100]. However, we observed no effect of the hydroxyl radical scavenger, mannitol/ethanol mixture, on iron-catalyzed tyrosine nitration of BSA (98). In addition, the Fenton reaction effectively produces hydroxyl radical in several seconds; while our study showed that 16-h incubation is needed for iron to induce protein nitration. These data cast doubts on the notion that heme/iron-mediated protein nitration proceed through the hydroxyl radical pathway.

Formation of the Fe(IV) hypervalent compound may be an essential intermediate in the process of heme- or iron-catalyzed nitration. This possibility was tested using cyanide (CN) and EDTA as iron chelators [98]. We found that chelation of iron ion with CN before the initiation of BSA nitration completely abolishes the tyrosine nitration. As a potent iron-chelating agent, CN^- can occupy all of the binding sites of iron in a 6:1 ratio and completely block the catalytic effect of iron on BSA nitration. Addition of CN^- after the initiation of reaction had no inhibitory effect. Addition of EDTA, which has a weaker interaction with iron ion, did not effect the nitriation of BSA, but partly reversed the inhibitory effect of CN^-. Of interest, the addition of cyanide into the iron-EDTA mixture markedly enhanced NO_2-Tyr formation. Weaker chelating properties of either EDTA or BSA [101] may facilitate the formation [Fe(IV)] redox status [102], which can be further stabilized by CN . Hemin or chelated iron reacts with H_2O_2 to form the ferryl π-cation radical complex, which then oxidizes both nitrite and tyrosine to form nitric dioxide radical ($^{\bullet}NO_2$) and tyrosine radical (Tyr$^{\bullet}$), respectively. These nitrating species nitrate tyrosyl residues in proteins to form NO_2-Tyr.

S-nitrosothiols (RSNO) formation – endocrinology of RNS. As early as 1981, Ignarro's group demonstrated that the bioactivities of certain pharmacological nitrogen oxide (NOx) donors are attributed to reactions with cellular thiols [103]. This observation was made several years before the discovery that NO is actually synthesized endogenously in mammalian cells. An increasing number of reports implicate S-nitrosothiol (SNO) as intermediate in NO-dependent and cGMP-independent signaling. Reactive protein thiols are more and more appreciated as major intracellular target of nitric oxide [104]. S-nitrosothiols are thio-esters with the general structure R-S-N=O (RSNO), in which R is an amino acid, polypeptide or protein. Naturally occurring examples of RSNO include S-nitrosocysteine, S-nitrosoglutathione and S-nitrosoalbumin (for detailed review see [105, 106]). S-nitrosothiols can be synthesized *in vitro* by a reaction between thiols and nitrous acid in extremely acidic condition ($pH < 3$). Decomposition of RSNOs releases NO and forms corresponding disulphide (RSSR). Such factors as heat-, UV light- and metal ion (e.g., Cu^{2+} or Hg^{2+}) contribute to RSNO decomposition. Ascorbate, thiols, high oxygen tension and pH shift to more basic values also promote decomposition. In addition, enzymatic decomposition by γ-glutamyl transpeptidase, glutathione peroxidase and xanthenes oxidize has also been demonstrated.

Although RSNOs, such as S-nitroso-N-acetyl-penicillamine (SNAP), N-acetyl-S-nitrosopenicillaminyl-, S-nitrosopenicillamine and S-nitrosoglutathione (GSNO), are commercially available, none of them has been used therapeutically due to the unpredictable rate of decomposition in the body. In contrast, an increasing number of proteins are demonstrated to undergo S-nitrosylation *in vivo*. In many instances, important roles in physiologic and pathologic processes are attributed to protein S-nitrosylation.

Hemoglobin (Hb) is one of the most well studied proteins in the human body. The function of S-nitrosohemoglobin is probably the best example illustrating the potential significance protein nitrosylation. In addition to the well recognized function as O_2/CO_2 transporter, Hb has been shown to capture and deliver a third gas, NO [107]. Hb exists in two alternative structures, known as R (for relaxed, high O_2 affinity) and T (for tense, low O_2 affinity). The R→T allosteric transition of Hb controls the reactivity of one pair of highly conserved cysteines (Cys93) that are located in a protected pocket within R conformation and are exposed in the T conformation. At high pO_2, endothelium-derived NO is captured by the heme iron in R-state Hb and is then transferred to Cys93 in R-state Hb, forming SNO-Hb [108]. After reaching regions with low pO_2, allosteric conversion of SNO-Hb to the T-state promotes trans-nitrosation. During trans-nitrosation NO is transferred to acceptor thiols, such as glutathione, or other proteins in red blood cells. This protein to protein trans-nitrosylation was best demonstrated on chloride/bicarbonate anion exchange protein AE1. AE1, a transmemebrane protein in red blood cells, contains two reactive cysteine residues, which can be oxidatively crosslinked through a disulphide bond to the reactive β-chain cysteine in Hb. It was proposed that transnitrosylation of a vicinal thiol in the cytoplasmic domain of AE1 is a strong candidate mechanism for the transfer of the NO group from the β-chain Cys93 of SNO-Hb to the red blood cells membrane [109, 110].

Conclusion

Nitric oxide exerts its functions in cellular signaling by activationg sGC in order to recruit several cGMP-regulated effector molecules. This affects a pleithora of cellular processes. In addition, nitric oxide regulates protein function through direct protein modification mediated by RNS. However, this is only a small subset of signaling processes associated with NO. NO-dependent modifications of nucleic acid, lipids and saccharides are also integral parts of complex cellular processes that occur in response to NOS-mediated synthesis of nitric oxide. Readers are encouraged to consult a great number of excellent reviews that appeared over the years on these subjects.

Acknowledgments

The financial support of this research was provided by the John S. Dunn, Harold and Leila Mathers, and Robert A. Welch Foundations (AU-1437); National Institute of General Medical Sciences Grant (GM-061731); The Department of Defense (T5 project); as well as The University of Texas and The Shanghai University of TCM. We also acknowledge the support fromthe National Natural Science Foundation (30570710); E-Institutes for Nitric Oxide and Inflammatory Medicine (E-04010) and Key Project (05ZZ11) from Shanghai Municipal Education Commission; Key Project of Shanghai Committee of Science and Technology (05JC14056).

References

[1] Lang G. Mossbauer spectroscopy of haem proteins. *Q. Rev. Biophys.* 1970;3(1):1-60.

[2] Wink DA, Vodovotz Y, Laval J, Laval F, Dewhirst MW, Mitchell JB. The multifaceted roles of nitric oxide in cancer. *Carcinogenesis* 1998;19(5):711-21.

[3] Wink DA, Mitchell JB. Chemical biology of nitric oxide: Insights into regulatory, cytotoxic, and cytoprotective mechanisms of nitric oxide. *Free Radic Biol. Med.* 1998;25(4-5):434-56.

[4] Wink DA, Miranda KM, Espey MG. Cytotoxicity related to oxidative and nitrosative stress by nitric oxide. *Exp. Biol. Med.* (Maywood) 2001;226(7):621-3.

[5] Abe K, Pan LH, Watanabe M, Kato T, Itoyama Y. Induction of nitrotyrosine-like immunoreactivity in the lower motor neuron of amyotrophic lateral sclerosis. *Neurosci. Lett.* 1995;199(2):152-4.

[6] Jang D, Murrell GA. Nitric oxide in arthritis. *Free Radic. Biol. Med.* 1998;24(9):1511-9.

[7] Kooy NW, Lewis SJ, Royall JA, Ye YZ, Kelly DR, Beckman JS. Extensive tyrosine nitration in human myocardial inflammation: evidence for the presence of peroxynitrite. *Crit. Care Med.* 1997;25(5):812-9.

[8] Iwashita E, Miyahara T, Hino K, Tokunaga T, Wakisaka H, Sawazaki Y. High nitric oxide synthase activity in endothelial cells in ulcerative colitis. *J. Gastroenterol.* 1995;30(4):551-4.

[9] Kelly CJ, Gold DP. Nitric oxide in interstitial nephritis and other autoimmune diseases. *Semin. Nephrol.* 1999;19(3):288-95.

[10] Chan NN, Vallance P, Colhoun HM. Nitric oxide and vascular responses in Type I diabetes. *Diabetologia* 2000;43(2):137-47.

[11] Schulz JB, Matthews RT, Klockgether T, Dichgans J, Beal MF. The role of mitochondrial dysfunction and neuronal nitric oxide in animal models of neurodegenerative diseases. *Mol. Cell Biochem.* 1997;174(1-2):193-7.

[12] Zhang J, Martasek P, Paschke R, Shea T, Siler Masters BS, Kim JJ. Crystal structure of the FAD/NADPH-binding domain of rat neuronal nitric-oxide synthase. Comparisons with NADPH-cytochrome P450 oxidoreductase. *J. Biol. Chem.* 2001;276(40):37506-13.

[13] Crane BR, Arvai AS, Gachhui R, Wu C, Ghosh DK, Getzoff ED, et al. The structure of nitric oxide synthase oxygenase domain and inhibitor complexes. *Science* 1997;278(5337):425-31.

[14] Raman CS, Li H, Martasek P, Kral V, Masters BS, Poulos TL. Crystal structure of constitutive endothelial nitric oxide synthase: a paradigm for pterin function involving a novel metal center. *Cell* 1998;95(7):939-50.

[15] Lucas KA, Pitari GM, Kazerounian S, Ruiz-Stewart I, Park J, Schulz S, et al. Guanylyl cyclases and signaling by cyclic GMP. *Pharmacol. Rev.* 2000;52(3):375-414.

[16] Kamisaki Y, Saheki S, Nakane M, Palmieri JA, Kuno T, Chang BY, et al. Soluble guanylate cyclase from rat lung exists as a heterodimer. *J. Biol. Chem.* 1986;261(16):7236-41.

[17] Zabel U, Weeger M, La M, Schmidt HH. Human soluble guanylate cyclase: functional expression and revised isoenzyme family. *Biochem. J.* 1998;335(Pt 1):51-7.

[18] Andreopoulos S, Papapetropoulos A. Molecular aspects of soluble guanylyl cyclase regulation. *Gen. Pharmacol*. 2000;34(3):147-57.

[19] Russwurm M, Behrends S, Harteneck C, Koesling D. Functional properties of a naturally occurring isoform of soluble guanylyl cyclase. *Biochem. J.* 1998;335(Pt 1):125-30.

[20] Yuen PS, Potter LR, Garbers DL. A new form of guanylyl cyclase is preferentially expressed in rat kidney. *Biochemistry* 1990;29(49):10872-8.

[21] Buechler WA, Nakane M, Murad F. Expression of soluble guanylate cyclase activity requires both enzyme subunits. *Biochem. Biophys. Res. Commun.* 1991;174(1):351-7.

[22] Zabel U, Hausler C, Weeger M, Schmidt HH. Homodimerization of soluble guanylyl cyclase subunits. Dimerization analysis using a glutathione s-transferase affinity tag. *J. Biol. Chem*. 1999;274(26):18149-52.

[23] Koglin M, Vehse K, Budaeus L, Scholz H, Behrends S. Nitric oxide activates the beta 2 subunit of soluble guanylyl cyclase in the absence of a second subunit. *J. Biol. Chem.* 2001;276(33):30737-43.

[24] Stone JR, Marletta MA. Soluble guanylate cyclase from bovine lung: activation with nitric oxide and carbon monoxide and spectral characterization of the ferrous and ferric states. *Biochemistry* 1994;33(18):5636-40.

[25] Martin E, Sharina I, Kots A, Murad F. A constitutively activated mutant of human soluble guanylyl cyclase (sGC): implication for the mechanism of sGC activation. *Proc. Natl. Acad. Sci. USA* 2003;100(16):9208-13.

[26] Koesling D. Studying the structure and regulation of soluble guanylyl cyclase. *Methods* 1999;19(4):485-93.

[27] Zhao Y, Brandish PE, Ballou DP, Marletta MA. A molecular basis for nitric oxide sensing by soluble guanylate cyclase. *Proc. Natl. Acad. Sci. USA* 1999;96(26):14753-8.

[28] Gerzer R, Bohme E, Hofmann F, Schultz G. Soluble guanylate cyclase purified from bovine lung contains heme and copper. *FEBS Lett.* 1981;132(1):71-4.

[29] Zhao Y, Schelvis JP, Babcock GT, Marletta MA. Identification of histidine 105 in the beta1 subunit of soluble guanylate cyclase as the heme proximal ligand. *Biochemistry* 1998;37(13):4502-9.

[30] Stone JR, Marletta MA. Spectral and kinetic studies on the activation of soluble guanylate cyclase by nitric oxide. *Biochemistry* 1996;35(4):1093-9.

[31] Stone JR, Sands RH, Dunham WR, Marletta MA. Electron paramagnetic resonance spectral evidence for the formation of a pentacoordinate nitrosyl-heme complex on soluble guanylate cyclase. *Biochem. Biophys. Res. Commun.* 1995;207(2):572-7.

[32] Tomita T, Ogura T, Tsuyama S, Imai Y, Kitagawa T. Effects of GTP on bound nitric oxide of soluble guanylate cyclase probed by resonance Raman spectroscopy. *Biochemistry* 1997;36(33):10155-60.

[33] Deinum G, Stone JR, Babcock GT, Marletta MA. Binding of nitric oxide and carbon monoxide to soluble guanylate cyclase as observed with Resonance raman spectroscopy. *Biochemistry* 1996;35(5):1540-7.

[34] Martin E, Berka V, Tsai AL, Murad F. Soluble Guanylyl Cyclase: The Nitric Oxide Receptor. *Methods Enzymol.* 2005;396PE:478-492.

[35] Martin E, Berka V, Bogatenkova E, Murad F, Tsai AL. Ligand selectivity of soluble guanylyl cyclase: effect of the hydrogen-bonding tyrosine in the distal heme pocket on binding of oxygen, nitric oxide, and carbon monoxide. *J. Biol. Chem.* 2006;281(38):27836-45.

[36] Russwurm M, Koesling D. NO activation of guanylyl cyclase. *Embo. J.* 2004;23(22):4443-50.

[37] Boon EM, Marletta MA. Ligand discrimination in soluble guanylate cyclase and the H-NOX family of heme sensor proteins. *Curr. Opin. Chem. Biol.* 2005;9(5):441-6.

[38] Boon EM, Huang SH, Marletta MA. A molecular mechanism for NO selectivity in soluble guanylate cyclase. *Nature Chem. Biol.* 2005;1(1):53-59.

[39] Francis SH, Corbin JD. Cyclic nucleotide-dependent protein kinases: intracellular receptors for cAMP and cGMP action. *Crit. Rev. Clin. Lab. Sci.* 1999;36(4):275-328.

[40] Francis SH, Woodford TA, Wolfe L, Corbin JD. Types I alpha and I beta isozymes of cGMP-dependent protein kinase: alternative mRNA splicing may produce different inhibitory domains. *Second Messengers Phosphoproteins* 1988;12(5-6):301-10.

[41] Wernet W, Flockerzi V, Hofmann F. The cDNA of the two isoforms of bovine cGMP-dependent protein kinase. *FEBS Lett*. 1989;251(1-2):191-6.

[42] Lohmann SM, Vaandrager AB, Smolenski A, Walter U, De Jonge HR. Distinct and specific functions of cGMP-dependent protein kinases. *Trends Biochem. Sci.* 1997;22(8):307-12.

[43] Pfeifer A, Klatt P, Massberg S, Ny L, Sausbier M, Hirneiss C, et al. Defective smooth muscle regulation in cGMP kinase I-deficient mice. *Embo. J.* 1998;17(11):3045-51.

[44] Feil R, Lohmann SM, de Jonge H, Walter U, Hofmann F. Cyclic GMP-dependent protein kinases and the cardiovascular system: insights from genetically modified mice. *Circ. Res*. 2003;93(10):907-16.

[45] Wall ME, Francis SH, Corbin JD, Grimes K, Richie-Jannetta R, Kotera J, et al. Mechanisms associated with cGMP binding and activation of cGMP-dependent protein kinase. *Proc. Natl. Acad. Sci. USA* 2003;100(5):2380-5.

[46] Chu DM, Corbin JD, Grimes KA, Francis SH. Activation by cyclic GMP binding causes an apparent conformational change in cGMP-dependent protein kinase. *J. Biol. Chem.* 1997;272(50):31922-8.

[47] Aitken A, Hemmings BA, Hofmann F. Identification of the residues on cyclic GMP-dependent protein kinase that are autophosphorylated in the presence of cyclic AMP and cyclic GMP. *Biochim. Biophys. Acta* 1984;790(3):219-25.

[48] Smith JA, Francis SH, Walsh KA, Kumar S, Corbin JD. Autophosphorylation of type Ibeta cGMP-dependent protein kinase increases basal catalytic activity and enhances allosteric activation by cGMP or cAMP. *J. Biol. Chem.* 1996;271(34):20756-62.

[49] Francis SH, Smith JA, Colbran JL, Grimes K, Walsh KA, Kumar S, et al. Arginine 75 in the pseudosubstrate sequence of type Ibeta cGMP-dependent protein kinase is critical for autoinhibition, although autophosphorylated serine 63 is outside this sequence. *J. Biol. Chem*. 1996;271(34):20748-55.

[50] Landgraf W, Hofmann F. The amino terminus regulates binding to and activation of cGMP-dependent protein kinase. *Eur. J. Biochem.* 1989;181(3):643-50.

[51] de Jonge HR, Rosen OM. Self-phosphorylation of cyclic guanosine 3':5'-monophosphate-dependent protein kinase from bovine lung. Effect of cyclic adenosine 3':5'-monophosphate, cyclic guanosine 3':5'-monophosphate and histone. *J. Biol. Chem.* 1977;252(8):2780-3.

[52] Gamm DM, Francis SH, Angelotti TP, Corbin JD, Uhler MD. The type II isoform of cGMP-dependent protein kinase is dimeric and possesses regulatory and catalytic properties distinct from the type I isoforms. *J. Biol. Chem.* 1995;270(45):27380-8.

[53] Butcher RW, Sutherland EW. Adenosine 3',5'-phosphate in biological materials. I. Purification and properties of cyclic 3',5'-nucleotide phosphodiesterase and use of this enzyme to characterize adenosine 3',5'-phosphate in human urine. *J. Biol. Chem.* 1962;237:1244-50.

[54] Rall TW, Sutherland EW. Formation of a cyclic adenine ribonucleotide by tissue particles. *J. Biol. Chem.* 1958;232(2):1065-76.

[55] Maurice DH, Palmer D, Tilley DG, Dunkerley HA, Netherton SJ, Raymond DR, et al. Cyclic nucleotide phosphodiesterase activity, expression, and targeting in cells of the cardiovascular system. *Mol. Pharmacol.* 2003;64(3):533-46.

[56] Corbin JD, Francis SH. Cyclic GMP phosphodiesterase-5: target of sildenafil. *J. Biol. Chem.* 1999;274(20):13729-32.

[57] Mullershausen F, Russwurm M, Thompson WJ, Liu L, Koesling D, Friebe A. Rapid nitric oxide-induced desensitization of the cGMP response is caused by increased activity of phosphodiesterase type 5 paralleled by phosphorylation of the enzyme. *J. Cell Biol.* 2001;155(2):271-8.

[58] Mullershausen F, Friebe A, Feil R, Thompson WJ, Hofmann F, Koesling D. Direct activation of PDE5 by cGMP: long-term effects within NO/cGMP signaling. *J. Cell Biol.* 2003;160(5):719-27.

[59] Cote RH. Characteristics of photoreceptor PDE (PDE6): similarities and differences to PDE5. *Int. J. Impot. Res.* 2004;16 Suppl 1:S28-33.

[60] Bradley J, Frings S, Yau KW, Reed R. Nomenclature for ion channel subunits. *Science* 2001;294(5549):2095-6.

[61] Fesenko EE, Kolesnikov SS, Lyubarsky AL. Induction by cyclic GMP of cationic conductance in plasma membrane of retinal rod outer segment. *Nature* 1985;313(6000):310-3.

[62] Haynes L, Yau KW. Cyclic GMP-sensitive conductance in outer segment membrane of catfish cones. *Nature* 1985;317(6032):61-4.

[63] Nakamura T, Gold GH. A cyclic nucleotide-gated conductance in olfactory receptor cilia. *Nature* 1987;325(6103):442-4.

[64] Dryer SE, Henderson D. A cyclic GMP-activated channel in dissociated cells of the chick pineal gland. *Nature* 1991;353(6346):756-8.

[65] Hofmann F, Biel M, Kaupp UB. International Union of Pharmacology. XLII. Compendium of voltage-gated ion channels: cyclic nucleotide-modulated channels. *Pharmacol. Rev.* 2003;55(4):587-9.

[66] Bradley J, Zhang Y, Bakin R, Lester HA, Ronnett GV, Zinn K. Functional expression of the heteromeric "olfactory" cyclic nucleotide-gated channel in the hippocampus: a potential effector of synaptic plasticity in brain neurons. *J. Neurosci.* 1997;17(6):1993-2005.

[67] Biel M, Altenhofen W, Hullin R, Ludwig J, Freichel M, Flockerzi V, et al. Primary structure and functional expression of a cyclic nucleotide-gated channel from rabbit aorta. *FEBS Lett.* 1993;329(1-2):134-8.

[68] Biel M, Zong X, Distler M, Bosse E, Klugbauer N, Murakami M, et al. Another member of the cyclic nucleotide-gated channel family, expressed in testis, kidney, and heart. *Proc. Natl. Acad. Sci. USA* 1994;91(9):3505-9.

[69] Biel M, Zong X, Ludwig A, Sautter A, Hofmann F. Molecular cloning and expression of the Modulatory subunit of the cyclic nucleotide-gated cation channel. *J. Biol. Chem.* 1996;271(11):6349-55.

[70] Distler M, Biel M, Flockerzi V, Hofmann F. Expression of cyclic nucleotide-gated cation channels in non-sensory tissues and cells. *Neuropharmacology* 1994;33(11):1275-82.

[71] Kingston PA, Zufall F, Barnstable CJ. Rat hippocampal neurons express genes for both rod retinal and olfactory cyclic nucleotide-gated channels: novel targets for cAMP/cGMP function. *Proc. Natl. Acad. Sci. USA* 1996;93(19):10440-5.

[72] Weyand I, Godde M, Frings S, Weiner J, Muller F, Altenhofen W, et al. Cloning and functional expression of a cyclic-nucleotide-gated channel from mammalian sperm. *Nature* 1994;368(6474):859-63.

[73] Varnum MD, Zagotta WN. Subunit interactions in the activation of cyclic nucleotide-gated ion channels. *Biophys. J.* 1996;70(6):2667-79.

[74] Liu DT, Tibbs GR, Siegelbaum SA. Subunit stoichiometry of cyclic nucleotide-gated channels and effects of subunit order on channel function. *Neuron.* 1996;16(5):983-90.

[75] Gordon SE, Zagotta WN. Subunit interactions in coordination of Ni2+ in cyclic nucleotide-gated channels. *Proc. Natl. Acad. Sci. USA* 1995;92(22):10222-6.

[76] Su Y, Dostmann WR, Herberg FW, Durick K, Xuong NH, Ten Eyck L, et al. Regulatory subunit of protein kinase A: structure of deletion mutant with cAMP binding domains. *Science* 1995;269(5225):807-13.

[77] Kurenny DE, Moroz LL, Turner RW, Sharkey KA, Barnes S. Modulation of ion channels in rod photoreceptors by nitric oxide. *Neuron* 1994;13(2):315-24.

[78] Koch KW, Lambrecht HG, Haberecht M, Redburn D, Schmidt HH. Functional coupling of a Ca2+/calmodulin-dependent nitric oxide synthase and a soluble guanylyl cyclase in vertebrate photoreceptor cells. *Embo. J.* 1994;13(14):3312-20.

[79] Liepe BA, Stone C, Koistinaho J, Copenhagen DR. Nitric oxide synthase in Muller cells and neurons of salamander and fish retina. *J. Neurosci.* 1994;14(12):7641-54.

[80] Savchenko A, Barnes S, Kramer RH. Cyclic-nucleotide-gated channels mediate synaptic feedback by nitric oxide. *Nature* 1997;390(6661):694-8.

[81] Broillet MC, Firestein S. Direct activation of the olfactory cyclic nucleotide-gated channel through modification of sulfhydryl groups by NO compounds. *Neuron* 1996;16(2):377-85.

[82] Brown RL, Snow SD, Haley TL. Movement of gating machinery during the activation of rod cyclic nucleotide-gated channels. *Biophys. J.* 1998;75(2):825-33.

[83] Komatsu H, Jin YH, L'Etoile N, Mori I, Bargmann CI, Akaike N, et al. Functional reconstitution of a heteromeric cyclic nucleotide-gated channel of Caenorhabditis elegans in cultured cells. *Brain Res.* 1999;821(1):160-8.

[84] Trivedi B, Kramer RH. Real-time patch-cram detection of intracellular cGMP reveals long-term suppression of responses to NO and muscarinic agonists. *Neuron* 1998;21(4):895-906.

[85] Breer H, Klemm T, Boekhoff I. Nitric oxide mediated formation of cyclic GMP in the olfactory system. *Neuroreport* 1992;3(11):1030-2.

[86] Lischka FW, Schild D. Effects of nitric oxide upon olfactory receptor neurones in Xenopus laevis. *Neuroreport* 1993;4(5):582-4.

[87] Inamura K, Kashiwayanagi M, Kurihara K. Effects of cGMP and sodium nitroprusside on odor responses in turtle olfactory sensory neurons. *Am. J. Physiol.* 1998;275(5 Pt 1):C1201-6.

[88] Kawai F, Sterling P. AMPA receptor activates a G-protein that suppresses a cGMP-gated current. *J. Neurosci.* 1999;19(8):2954-9.

[89] Kishimoto J, Keverne EB, Hardwick J, Emson PC. Localization of nitric oxide synthase in the mouse olfactory and vomeronasal system: a histochemical, immunological and in situ hybridization study. *Eur. J. Neurosci.* 1993;5(12):1684-94.

[90] Bredt DS, Snyder SH. Nitric oxide: a physiologic messenger molecule. *Annu. Rev. Biochem.* 1994;63:175-95.

[91] Kulkarni AP, Getchell TV, Getchell ML. Neuronal nitric oxide synthase is localized in extrinsic nerves regulating perireceptor processes in the chemosensory nasal mucosae of rats and humans. *J. Comp. Neurol.* 1994;345(1):125-38.

[92] Roskams AJ, Bredt DS, Dawson TM, Ronnett GV. Nitric oxide mediates the formation of synaptic connections in developing and regenerating olfactory receptor neurons. *Neuron* 1994;13(2):289-99.

[93] Bian K, Ke Y, Kamisaki Y, Murad F. Proteomic modification by nitric oxide. *J. Pharmacol. Sci.* 2006;101(4):271-9.

[94] Bian K, Murad F. Nitric oxide (NO)--biogeneration, regulation, and relevance to human diseases. *Front Biosci.* 2003;8:d264-78.

[95] Turko IV, Murad F. Protein nitration in cardiovascular diseases. *Pharmacol. Rev.* 2002;54(4):619-34.

[96] Poulos TL, Li H, Raman CS. Heme-mediated oxygen activation in biology: cytochrome c oxidase and nitric oxide synthase. *Curr. Opin. Chem. Biol.* 1999;3(2):131-7.

[97] Ponka P. Cell biology of heme. *Am. J. Med. Sci.* 1999;318(4):241-56.

[98] Bian K, Gao Z, Weisbrodt N, Murad F. The nature of heme/iron-induced protein tyrosine nitration. *Proc. Natl. Acad. Sci. USA* 2003;100(10):5712-7.

[99] Isogai Y, Iizuka T, Shiro Y. The mechanism of electron donation to molecular oxygen by phagocytic cytochrome b558. *J. Biol. Chem.* 1995;270(14):7853-7.

[100] Thomas DD, Espey MG, Vitek MP, Miranda KM, Wink DA. Protein nitration is mediated by heme and free metals through Fenton-type chemistry: an alternative to the NO/O2- reaction. *Proc. Natl. Acad. Sci. USA* 2002;99(20):12691-6.

[101] Lovstad RA. A kinetic study of the coupled iron-ceruloplasmin catalyzed oxidation of ascorbate in the presence of albumin. *Biometals* 1995;8(4):328-31.
[102] Welch KD, Davis TZ, Aust SD. Iron autoxidation and free radical generation: effects of buffers, ligands, and chelators. *Arch. Biochem. Biophys*. 2002;397(2):360-9.
[103] Ignarro LJ, Kadowitz PJ, Baricos WH. Evidence that regulation of hepatic guanylate cyclase activity involves interactions between catalytic site -SH groups and both substrate and activator. *Arch. Biochem. Biophys*. 1981;208(1):75-86.
[104] Gow AJ, Chen Q, Hess DT, Day BJ, Ischiropoulos H, Stamler JS. Basal and stimulated protein S-nitrosylation in multiple cell types and tissues. *J. Biol. Chem*. 2002;277(12):9637-40.
[105] Hogg N. The biochemistry and physiology of S-nitrosothiols. *Annu. Rev. Pharmacol.* Toxicol 2002;42:585-600.
[106] Richardson G, Benjamin N. Potential therapeutic uses for S-nitrosothiols. *Clin. Sci.* (Lond) 2002;102(1):99-105.
[107] Patel RP. Biochemical aspects of the reaction of hemoglobin and NO: implications for Hb-based blood substitutes. *Free Radic. Biol. Med.* 2000;28(10):1518-25.
[108] McMahon TJ, Moon RE, Luschinger BP, Carraway MS, Stone AE, Stolp BW, et al. Nitric oxide in the human respiratory cycle. *Nat. Med.* 2002;8(7):711-7.
[109] Pawloski JR, Hess DT, Stamler JS. Export by red blood cells of nitric oxide bioactivity. *Nature* 2001;409(6820):622-6.
[110] Joshi MS, Ferguson TB, Jr., Han TH, Hyduke DR, Liao JC, Rassaf T, et al. Nitric oxide is consumed, rather than conserved, by reaction with oxyhemoglobin under physiological conditions. *Proc. Natl. Acad. Sci. USA* 2002;99(16):10341-6.

In: Oxidative Stress: Clinical and Biomedical Implications ISBN: 978-1-60021-888-0
Editors: B. M. Matata, M. M. Elahi, pp. 245-258

Chapter XV

The Role of Nitric Oxide in Cardiovascular Diseases

Khalid Naseem *

Centre for Atherothrombosis Research, Medical Biosciences,
University of Bradford, Richmond Road,
Bradford, BD7 1DP

Abstract

Nitric oxide (NO), a gaseous intracellular messenger, maintains vascular homeostasis by regulating blood pressure, vascular tone and haemostasis, but is also intimately involved in the development of cardiovascular disease. The pathology underlying cardiovascular disease is atherothrombosis, which is associated with hypercholesterolaemia, diabetes and hypertension. The atherothrombotic process is characterised by endothelial dysfunction and the loss of NO-mediated vasodilatation. NO plays a number of cardioprotective roles including regulation of platelet and leukocyte adhesion to the endothelium, prevention of smooth muscle cell proliferation and stimulation of angiogenesis. Reduced bioavailability of NO, probably through the inactivation of NO by superoxide anions, plays an important role in the disease process. Diminished NO availability leads to the loss of the cardioprotective roles and in certain situations may act to accelerate disease progression.

Introduction

Nitric oxide is an important modulator of vascular function and as such plays a key role in the aetiology of cardiovascular disease (CVD). Cardiovascular disease encompasses a number of different diseases including coronary heart disease, stroke and peripheral vascular

* Correspondence to Khalid Naseem PhD, Tel: +44 1274 236215 ; Fax: +44 1274 309742 ; Email:k.m.naseem@bradford.ac.uk

disease, and is the leading cause of morbidity and mortality in industrialised nations. Endothelial-derived NO plays a crucial cardioprotective role and a reduction in the bioavailability of NO leads to accelerated atherothrombosis, a pathology that underlies cardiovascular disease. This review provides an overview of the important biochemical and molecular mechanisms that regulate NO availability, and how these mechanisms are compromised in cardiovascular diseases.

The Role of NO in Atherothrombosis

Atherothrombosis is a chronic inflammatory process, which is multicellular in nature and involves the structural elements of the arterial wall, circulating platelets and leukocytes and a number of inflammatory cells particularly monocytes/macrophages. The pathobiology of atherothrombosis can be separated into three different phases termed inflammatory, reparative and thrombotic. Central to the disease process is endothelial dysfunction, which precedes clinical identification of disease [1]. Here endothelial cell function is compromised, which in turn facilitates altered activity of platelets, increased the permeability to inflammatory cells such as leukocytes and results in phenotypic changes in vascular smooth muscle cells (VSMC) [2]. Low-density lipoproteins (LDL) migrate into the subendothelial space, where they subjected to free radical oxidative attack resulting in the formation of cytotoxic oxidised LDL (oxLDL) [3]. The dysfunctional endothelium releases growth factors, cytokines and expresses cell adhesion molecules that recruit inflammatory cells such as monocytes. This process is aggravated by different forms of oxLDL present in the subendothelial space. Minimally modified LDL increases the release of endothelial-derived chemoattractants monocytes chemoattractant protein-1 (MCP-1) and macrophage colony stimulating factor (MCSF), while more fully oxLDL act as a chemotactic agents leading to increased recruitment of monocytes and T-lymphocytes(2). These inflammatory cells, which are initially recruited to remove oxLDL, are transformed into lipid-laden foam cells by the unregulated uptake of oxLDL, and exacerbate the inflammatory response by releasing further cytokines. The pro inflammatory cytokines such as interferon-γ□□□ and interleukin-12 [5] induce migration of VSMC from the media to the intima of the vessel and cause phenotypic changes leading to increased deposition of matrix proteins such as collagen. Thus, over time the core of the atherosclerotic plaque consisting of extracellular lipid and foam cells is covered by a fibrous cap, causing the plaque to protrude into the blood vessel lumen. The thick fibrous cap provides the plaque with mechanical stability but also reduces the supply of oxygen and nutrients to the foam cells, which begin to necrose. To counteract this, angiogenesis is induced to improve nutrient supply, although at the same time this can slowly allow the plaque to enlarge. In the final thrombotic phase, the complex atherosclerotic plaque ruptures and activates the haemostatic system. If the resultant thrombus does not occlude the vessel, it is incorporated into the plaque, while a more aggressive thrombotic event can completely occlude the blood vessel. It is the thrombotic events after plaque rupture that leads to clinical events such as acute coronary syndromes, myocardial infarction (MI) and stroke depending on the vascular bed. For a more detailed understanding of the molecular pathogenesis of atherothrombosis the reader is directed to two excellent reviews [2, 6].

The roles of NO in CVD can be viewed as a double-edged sword. The low concentrations of NO released under physiological conditions, derived primarily from endothelial nitric oxide synthase (eNOS), are essential to prevent atherothrombosis [7]. While under pathological conditions increased levels of NO from inducible NOS (iNOS) may increase disease by favouring formation of toxic secondary oxidants such as peroxynitrite. Numerous studies have demonstrated the loss of NO-mediated vasodilatation in patients with atherosclerosis. Furthermore, it seems that the pathobiology is more aggressive if other risk factors are present that limits the bioavailability of NO [8, 9]. For example, hypercholesterolaemic mice show accelerated atherosclerosis if NO synthases are inhibited. Endothelial-derived NO has a number of vasoprotective properties that prevent the initiation and development of atherothrombosis. At the luminal interface NO inhibits the recruitment, rolling and diapedesis of leukocytes into the subendothelial space [10, 11]. This is achieved by limiting the expression of VCAM-1 and ICAM-1, and the release of endothelial-derived MCP-1 [12, 13]. In addition to regulating endothelium-leukocyte interactions, NO is a potent inhibitor of platelet adhesion to the extracellular matrix and their subsequent aggregation [14], and therefore is a key anti-thrombotic agent [15]. A recent study has demonstrated that the sensitivity of platelets to inhibition by NO *in vitro*, is an independent marker of acute coronary syndromes [16]. Interestingly platelets also release NO upon stimulation, the more localised release of NO acts to limit the recruitment of platelets into a growing thrombus [17]; the loss of platelet NO biosynthesis is associated with increased likelihood of MI [18]. The activation of platelets at the sites of vascular injury leads to the release of a battery of growth factors, including platelet-derived growth factor and vascular endothelial cell growth factor (VEGF) [6]. By inhibiting the release of these agents NO prevents the underlying smooth muscle cells from being exposed to potent proliferating agents. NO also inhibits smooth muscle cell proliferation directly by a cGMP-dependent mechanism [19-21]. These actions of NO are critical to disease prevention since the proliferation of VSMC and their over production of extracellular matrix is one of the key events in the reparative phase of atherogenesis [2]. It is proposed that NO acts as an antioxidant by terminating lipid peroxyl radicals thereby reducing LDL oxidation[22], and thereby limiting the deleterious effects of toxic lipids on the surround cells. The thrombotic phase is characterised by a haemostatic imbalance resulting from inappropriate platelet activity and coagulation. Atherosclerotic plaques are rich in the prothrombotic protein tissue factor (TF)[23] which is plays an integral role in thrombus formation[24, 25]. In an *in vitro* setting NO inhibits the expression of TF on both endothelial cells and VSMC [26, 27]. Thus, NO regulates primary haemostasis through inhibition of platelet aggregation, and secondary haemostasis by reduction of TF exposure. NO has a number of important vasoprotective roles, and it is obvious that any loss in the availability of NO could significantly increase the progression of vascular diseases.

Factors That Affect the Bioavailability of NO in the Cardiovascular System

Since NO plays a central vasoprotective role, clearly any changes in NO metabolism will have major consequences for disease progression. The normal function of eNOS and the tonic production of NO are associated with a fully functional and compliant endothelium, while the reduced bioavailability of NO is thought to accelerate the atherothrombosis process. Certainly, eNOS -/- mice are hypertensive and have more accelerated and aggressive atherothrombotic pathology [28]. To date eNOS deficiency in humans has not been reported, although a number of eNOS gene polymorphisms have been reported [29, 30]. Variations in the eNOS gene could have a plethora of effects on the enzyme including altered protein stability, altered post-translational processing such as acylation or phosphorylation, altered intracellular distribution or altered cofactor association: all of which could influence the enzymatic activity. Unfortunately, linkage studies analysing the relationship between eNOS polymorphisms and incidence of cardiovascular diseases have been inconclusive [29]. However, recent evidence suggests that nNOS also plays a significant role in protecting against atherothrombosis [31], and so it may be the relative contribution of the different isoforms of the enzyme which determine disease progression.

A number of factors which could potentially affect either the production of NO, or the ability of NO to diffuse to its cellular targets, including co-factor availability, endogenous NOS inhibitors, the levels of proatherogenic factors circulating in the blood and reactive oxygen species (ROS). A number of circulating factors known to influence CVD modulate the cellular location and therefore activity of eNOS. In endothelial cells eNOS is located in specialized membrane rafts termed caveolae, and correct targeting of eNOS to caveolae via acylation is required for effective functioning of the enzyme. Caveolin-1, the predominant protein present in caveolae, negatively regulates eNOS activity by preventing calmodulin binding. OxLDL increases the expression of caveolin-1 and causes the depletion of cholesterol from membrane caveolae via the CD36 scavenger receptor [32], resulting in diminished eNOS activity. Furthermore, in apoE knockout mice where eNOS is not present in caveolae, acetylcholine (Ach) is unable to reduce blood pressure [33]. In contrast to oxLDL, HDL maintains the cholesterol content of caveolae leading to retention of eNOS in lipid raft, normal responses to Ach and reverses the detrimental effects of oxLDL on eNOS localisation to caveolae [33]. Thus, the ratio of pro and anti-atherogenic lipoproteins may have profound effects on eNOS function and therefore NO bioavailability.

eNOS activity is determined by availability of key substrates and co-factors. It is highly unlikely that L-arginine becomes a rate-limiting factor since the Km of eNOS is approximately 2.9μM [34], while the cytoplasmic levels of arginine are in the region of 100μM [35]. In addition the plasma concentration of arginine is 500μM [36], suggesting an adequate supply of arginine for optimal enzyme activity. However, the circulating level of asymmetrical dimethylarginine (ADMA), an endogenous NO inhibitor, is increased in a number of vascular disorders including hypercholesterolaemia and hypertension, and has been proposed as a strong predictor of the risk of acute coronary events [37, 38]. Tetrahydrobiopterin (BH_4), an essential co-factor for NOS, has profound effects on the structure and function of NOS including stabilisation of the dimeric structure, and facilitating and enhancing the binding of L-arginine [23]. *In vitro* studies have demonstrated that reduced availability of BH_4 results in the 'uncoupling' of NOS, leading to the production of superoxide anion ($O_2\bullet^-$) and hydrogen peroxide (H_2O_2) [39]. Hence it is conceivable that NOS itself could contribute to the oxidative stress associated with CVD. Certainly short-term administration of BH_4 led to improvements in endothelial cell function [40-42]. The mechanism is uncertain although the authors suggest that an increased formation of NO is the most likely explanation.

Reactive oxygen species (ROS) including $O_2\bullet^-$, hydroxyl radicals and H_2O_2 are produced through the incomplete reduction of oxygen. A number of vascular pathologies are associated with increased ROS production, which in turn has a profound effect on the availability of NO. The reaction of NO with $O_2\bullet^-$ is diffusion-limited for NO and is approximately six-times faster than the dismutation of $O_2\bullet^-$ by superoxide dismutase (SOD) [43]. This reaction has the dual effect of scavenging NO and thereby reducing its availability, but also of producing the potent oxidant peroxynitrite [Figure 1]. Certainly, the loss of NO would abrogate its anti-platelet and anti-leukocyte actions, and the antioxidant effects of NO on LDL oxidation. However, the pro-oxidant effects of peroxynitrite formation are far more damaging. Peroxynitrite, but not NO, has the capacity to induce the oxidation of LDL, a critical event in the development of atherosclerotic plaques [44]. Unlike other reactive species, peroxynitrite has the capacity to modify both lipid and protein moieties of LDL. Peroxynitrite can deplete LDL of the antioxidants, α-tocopherol and β-carotene, and induce the oxidation of lipids to produce lipid hydroperoxides and isoprostanes [45].

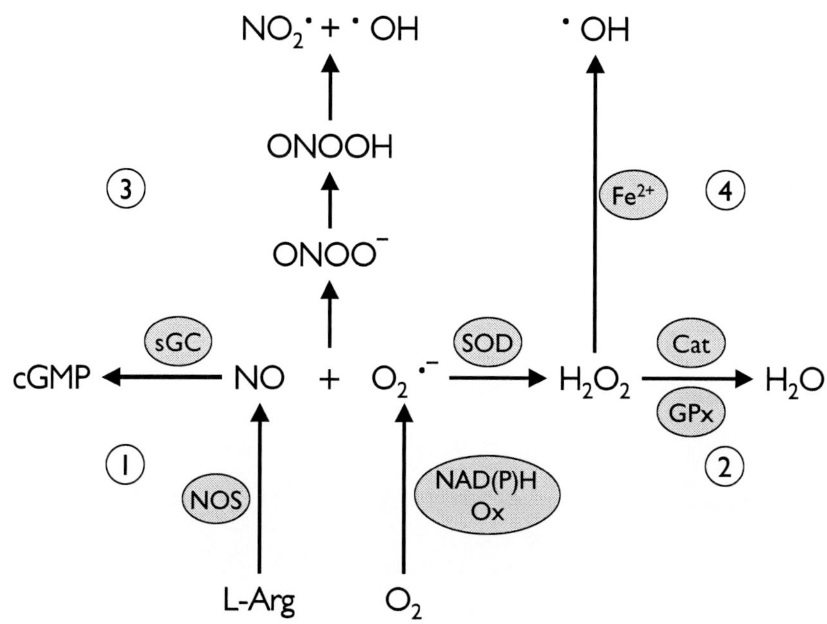

Figure 1. The formation of peroxynitrite by the reaction of nitric oxide and superoxide anions. Endothelial cells have the capacity to produce NO and superoxide, the reactants required for the production of peroxynitrite, in both extra and intracellular compartments. Under normal physiological conditions NO produced from L-arginine diffuses to its target soluble guanylyl cyclase (sGC) (1), while the levels of superoxide anions ($O_2^{\bullet-}$) are kept low by conversion to hydrogen peroxide (H_2O_2) and then water (2). However the levels of $O_2^{\bullet-}$ increase such that the interaction with NO becomes the favoured reaction (3), leading to the production of peroxynitrite and the potent nitrogen dioxide ($NO_2^{\bullet}$) and hydroxyl radicals ($^{\bullet}OH$). Under certain conditions hydroxyl radicals can be formed from hydrogen peroxide through the Fenton reaction (4).

Protein moiety of LDL, apolipoprotein B_{100}, is modified by both direct and indirect mechanisms. Peroxynitrite directly oxidises cysteine and tryptophan residues, while modification of lysine and arginine probably occurs via secondary reactions with lipid hydroperoxide radicals. The modification of LDL converts it to a highly atherogenic particle, which promotes SMC dysfunction and foam cell formation [44].

Interestingly peroxynitrite also causes the oxidation of BH_4 rendering it inactive [46]. It has been suggested that the beneficial effects of BH_4 supplementation in atherosclerosis may be by replacing lost BH_4. Peroxynitrite can potentially lead to cellular dysfunction through the modulation of important signaling cascades. Once formed, peroxynitrite can chemically modify amino acids, nucleic acids and thiol containing proteins and peptides [47]. The $NO_2\bullet$ attacks phenol groups to produce nitrophenols [48]. In biological systems this leads to modification of tyrosine residues to produce 3-nitrotyrosine [Figure 2]. The formation of 3-nitrotyrosine is viewed as a stable biological marker for the formation of peroxynitrite previously shown to be elevated in a number of cardiovascular diseases [48]. Furthermore the nitration of important tyrosine residues leads to altered enzyme activity and cell signaling pathways [48]. Thus, the increased production of oxidants by vascular cells contributes to the removal of NO. This not only reduces NO bioavailability but also produces highly reactive secondary oxidants that can further compromise cell function.

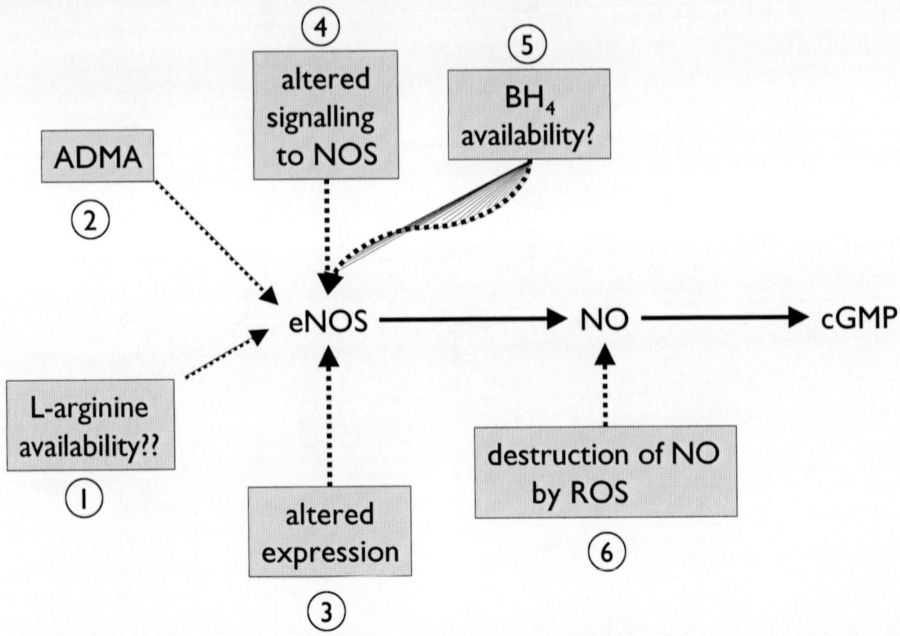

Figure 2. Factors that may contribute to a reduction in nitric oxide bioavailability. One of the major contributing factors to the progression of cardiovascular diseases is a loss of NO-dependent actions. It is proposed that this is due to a reduction in the bioavailability of NO. There are a variety of factors which could reduce NO availability including, (1) a reduction in the availability of substrate L-arginine; (2) increased concentration of circulating inhibitors such as ADMA; (3) altered levels of eNOS expression; (4) perturbed signal transduction reducing agonist-induced eNOS activation; (5) reduced availability of tetrahydrobiopterin (BH_4) an essential co-factor; or (6) the destruction of NO by other free radical species.

Nitric Oxide and Disease

Atherosclerosis and Hypercholesterolaemia

The reduced bioavailability of NO occurs early in the development of atherosclerosis, since hypercholesterolaemic subjects show impaired acetylcholine-induced vasorelaxation prior to the development of atherosclerosis as visualised by angiography [1]. Disease is therefore associated with either diminished NO production or increased NO scavenging or both. Currently there is little evidence of impaired production of NO, although there is increasing evidence to suggest that the scavenging of NO by reactive oxygen species plays a key role in disease. The dysfunctional arteries found in atherosclerotic and hypercholesterolaemic animals produce larger quantities of $O_2•^-$ and probably other ROS, than do healthy vessels [49]. Animal studies infusing polyethylene glycated SOD into atherosclerotic rabbits resulted in an improved endothelium-dependent relaxation [50, 51]. Animal models of early atherosclerosis have shown that hypercholesterolaemia causes a 2-fold increase in NADPH derived $O_2•^-$ [52], while in late atherosclerosis this increased to 3-fold [53]. This was reversed by endothelial denudation, implicating the endothelium as the primary source of $O_2•^-$. NADPH oxidase is stimulated by a variety of circulating factors such as proinflammatory cytokines and angiotensin II [54]. Furthermore, enzyme activity is elevated in CVD, probably due to the increased circulating plasma–derived risk factors. The increased $O_2•^-$ production is compounded by a diminished SOD activity in patients with CVD [55]. The elevated levels of both NO and $O_2•^-$ would promote formation of peroxynitrite, which would exacerbate disease progression through increased oxidation of LDL and destruction of BH_4. Certainly there is strong evidence that peroxynitrite is formed in the location of atherosclerotic plaques. Seminal work by Beckman et al demonstrated a high degree of staining for nitrotyrosine, in atherosclerotic plaques at post-mortem, suggesting that increased levels of peroxynitrite in the areas of atherosclerosis [56].

L-arginine supplementation has profound positive effects on a range of anti-atherosclerotic actions of NO. In hypercholesterolaemic rabbits L-arginine has a restorative effect on vasodilator function [57], inhibits platelet function and leukocyte adhesion [58]. Similar effects have also been observed in studies on patients with atherosclerosis [59]. Initially it was proposed that the excess arginine facilitated an increased production of NO, although this was never confirmed *in vivo*. At a cellular level this mechanism seems unlikely, since both cellular and plasma levels are well in excess of the Km of eNOS. Thus, the mechanism by which L-arginine supplementation improves NO-mediated actions remains unclear. eNOS activity may also be affected by ADMA since increased circulating levels of this endogenous inhibitor have been demonstrated in patients with coronary atherosclerosis, peripheral vascular disease and hypercholesterolaemic rabbits [38]. It is interesting to speculate that excess L-arginine could effectively compete with ADMA for NOS binding thereby abrogating the effects of ADMA.

Diabetes

Diabetes mellitus is associated with increased rates of morbidity and mortality caused primarily by the accelerated development of atherosclerotic disease [60]. Similar to other atherosclerotic pathologies diabetic vascular disease is characterised by endothelial dysfunction [61], reduced endothelium-dependent relaxation [62, 63], diminished cGMP formation in arterial vessels, increased superoxide [64] and peroxynitrite formation [65] are all indicative of a reduced bioavailability of NO which would act to accelerate disease progression. In human subjects both reduced and normal levels of arginine have been reported, although L-arginine supplementation in diabetic patients has been shown to increase plasma cGMP, L-citrulline levels, and endothelium-dependent relaxation in type 1 diabetes [60]. The effects of increasing the L-arginine concentration *in vivo* should be viewed cautiously, since the amino acid can also increase insulin secretion.

The relationship between insulin and NO in diabetes is further complicated by the ability of insulin to stimulate NO synthesis. Insulin induces NO-dependent relaxation of blood vessels [66] and activates eNOS in both cultured endothelial cells and blood platelets [67, 68]. However, more recent data indicate that NO availability and insulin resistance are intimately linked. Insulin stimulates blood flow through a NO-dependent mechanism, thereby increasing substrate delivery to muscle tissue. Indeed insulin resistance is associated with reduced endothelium-dependent vasodilatation [69]. Animal models have provided the biochemical and molecular evidence for the mechanism by which these interactions occur. NO donors stimulate glucose uptake [70], while NOS inhibitors reduce insulin-dependent glucose uptake [71]. Furthermore, eNOS knockout mice are insulin-resistant both in terms of glucose uptake and insulin stimulated blood flow [69]. Therefore, the loss of NO may further aggravate diabetes by preventing the biological actions of insulin.

Nitric Oxide and Hypertension

Impaired NO-mediated vasodilatation has been reported in brachial [72], coronary [73] and renal arteries [74] in patients with essential hypertension compared to controls. The reason for reduced NO dependent vasoactivity is unclear. The diminished NO responses are genetically determined, since basal NO production is impaired in offspring of patients with essential hypertension [75], and implies his that impaired NO production does not occur simply as a consequence of the disease but may be integral to disease development. Essential hypertension is associated with a number of eNOS polymorphisms associated with reduced NO synthesis. Essential hypertension is a multifactorial disease where the interactions between neuronal, hormonal and cellular signalling processes all contribute to the pathogenesis. ENOS gene polymorphisms probably led to only very subtle changes in NO production, and thus the contribution of these polymorphisms may only become important in the presence of other contributing factors [29]. Therefore, the group of patients used for each study is a critical determinant of the results.

Vasodilatation in response to the NO-donor sodium nitroprusside, was identical in both hypertensive and normotensives [76], suggesting that the cGMP-signalling pathway in smooth muscle cells is normal. Interestingly, hypertensive subjects were found to have defect vascular responses to acetylcholine, but normal responses to β-adrenergic stimulation by bradykinin [77]. These findings suggested that endothelial dysfunction associated with essential hypertension is due to a selective abnormality of the signalling pathways leading to eNOS activation. Indeed, the hypotensive effects of angiotensin-converting enzyme (ACE) inhibitors in hypertensive subjects, has been shown to occur through increased formation of bradykinin and leading to enhanced NO formation [77].

Conclusion

Endothelial-derived NO plays several key cardioprotective roles that prevent initiation and progression of atherothrombosis, the pathology that underlies cardiovascular diseases. However, hypercholesterolaemia, diabetes, hypertension and heart failure are all associated with reduced bioavailability of NO. Current evidence suggests that it is the balance between NO and ROS that determines NO availability and that atherothrombosis is associated with elevated ROS. Future therapeutic interventions that increase NO bioavailability to treat CVD are required. These strategies may include the use of agents to enhance of endogenous NO production, more selective NO donors and more effective scavengers of ROS.

References

[1] Forstermann U, Mugge A, Alheid U, Haverich A, Frolich JC. Selective attenuation of endothelium-mediated vasodilation in atherosclerotic human coronary arteries. *Circ. Res.* 1988 Feb;62(2):185-90.

[2] Lusis AJ. Atherosclerosis. *Nature*. 2000 Sep 14;407(6801):233-41.

[3] Witztum JL, Steinberg D. Role of oxidized low density lipoprotein in atherogenesis. *J. Clin. Invest.* 1991 Dec;88(6):1785-92.

[4] Gupta S, Pablo AM, Jiang X, Wang N, Tall AR, Schindler C. IFN-gamma potentiates atherosclerosis in ApoE knock-out mice. *J. Clin. Invest.* 1997 Jun 1;99(11):2752-61.

[5] Lee TS, Yen HC, Pan CC, Chau LY. The role of interleukin 12 in the development of atherosclerosis in ApoE-deficient mice. *Arterioscler Thromb. Vasc. Biol.* 1999 Mar;19(3):734-42.

[6] Ruggeri ZM. Platelets in atherothrombosis. *Nat. Med.* 2002 Nov;8(11):1227-34.

[7] Loscalzo J. Nitric oxide insufficiency, platelet activation, and arterial thrombosis. *Circ. Res.* 2001 Apr 27;88(8):756-62.

[8] Cayatte AJ, Palacino JJ, Horten K, Cohen RA. Chronic inhibition of nitric oxide production accelerates neointima formation and impairs endothelial function in hypercholesterolemic rabbits. *Arterioscler Thromb.* 1994 May;14(5):753-9.

[9] Naruse K, Shimizu K, Muramatsu M, Toki Y, Miyazaki Y, Okumura K, et al. Long-term inhibition of NO synthesis promotes atherosclerosis in the hypercholesterolemic rabbit thoracic aorta. PGH2 does not contribute to impaired endothelium-dependent relaxation. *Arterioscler Thromb.* 1994 May;14(5):746-52.

[10] Kubes P, Suzuki M, Granger DN. Nitric oxide: an endogenous modulator of leukocyte adhesion. *Proc. Natl. Acad. Sci. USA*. 1991 Jun 1;88(11):4651-5.

[11] Lefer DJ, Jones SP, Girod WG, Baines A, Grisham MB, Cockrell AS, et al. Leukocyte-endothelial cell interactions in nitric oxide synthase-deficient mice. *Am. J. Physiol.* 1999 Jun;276(6 Pt 2):H1943-50.

[12] Zeiher AM, Fisslthaler B, Schray-Utz B, Busse R. Nitric oxide modulates the expression of monocyte chemoattractant protein 1 in cultured human endothelial cells. *Circ. Res*. 1995 Jun;76(6):980-6.

[13] Tomita H, Egashira K, Kubo-Inoue M, Usui M, Koyanagi M, Shimokawa H, et al. Inhibition of NO synthesis induces inflammatory changes and monocyte chemoattractant protein-1 expression in rat hearts and vessels. *Arterioscler Thromb. Vasc. Biol*. 1998 Sep;18(9):1456-64.

[14] de Graaf JC, Banga JD, Moncada S, Palmer RM, de Groot PG, Sixma JJ. Nitric oxide functions as an inhibitor of platelet adhesion under flow conditions. *Circulation*. 1992 Jun;85(6):2284-90.

[15] Freedman JE, Loscalzo J. Nitric oxide and its relationship to thrombotic disorders. *J. Thromb Haemost*. 2003 Jun;1(6):1183-8.

[16] Willoughby SR, Stewart S, Holmes AS, Chirkov YY, Horowitz JD. Platelet nitric oxide responsiveness: a novel prognostic marker in acute coronary syndromes. *Arterioscler Thromb. Vasc. Biol*. 2005 Dec;25(12):2661-6.

[17] Freedman JE, Loscalzo J, Barnard MR, Alpert C, Keaney JF, Michelson AD. Nitric oxide released from activated platelets inhibits platelet recruitment. *J. Clin. Invest.* 1997 Jul 15;100(2):350-6.

[18] Ikeda H, Takajo Y, Murohara T, Ichiki K, Adachi H, Haramaki N, et al. Platelet-derived nitric oxide and coronary risk factors. *Hypertension*. 2000 Apr;35(4):904-7.

[19] Garg UC, Hassid A. Inhibition of rat mesangial cell mitogenesis by nitric oxide-generating vasodilators. *Am. J. Physiol.* 1989 Jul;257(1 Pt 2):F60-6.

[20] Garg UC, Hassid A. Nitric oxide-generating vasodilators and 8-bromo-cyclic guanosine monophosphate inhibit mitogenesis and proliferation of cultured rat vascular smooth muscle cells. *J. Clin. Invest.* 1989 May;83(5):1774-7.

[21] Tanner FC, Meier P, Greutert H, Champion C, Nabel EG, Luscher TF. Nitric oxide modulates expression of cell cycle regulatory proteins: a cytostatic strategy for inhibition of human vascular smooth muscle cell proliferation. *Circulation.* 2000 Apr 25;101(16):1982-9.

[22] Hogg N, Kalyanaraman B, Joseph J, Struck A, Parthasarathy S. Inhibition of low-density lipoprotein oxidation by nitric oxide. Potential role in atherogenesis. *FEBS Lett.* 1993 Nov 15;334(2):170-4.

[23] Cosentino F, Luscher TF. Tetrahydrobiopterin and endothelial nitric oxide synthase activity. *Cardiovasc Res*. 1999 Aug 1;43(2):274-8.

[24] Furie B, Furie BC. Thrombus formation in vivo. *J. Clin. Invest*. 2005 Dec;115(12):3355-62.

[25] Falati S, Gross P, Merrill-Skoloff G, Furie BC, Furie B. Real-time in vivo imaging of platelets, tissue factor and fibrin during arterial thrombus formation in the mouse. *Nat. Med*. 2002 Oct;8(10):1175-81.

[26] Kibbe MR, Johnnides C, Gleixner S, Kovesdi I, Lizonova A, Zuckerbraun B, et al. Regulation of tissue factor expression in smooth muscle cells with nitric oxide. *J. Vasc. Surg.* 2003 Mar;37(3):650-9.

[27] Yang Y, Loscalzo J. Regulation of tissue factor expression in human microvascular endothelial cells by nitric oxide. *Circulation.* 2000 May 9;101(18):2144-8.

[28] Kuhlencordt PJ, Gyurko R, Han F, Scherrer-Crosbie M, Aretz TH, Hajjar R, et al. Accelerated atherosclerosis, aortic aneurysm formation, and ischemic heart disease in apolipoprotein E/endothelial nitric oxide synthase double-knockout mice. *Circulation.* 2001 Jul 24;104(4):448-54.

[29] Wang XL, Wang J. Endothelial nitric oxide synthase gene sequence variations and vascular disease. *Mol. Genet. Metab.* 2000 Aug;70(4):241-51.

[30] Agema WR, de Maat MP, Zwinderman AH, Kastelein JJ, Rabelink TJ, van Boven AJ, et al. An integrated evaluation of endothelial constitutive nitric oxide synthase polymorphisms and coronary artery disease in men. *Clin. Sci.* (Lond). 2004 Sep;107(3):255-61.

[31] Kuhlencordt PJ, Hotten S, Schodel J, Rutzel S, Hu K, Widder J, et al. Atheroprotective effects of neuronal nitric oxide synthase in apolipoprotein e knockout mice. *Arterioscler Thromb Vasc. Biol.* 2006 Jul;26(7):1539-44.

[32] Blair A, Shaul PW, Yuhanna IS, Conrad PA, Smart EJ. Oxidized low density lipoprotein displaces endothelial nitric-oxide synthase (eNOS) from plasmalemmal caveolae and impairs eNOS activation. *J. Biol. Chem.* 1999 Nov 5;274(45):32512-9.

[33] Uittenbogaard A, Shaul PW, Yuhanna IS, Blair A, Smart EJ. High density lipoprotein prevents oxidized low density lipoprotein-induced inhibition of endothelial nitric-oxide synthase localization and activation in caveolae. *J. Biol. Chem.* 2000 Apr 14;275(15):11278-83.

[34] Stuehr DJ, Cho HJ, Kwon NS, Weise MF, Nathan CF. Purification and characterization of the cytokine-induced macrophage nitric oxide synthase: an FAD- and FMN-containing flavoprotein. *Proc. Natl. Acad. Sci. USA.* 1991 Sep 1;88(17):7773-7.

[35] Baydoun AR, Emery PW, Pearson JD, Mann GE. Substrate-dependent regulation of intracellular amino acid concentrations in cultured bovine aortic endothelial cells. *Biochem. Biophys. Res. Commun.* 1990 Dec 31;173(3):940-8.

[36] Cynober LA. Plasma amino acid levels with a note on membrane transport: characteristics, regulation, and metabolic significance. *Nutrition.* 2002 Sep;18(9):761-6.

[37] Leiper J, Vallance P. New tricks from an old dog: nitric oxide-independent effects of dimethylarginine dimethylaminohydrolase. *Arterioscler Thromb. Vasc. Biol.* 2006 Jul;26(7):1419-20.

[38] Vallance P, Leiper J. Cardiovascular biology of the asymmetric dimethylarginine:dimethylarginine dimethylaminohydrolase pathway. Arterioscler *Thromb. Vasc. Biol.* 2004 Jun;24(6):1023-30.

[39] Stroes E, Hijmering M, van Zandvoort M, Wever R, Rabelink TJ, van Faassen EE. Origin of superoxide production by endothelial nitric oxide synthase. *FEBS Lett.* 1998 Nov 6;438(3):161-4.

[40] Maier W, Cosentino F, Lutolf RB, Fleisch M, Seiler C, Hess OM, et al. Tetrahydrobiopterin improves endothelial function in patients with coronary artery disease. *J. Cardiovasc. Pharmacol.* 2000 Feb;35(2):173-8.

[41] Heitzer T, Brockhoff C, Mayer B, Warnholtz A, Mollnau H, Henne S, et al. Tetrahydrobiopterin improves endothelium-dependent vasodilation in chronic smokers: evidence for a dysfunctional nitric oxide synthase. *Circ. Res.* 2000 Feb 4;86(2):E36-41.

[42] Stroes E, Kastelein J, Cosentino F, Erkelens W, Wever R, Koomans H, et al. Tetrahydrobiopterin restores endothelial function in hypercholesterolemia. *J. Clin. Invest.* 1997 Jan 1;99(1):41-6.

[43] Beckman JS, Ischiropoulos H, Zhu L, van der Woerd M, Smith C, Chen J, et al. Kinetics of superoxide dismutase- and iron-catalyzed nitration of phenolics by peroxynitrite. *Arch. Biochem. Biophys.* 1992 Nov 1;298(2):438-45.

[44] Graham A, Hogg N, Kalyanaraman B, O'Leary V, Darley-Usmar V, Moncada S. Peroxynitrite modification of low-density lipoprotein leads to recognition by the macrophage scavenger receptor. *FEBS Lett.* 1993 Sep 13;330(2):181-5.

[45] Patel RP, Levonen A, Crawford JH, Darley-Usmar VM. Mechanisms of the pro- and anti-oxidant actions of nitric oxide in atherosclerosis. *Cardiovasc Res.* 2000 Aug 18;47(3):465-74.

[46] Kuzkaya N, Weissmann N, Harrison DG, Dikalov S. Interactions of peroxynitrite, tetrahydrobiopterin, ascorbic acid, and thiols: implications for uncoupling endothelial nitric-oxide synthase. *J. Biol. Chem.* 2003 Jun 20;278(25):22546-54.

[47] Beckman JS, Koppenol WH. Nitric oxide, superoxide, and peroxynitrite: the good, the bad, and ugly. Am J Physiol. 1996 Nov;271(5 Pt 1):C1424-37.

[48] Greenacre SA, Ischiropoulos H. Tyrosine nitration: localisation, quantification, consequences for protein function and signal transduction. *Free Radic Res.* 2001 Jun;34(6):541-81.

[49] Ohara Y, Peterson TE, Harrison DG. Hypercholesterolemia increases endothelial superoxide anion production. *J. Clin. Invest.* 1993 Jun;91(6):2546-51.

[50] Mugge A, Elwell JH, Peterson TE, Hofmeyer TG, Heistad DD, Harrison DG. Chronic treatment with polyethylene-glycolated superoxide dismutase partially restores endothelium-dependent vascular relaxations in cholesterol-fed rabbits. *Circ. Res.* 1991 Nov;69(5):1293-300.

[51] Hattori Y, Kawasaki H, Abe K, Kanno M. Superoxide dismutase recovers altered endothelium-dependent relaxation in diabetic rat aorta. *Am. J. Physiol.* 1991 Oct;261(4 Pt 2):H1086-94.

[52] Warnholtz A, Nickenig G, Schulz E, Macharzina R, Brasen JH, Skatchkov M, et al. Increased NADH-oxidase-mediated superoxide production in the early stages of atherosclerosis: evidence for involvement of the renin-angiotensin system. *Circulation.* 1999 Apr 20;99(15):2027-33.

[53] Miller FJ, Jr., Gutterman DD, Rios CD, Heistad DD, Davidson BL. Superoxide production in vascular smooth muscle contributes to oxidative stress and impaired relaxation in atherosclerosis. *Circ. Res.* 1998 Jun 29;82(12):1298-305.

[54] Guzik TJ, West NE, Black E, McDonald D, Ratnatunga C, Pillai R, et al. Vascular superoxide production by NAD(P)H oxidase: association with endothelial dysfunction and clinical risk factors. *Circ. Res.* 2000 May 12;86(9):E85-90.

[55] Landmesser U, Merten R, Spiekermann S, Buttner K, Drexler H, Hornig B. Vascular extracellular superoxide dismutase activity in patients with coronary artery disease: relation to endothelium-dependent vasodilation. *Circulation.* 2000 May 16;101(19):2264-70.

[56] Beckmann JS, Ye YZ, Anderson PG, Chen J, Accavitti MA, Tarpey MM, et al. Extensive nitration of protein tyrosines in human atherosclerosis detected by immunohistochemistry. *Biol. Chem. Hoppe Seyler.* 1994 Feb;375(2):81-8.

[57] Girerd XJ, Hirsch AT, Cooke JP, Dzau VJ, Creager MA. L-arginine augments endothelium-dependent vasodilation in cholesterol-fed rabbits. *Circ. Res.* 1990 Dec;67(6):1301-8.

[58] Rossitch E, Jr., Alexander E, 3rd, Black PM, Cooke JP. L-arginine normalizes endothelial function in cerebral vessels from hypercholesterolemic rabbits. *J. Clin. Invest.* 1991 Apr;87(4):1295-9.

[59] Creager MA, Gallagher SJ, Girerd XJ, Coleman SM, Dzau VJ, Cooke JP. L-arginine improves endothelium-dependent vasodilation in hypercholesterolemic humans. *J. Clin. Invest.* 1992 Oct;90(4):1248-53.

[60] Beckman JA, Creager MA, Libby P. Diabetes and atherosclerosis: epidemiology, pathophysiology, and management. *Jama.* 2002 May 15;287(19):2570-81.

[61] Hink U, Li H, Mollnau H, Oelze M, Matheis E, Hartmann M, et al. Mechanisms underlying endothelial dysfunction in diabetes mellitus. *Circ. Res.* 2001 Feb 2;88(2):E14-22.

[62] Johnstone MT, Creager SJ, Scales KM, Cusco JA, Lee BK, Creager MA. Impaired endothelium-dependent vasodilation in patients with insulin-dependent diabetes mellitus. *Circulation.* 1993 Dec;88(6):2510-6.

[63] McVeigh GE, Brennan GM, Johnston GD, McDermott BJ, McGrath LT, Henry WR, et al. Impaired endothelium-dependent and independent vasodilation in patients with type 2 (non-insulin-dependent) diabetes mellitus. *Diabetologia.* 1992 Aug;35(8):771-6.

[64] Cosentino F, Hishikawa K, Katusic ZS, Luscher TF. High glucose increases nitric oxide synthase expression and superoxide anion generation in human aortic endothelial cells. *Circulation.* 1997 Jul 1;96(1):25-8.

[65] Ceriello A, Mercuri F, Quagliaro L, Assaloni R, Motz E, Tonutti L, et al. Detection of nitrotyrosine in the diabetic plasma: evidence of oxidative stress. *Diabetologia.* 2001 Jul;44(7):834-8.

[66] Steinberg HO, Brechtel G, Johnson A, Fineberg N, Baron AD. Insulin-mediated skeletal muscle vasodilation is nitric oxide dependent. A novel action of insulin to increase nitric oxide release. *J. Clin. Invest.* 1994 Sep;94(3):1172-9.

[67] Gao F, Gao E, Yue TL, Ohlstein EH, Lopez BL, Christopher TA, et al. Nitric oxide mediates the antiapoptotic effect of insulin in myocardial ischemia-reperfusion: the roles of PI3-kinase, Akt, and endothelial nitric oxide synthase phosphorylation. *Circulation.* 2002 Mar 26;105(12):1497-502.

[68] Fleming I, Schulz C, Fichtlscherer B, Kemp BE, Fisslthaler B, Busse R. AMP-activated protein kinase (AMPK) regulates the insulin-induced activation of the nitric oxide synthase in human platelets. *Thromb. Haemost*. 2003 Nov;90(5):863-71.

[69] Duplain H, Burcelin R, Sartori C, Cook S, Egli M, Lepori M, et al. Insulin resistance, hyperlipidemia, and hypertension in mice lacking endothelial nitric oxide synthase. *Circulation.* 2001 Jul 17;104(3):342-5.

[70] Higaki Y, Hirshman MF, Fujii N, Goodyear LJ. Nitric oxide increases glucose uptake through a mechanism that is distinct from the insulin and contraction pathways in rat skeletal muscle. *Diabetes.* 2001 Feb;50(2):241-7.

[71] Roy D, Perreault M, Marette A. Insulin stimulation of glucose uptake in skeletal muscles and adipose tissues in vivo is NO dependent. *Am. J. Physiol.* 1998 Apr;274(4 Pt 1):E692-9.

[72] Panza JA, Garcia CE, Kilcoyne CM, Quyyumi AA, Cannon RO, 3rd. Impaired endothelium-dependent vasodilation in patients with essential hypertension. Evidence that nitric oxide abnormality is not localized to a single signal transduction pathway. *Circulation.* 1995 Mar 15;91(6):1732-8.

[73] Treasure CB, Klein JL, Vita JA, Manoukian SV, Renwick GH, Selwyn AP, et al. Hypertension and left ventricular hypertrophy are associated with impaired endothelium-mediated relaxation in human coronary resistance vessels. *Circulation.* 1993 Jan;87(1):86-93.

[74] Higashi Y, Oshima T, Ozono R, Watanabe M, Matsuura H, Kajiyama G. Effects of L-arginine infusion on renal hemodynamics in patients with mild essential hypertension. *Hypertension.* 1995 Apr;25(4 Pt 2):898-902.

[75] Kelm M. The L-arginine-nitric oxide pathway in hypertension. *Curr. Hypertens. Rep.* 2003 Feb;5(1):80-6.

[76] Calver A, Collier J, Moncada S, Vallance P. Effect of local intra-arterial NG-monomethyl-L-arginine in patients with hypertension: the nitric oxide dilator mechanism appears abnormal. *J. Hypertens*. 1992 Sep;10(9):1025-31.

[77] Ignjatovic T, Stanisavljevic S, Brovkovych V, Skidgel RA, Erdos EG. Kinin B1 receptors stimulate nitric oxide production in endothelial cells: signaling pathways activated by angiotensin I-converting enzyme inhibitors and peptide ligands. *Mol. Pharmacol.* 2004 Nov;66(5):1310-6.

In: Oxidative Stress: Clinical and Biomedical Implications ISBN: 978-1-60021-888-0
Editors: B. M. Matata, M. M. Elahi, pp. 259-276

Chapter XVI

Airways Inflammation and Reactive Oxygen/Nitrogen Species in Pulmonary Hypertension

Weiling Xu and Serpil C. Erzurum*
Departments of Pathobiology,
Pulmonary, Allergy and Critical Care Medicine,
Lerner Research Institute
Cleveland Clinic
Cleveland, Ohio 44195, USA

Abstract

Pulmonary arterial hypertension (PAH) is a fatal disease characterized by impaired regulation of pulmonary hemodynamics and vascular growth, and is associated with low levels of pulmonary nitric oxide (NO). The mechanisms for NO deficiency include both alterations in the pathways that affect NO biosynthesis, and increased oxidative consumption of NO. Studies in animal models of pulmonary hypertension and patients with PAH confirm that inflammation, excess reactive oxygen species generation, loss of antioxidant activity, and mitochondrial abnormalities play a role in the pathogenesis of PAH. Recent studies suggest that strategies to enhance NO production and to augment antioxidant capacity will be successful treatments for pulmonary hypertension.

* Corresponding author: Mailing Address: Serpil C. Erzurum, M.D. Cleveland Clinic, Lerner Research Institute, 9500 Euclid Ave./ NC22. Cleveland, Ohio 44195, USA. Telephone: 216-445-7191. Fax: 216-636-0104. E-mail: erzurus@ccf.org

Abbreviations

BMPR2	bone morphogenetic protein receptor type II;
COPD	chronic obstructive pulmonary disease;
FPAH	familial pulmonary arterial hypertension;
GPx	glutathione peroxidase;
cGMP	guanosine 3', 5' -cyclic monophosphate;
H_2O_2	hydrogen peroxide;
IPAH	idiopathic pulmonary arterial hypertension;
NADPH	nicotinamide adenine dinucleotide phosphate;
NO_3^-	nitrate;
NO	nitric oxide;
NOS	nitric oxide synthases;
NO_2^-	nitrite;
$ONOO^-$	peroxynitrite;
PAH	pulmonary arterial hypertension;
APAH	pulmonary arterial hypertension related to risk factors or associated conditions;
RNS	reactive nitrogen species;
ROS	reactive oxygen species;
GSH	reduced glutathione;
RANTES	regulated on activation, normal T cell expressed and secreted;
GSNO	S-nitrosoglutathione;
SNO	S-nitrosothiols;
$O_2^{\bullet -}$	superoxide;
SOD	superoxide dismutase;
BH_4	tetrahydrobiopterin;
TGF-β	transforming growth factor-beta;

A. Introduction

Pulmonary arterial hypertension (PAH) is a fatal disease of unknown etiology characterized by a progressive increase in pulmonary artery pressure and vascular remodeling [1, 2]. Secondary forms of PAH are associated with known diseases, such as collagen vascular diseases or portal hypertension, but in the absence of an identifiable etiology are classified as idiopathic (IPAH) [3]. Abnormalities in vasodilators, specifically nitric oxide (NO), and mitochondrial abnormalities have been implicated in the pathogenesis of IPAH [1, 2, 4-7]. The mechanisms for NO deficiency are complex, but inflammation and consequent oxidative stress contribute to the consumption of NO, and ultimately to the pathobiology of PAH [8-10].

B. Pulmonary Arterial Hypertension

Pulmonary arterial hypertension (PAH) is defined by a mean pulmonary artery pressure > 25 mm Hg at rest or > 30 mm Hg with exercise [11]. Current classification of PAH is a result of extensive discussion at the 2003 Third World Symposium held in Venice, Italy, which classified PAH into five categories, Group 1 pulmonary arterial hypertension, Group 2 pulmonary hypertension with left heart disease, Group 3 pulmonary hypertension associated with lung diseases and/or hypoxemia, Group 4 pulmonary hypertension due to chronic thrombotic and/or embolic disease and Group 5 pulmonary hypertension related to other miscellaneous conditions such as sarcoidosis, histiocytosis X or lymphangiomatosis [3]. Group 1 PAH consists of three main subgroups, idiopathic pulmonary arterial hypertension (IPAH), familial pulmonary arterial hypertension (FPAH), and pulmonary arterial hypertension related to associated conditions (APAH) [3]. IPAH is characterized pathologically by pulmonary vascular remodeling including medial hypertrophy and neointima formation in small arteries and plexiform lesions which are capillary-like channels comprised of proliferative monoclonal endothelial cells supported by stroma of myofibroblasts [1, 2, 12] [Figure 1]. Heterozygous germline mutations in the bone morphogenetic protein receptor type II (BMPR2) gene, localized to chromosome 2q33, have been linked to approximately 50% of cases of FPAH [13, 14] and identified in up to 26% of sporadic cases of IPAH [15]. As a member of the transforming growth factor-beta (TGF-β) superfamily of cytokines, BMPs transduce signals and induce growth arrest by binding to heteromeric complexes of type I and II receptors, e.g. BMPR2 [16, 17]. Thus, it is speculated that BMPR2 mutations may result in enhanced cell proliferation, leading to pulmonary artery remodeling, or alternatively that loss of BMP dependent survival signals may lead to endothelial cell apoptosis. IPAH pathogenesis is thus generally proposed to result from an imbalance between pulmonary vascular cell proliferation and susceptibility to apoptosis.

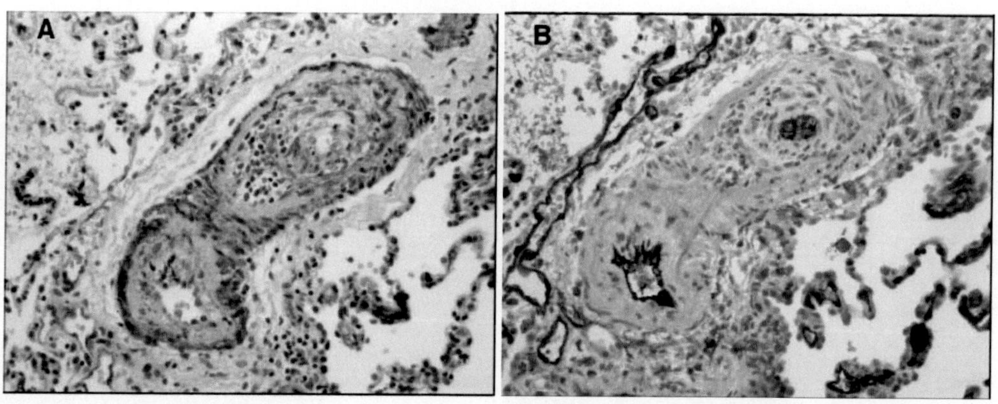

Figure 1. NOS III immunostaining in idiopathic pulmonary arterial hypertension (IPAH) lung. [A] IPAH vascular lesion with cellular proliferation of intima and concentric laminar intimal fibrosis shows strong positive immunoreactivity for NOS III on endothelial cells within the cellular intimal fibroplasia and on the smooth muscle cells of the concentric laminar intimal fibrosis. Immunoreactivity for NOS III is present on the endothelium and smooth muscle cells of an adjacent small artery with medial hypertrophy in IPAH lung. [B] Immunohistochemical staining for CD 31 in the laminar lining cells confirm their endothelial origin. Original magnification: × 400.

C. Reactive Oxygen and Nitrogen Species

Reactive oxygen species (ROS) including hydroxyl radicals (•OH), superoxide ($O_2\bullet^-$) and hydrogen peroxide (H_2O_2) [18, 19] are by-products of normal cellular metabolism. Mitochondria are the largest producer of ROS through the process of cellular respiration in which mitochondria catalyze one-electron reduction of oxygen to superoxide radical followed by formation of hydrogen peroxide, in addition to the four-electron reduction of oxygen to water [20]. Under physiological conditions, ROS are present in the picomolar-to-low nanomolar range and act as essential participants in cell signaling through oxidative modification of proteins. However, increased production of ROS and reactive nitrogen species (RNS) is injurious to tissues, and frequently serves as the final common mediator of pathology in the lung [18, 19].

RNS include reactive molecules such as nitric oxide (NO) and its derivative peroxynitrite ($ONOO^-$). NO is synthesized endogenously by NO synthases (NOS) (EC 1.14.13.39). Three isoforms of NOS, which are products of separate genes [21, 22], have been identified, including neuronal (nNOS or NOS I), endothelial (eNOS or NOS III) and inducible (iNOS or NOS II) [22-24]. NOS I and III, originally identified in neuronal and endothelial cells respectively, depend on increases in calcium to bind calmodulin which results in enzyme activation and picomolar levels of NO production. NOS II is calcium independent, usually expressed after exposure of cells to specific cytokines and/or endotoxin [22, 25-27], binding calmodulin even at low calcium concentrations intracellularly, and produces nanomolar levels of NO [22, 28]. In the healthy human lung, NOS II is continuously expressed in airway epithelial cells and active for high levels NO synthesis for host defense [26]. Once produced, NO is freely diffusible and enters target cells

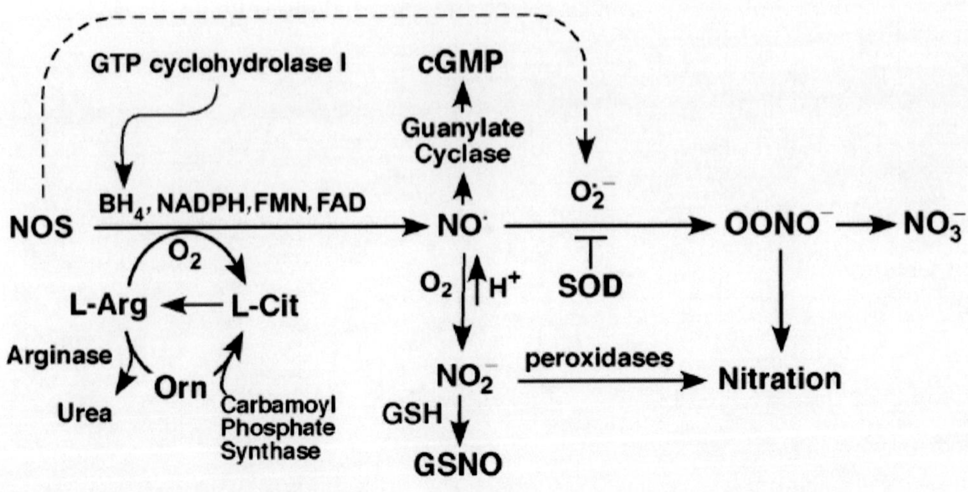

Figure 2. Pathways in NO Metabolism. NOS, nitric oxide synthases; $NO^\bullet$, nitric oxide; NO_2^-, nitrite; NO_3^-, nitrate; L-Arg, L-arginine; L-Cit, L-citrulline; BH_4, tetrahydrobiopterin; NADPH, nicotinamide adenine dinucleotide phosphate; cGMP, guanosine 3', 5'–cyclic monophosphate; $O2^{\bullet-}$, superoxide; $OONO^-$, peroxynitrite; SOD, superoxide dismutase; GSH, reduced glutathione; GSNO, S-nitrosoglutathione.

Vasodilatory effects of NO are mediated by signal transduction primarily through activation of soluble guanylate cyclase to produce guanosine 3', 5' -cyclic monophosphate (cGMP) [29]. NO is also readily oxidized to the metabolic end-products nitrite (NO_2^-) and nitrate (NO_3^-). In particular, NO and superoxide rapidly form the reactive nitrogen oxide, peroxynitrite ($ONOO^-$), which mediates effects through oxidation and nitration of amino acid residues of proteins or guanine of DNA, lipid peroxidation, and DNA cleavage. These modifications can cause cellular dysfunction and injury leading to cell death [30-35], but also may serve in essential functions for immune surveillance, such as host defense against microbes [33, 34, 36, 37] [Figure 2]. The lung avoids the adverse effects of ROS and RNS by minimizing oxidant damage to biological molecules through an integrated and abundant antioxidant system of enzymatic and expendable soluble antioxidants that are present intra- and extracellularly [38, 39]. Antioxidant enzymes include the families of superoxide dismutases (SOD), catalase, glutathione peroxidases (GPx), glutathione S-transferase, and thioredoxin. Nevertheless, despite these antioxidant systems, evidence indicates that excessive ROS contributes to the pathophysiology of PAH [1, 8-10, 40-46].

D. Reactive Oxygen/Nitrogen Species in PAH

Animal Models of PAH with Increased ROS and Inflammation

There is abundant evidence from animal models that proinflammatory cytokines and chemokines, which lead to ROS and RNS production, are involved in the pathogenesis of PAH. The monocrotaline animal model of PAH is well-described and characterized by high levels of pro-inflammatory cytokines and chemokines, e.g. IL-1, IL-6 and TNFα produced by macrophages [47, 48]. In this model, inflammatory cells increase in airway epithelial lining fluid, and inflammatory mediators increase in circulation, suggesting a contributory role of ROS to the development of PAH [48-50]. In the spontaneous idiopathic pulmonary hypertension of avian species, site-specific defects in the electron transport chain in the mitochondria have been identified, which lead to reduced respiratory chain coupling, inefficient use of oxygen and greater generation of ROS that is directly linked to the development of the pulmonary hypertension [51-53]. Similarly, Fawn Hooded rats, a spontaneously pulmonary hypertensive rodent strain, have dysmorphic mitochondria with reduced expression of electron transport chain components and the mitochondrial superoxide dismutase-2 (SOD2). Dichloroacetate, a mitochondrial pyruvate dehydrogenase kinase inhibitor, prevents and reverses the pulmonary hypertension of Fawn Hooded rats, indicating a relation between mitochondria function, ROS generation and/or clearance by antioxidant SOD and the pathogenesis of pulmonary hypertension [54, 55]. Interestingly, TNFα, a proinflammatory cytokine with potent modulatory effects on the pulmonary circulation, [45] [56], impairs both mitochondrial biogenesis and function by downregulating NOS III expression [46]. In animal studies, TNFα increases pulmonary vascular reactivity [45, 57, 58], decreases prostacyclin production in pulmonary artery smooth muscle cells, and potentiates platelet-activating factor (PAF)-induced pulmonary vasoconstriction [42]. Acute and chronic administration of TNFα increases pulmonary vascular reactivity in isolated rat lungs [58], and severe PAH develops in transgenic mice overexpressing TNFα, whereas TNF receptor-deficient mice are protected against PAH [45, 59].

Increased ROS and Inflammation in Human PAH

Multiple studies show that ROS levels are increased in human PAH, whether idiopathic or associated with other lung diseases, and that ROS play a role in the vascular remodeling process [4, 8, 60-63]. For example, PAH is a common complication in patients with chronic obstructive pulmonary disease (COPD), a disease characterized by chronic inflammation of the airways [64], even in those patients without significant arterial hypoxemia [40, 65]. Several studies have shown abnormalities in the structure of pulmonary muscular arteries in lungs of patients with COPD [43], and endothelial dysfunction in pulmonary arteries may be present in end-stage disease or in early mild airway disease [40, 66, 67]. The intensity of the intimal vascular thickening appears to correlate with the severity of the inflammatory infiltrate in the airways [43, 68], and greater number of inflammatory cells, especially CD8-positive lymphocytes are found in the adventitia of small pulmonary arteries in patients with COPD as compared with controls [42, 43]. Furthermore, the intensity of the inflammatory infiltrate in the arteries is associated with an impaired endothelium-dependent vascular relaxation and the thickening of the intimal layer [43], which suggests that an inflammatory process is involved in the pathogenesis of the structural and functional alterations of the vascular bed in COPD. Studies also show that serum C-reactive protein and TNFα levels in COPD patients with PAH are significantly higher than those patients without pulmonary hypertension, further supporting a role for inflammation in the pathogenesis of PAH [42, 69, 70].

Increased amounts of oxidatively modified products, including nitrotyrosine, 8-hydroxyguanosine and isoprostanes, in conjunction with decreased amounts of Mn-SOD protein and activity, are found in the lung tissues of patients with severe PAH, attesting to the presence of oxidant stress and antioxidant deficiency [61]. Isoprostanes, non-enzymatic products of lipid peroxidation of arachidonic acid, likely contribute to the disease process as they cause vasoconstriction, platelet aggregation, cellular proliferation, and induce endothelin synthesis [62, 71]. In support of the concept that oxidized metabolites participate in PAH pathogenesis, urinary levels of isoprostaglandin $F_{2\alpha}$, a marker of lipid peroxidation, are increased in PAH patients and correlate inversely to vasoreactivity to exogenously inhaled NO [7].

Studies also indicate that inflammation plays a role in the pathogenesis of IPAH. Immunological disturbances are present in IPAH patients [72], and a subset of IPAH patients have circulating autoantibodies, including antinuclear antibodies [73], and elevated serum levels of pro-inflammatory cytokines IL-1, IL-6, RANTES (regulated on activation, normal T cell expressed and secreted) and fractalkine [10, 41, 44, 74]. Lung histology reveals an inflammatory necrotizing arteritis, and infiltration of macrophages and lymphocytes into the pulmonary vasculature [9]. Similar to avian PAH and the fawn-hooded rat PAH, evaluation of pulmonary artery endothelial cells from human IPAH lungs reveal that mitochondrial function is reduced, primarily related to lower mitochondrial numbers per cell, and that glucose metabolism subserves the primary role for energy-requirements of IPAH cells [75]. Positron emission tomography scan with [^{18}F] fluoro-deoxy-D-glucose (FDG) performed on IPAH patients and healthy controls confirm significantly higher uptake of FDG in IPAH lungs as compared to controls, supporting that there are substantial changes in bioenergetics of IPAH endothelial cells, which likely has consequences on ROS and RNS generation [75].

Antioxidants and PAH

Perhaps some of the strongest data in support of ROS and RNS in the pathogenesis of PAH is derived from studies in which antioxidants prevent or abrogate PAH [4, 8, 60-63]. For example, the antioxidant dimethythiourea inhibits the right heart dysfunction induced by a 10-day exposure to hypoxia [10% oxygen] in rats [76-78], and pulmonary arterial pressure is lower in hypoxic rats treated with dimethylthiourea than in not-treated hypoxic rats [78]. PAH syndrome in avian species with lower respiratory chain coupling, inefficient use of oxygen and higher generation of ROS [51-53], is successfully treated by vitamin E implants applied after hatching [79]. The incidence of avian PAH syndrome is decreased, and antioxidant capacity increased, in the vitamin E-treated group of chickens [79]. N-acetyl-L-cysteine reduces pulmonary hypertension, right ventricular hypertrophy, and pulmonary vascular media thickening in rats caused by 3 weeks of normobaric hypoxic exposure [80]. In a human study, SOD supplementation improves oxygenation and reduces oxidation in neonatal PAH [63]. Altogether, these findings support a role for oxidant excess and antioxidant insufficiency in pulmonary hypertension.

E. Decreased NO in PAH

In addition to direct effects on biological compounds, excessive ROS may lead to PAH through effects on the bioavailability of the vasodilator substance NO [8, 61, 81-83] [Figure 2]. There is conclusive evidence from mice genetically deficient in NOS, and complementation studies with gene transfer of NOS, for the concept that NO is a critical determinant of pulmonary vascular tone [84-87]. Studies of mice genetically deficient in NOS suggest that NOS III is the primary regulator of basal pulmonary vascular tone, with a significant contributory role of NOS II [84-88]. Pulmonary and total body NO levels are lower in PAH patients as compared to healthy controls [4, 7, 89-92]. Furthermore, exhaled NO measured in PAH patients in conjunction with hemodynamic parameters over 2 years shows that NO is inversely related to pulmonary arterial pressures over time [7]. When compared with control individuals, IPAH patients have decreased nitrate levels in exhaled breath condensate [7]. Formation of nitrate occurs in part through decomposition of peroxynitrite generated by reaction of NO and superoxide. Thus, the finding of low nitrate in IPAH calls into question whether or not RNS significantly contribute to PAH [Figure 2]. However, nitrate levels in breath condensate of PAH patients are directly correlated with pulmonary arterial pressure, although NO levels are inversely related to pulmonary pressures, indicating that both the low levels of NO and greater RNS formation contribute to pulmonary hypertension [7].

Mechanisms of Low NO in PAH

The concentration of NO in any biologic system is the sum of its rate of enzymatic formation and consumption by other biological molecules, such as in reactions with ROS [8, 61, 81-83]. Decreased NO production in PAH results from both impaired NO synthase activity and increased oxidative consumption of NO.

NO biosynthesis is regulated at multiple levels in cells, i.e. NOS gene transcription, mRNA processing, protein expression and dimerization, and enzyme reaction kinetics [22]. All NOS convert L-arginine to NO and L-citrulline in a reaction that requires oxygen, NADPH, and cofactors FAD, FMN, tetrahydrobiopterin (BH_4), calmodulin and iron protoporphyrin IX [Figure 2]. Pathways that alter substrate L-arginine availability or co-factor levels are key mechanisms for regulation of NO synthesis [89, 93, 94]. Most previous studies indicate that NOS III is present in IPAH lungs, with no significant decrease in expression as compared with normal lung [28, 95-97] [Figure 1]. Similarly, NOS II expression in airway epithelial cells is similar among PAH and controls by western analyses [89]. Despite normal NOS protein expression, NOS biosynthetic activity is decreased in IPAH lung cells and tissues [98, 99].

Decreased NOS III activity in PAH

NOS activity is regulated by availability of its substrate and co-factors. In particular, the cofactor BH_4 is an important regulator of NOS enzymatic activity [100] [Figure 2], and under conditions of limiting BH_4, NO synthesis is reduced and superoxide production by NOS enzymes predominates as a result of the loss of enzymatic coupling between the reduction in molecular oxygen and oxidation of L-arginine [99, 101]. The hyperphenylalaninemic mutant mouse, which has deficient BH_4 due to reduced activity of the rate limiting enzyme in BH_4 synthesis GTP-cyclohydrolase I, provides proof for the concept that deficiency of this cofactor leads to reduction of NO synthesis and to the development of pulmonary hypertension. BH_4 bioavailability clearly regulates pulmonary vascular pressures and vascular structural remodeling in the mutant mouse under normoxia or hypoxia. Importantly, crossing the hyperphenylalaninemic mutant mice to mice with endothelial-targeted over-expression of GTP-cyclohydrolase I restores NO synthesis and rescues the progeny from developing pulmonary hypertension, confirming the essential role for endothelial BH_4 in the synthesis of NO and regulation of the pulmonary circulation [98, 99, 102].

Substrate Availability and NO Synthesis

Decrease of NO has also been linked to limitation of arginine bioavailability in PAH [89]. For example, abnormalities in arginine metabolism have been implicated in neonatal PAH. Infants with PAH have decreased NO synthesis with half-normal levels of plasma arginine, due to decreased activity of a variant form of carbamoyl-phosphate synthetase, the rate-limiting enzyme of the urea cycle that leads to arginine synthesis [103] [Figure 2]. Arginase, another critical enzyme in the urea cycle, converts arginine to ornithine and urea , and also plays a regulatory role in NO synthesis by modulating the availability of substrate arginine for NOS [104, 105]. Products of separate genes, arginase I on chromosome 6q23, and arginase II on 14q24, arginase I is expressed almost exclusively in red blood cells and cytosol of liver cells, and arginase II is located in the mitochondrial matrix [106-110]. Arginase activity is greater in IPAH serum than in control serum, and pulmonary artery endothelial cells derived from IPAH lung have greater arginase II expression and produce lower NO than control cells [89]. Arginase activity is also increased 2-fold in serum of sickle cell patients who have associated secondary PAH [111]. In support of a contributory role for arginine in PAH, supplemental arginine uniformly improves endothelium dependent vasodilation, and NO synthesis [112-115]. For example, arginine supplementation has been used in treatment of PAH, i.e. hypoxia-induced PAH in rodents is nearly abolished by daily injections of arginine [6], and oral or intravenous arginine reduces pulmonary pressure in PAH patients [111, 116, 117].

Oxidant Consumption and Reduced Bioavailable NO

In addition to the reduced NO biosynthesis, increased consumption of NO through oxidative reactions contributes to the low levels of NO in PAH. S-nitrosothiols (SNO) in human plasma and in the airway epithelial lining fluid are the physiologic bioactive pool that normally serves as a storage form and source of NO [118-120]. NO reacts with reduced glutathione (GSH), a water-soluble antioxidant present at abundant levels in human airway epithelial lining fluid, to form S-nitrosoglutathione (GSNO) [118] [Figure 2]. Similar to NO, GSNO levels are lower in PAH as compared to control lungs, and levels of S-nitrosohemoglobin in the systemic circulation are lower in PAH than in controls [120-122]. The successful therapy of neonatal PAH with GSNO highlights the potential importance of GSNO in pulmonary vasoregulation [120]. On the other hand, the increased presence of nitrotyrosine within IPAH lungs [61] indicates that NO is increasingly consumed in reactions with superoxide which results in peroxynitrite formation and subsequent nitration of tyrosine residues in proteins, as opposed to the formation of bioactive NO storage forms, e.g. GSNO [8, 61, 81-83] [Figure 2]. The greater formation of peroxynitrite likely also exacerbates oxidative stress through inactivation of antioxidant proteins [61, 123, 124]. For example, nitration of MnSOD by peroxynitrite leads to profound loss of the enzyme activity [123]. In addition, since NO is required for expression of the extracellular SOD gene (ecSOD) [125], loss of bioavailable NO may also lead to lower total SOD activity through reduced ecSOD expression. Overall, ROS consumption of NO contributes to the lower levels of bioactive NO and GSNO, and to the reduced antioxidant defenses in PAH, either through post-translational modifications and/or transcriptional effects on SODs, which further amplifies oxidant stress and NO consumption.

F. Conclusion

PAH is characterized by impaired regulation of pulmonary hemodynamics and vascular growth, and is associated with low levels of pulmonary NO. The mechanisms for NO deficiency include [1] alterations in the pathways that affect substrate L-arginine or co-factor BH_4 bioavailability for NOS biosynthetic functions, and [2] increased oxidative consumption of NO. There is strong support for inflammation, excess ROS generation, loss of antioxidant activity, and mitochondrial abnormalities in the pathogenesis of PAH. Taken together, the data suggest that strategies to enhance NO production through arginine, BH_4 and/or GSNO supplementation and to augment antioxidant levels might be successful and synergistic treatments for pulmonary hypertension.

Acknowledgments

WX and SCE supported by AI70649, HL60917, and M01 RR018390 from the National Center for Research Resources.

References

[1] Jeffery TK, Morrell NW. Molecular and cellular basis of pulmonary vascular remodeling in pulmonary hypertension. *Prog. Cardiovasc. Dis.* 2002;45:173-202.

[2] Archer S, Rich S. Primary pulmonary hypertension: A vascular biology and translational research "Work in progress". *Circulation* 2000;102:2781-2791.

[3] Simonneau G, Galie N, Rubin LJ, Langleben D, Seeger W, Domenighetti G, Gibbs S, Lebrec D, Speich R, Beghetti M, et al. Clinical classification of pulmonary hypertension. *J. Am. Coll Cardiol.* 2004;43:5S-12S.

[4] Kaneko FT, Arroliga AC, Dweik RA, Comhair SA, Laskowski D, Oppedisano R, Thomassen MJ, Erzurum SC. Biochemical reaction products of nitric oxide as quantitative markers of primary pulmonary hypertension. *Am. J. Respir. Crit Care Med.* 1998;158:917-923.

[5] Azeka E, Costa Auler JO, Jr., Kajita L, Alliman AC, Franchini Ramires JA, Ebaid M. Effects of low doses of inhaled nitric oxide combined with oxygen for the evaluation of pulmonary vascular reactivity in patients with pulmonary hypertension. *Pediatr. Cardiol.* 2002;23:20-26.

[6] Cooke JP, Dzau VJ. Derangements of the nitric oxide synthase pathway, l-arginine, and cardiovascular diseases. *Circulation* 1997;96:379-382.

[7] Machado RF, Londhe Nerkar MV, Dweik RA, Hammel J, Janocha A, Pyle J, Laskowski D, Jennings C, Arroliga AC, Erzurum SC. Nitric oxide and pulmonary

arterial pressures in pulmonary hypertension. *Free Radic. Biol. Med.* 2004;37:1010-1017.

[8] Wedgwood S, Black SM. Role of reactive oxygen species in vascular remodeling associated with pulmonary hypertension. *Antioxid Redox. Signal.* 2003;5:759-769.

[9] Tuder RM, Groves B, Badesch DB, Voelkel NF. Exuberant endothelial cell growth and elements of inflammation are present in plexiform lesions of pulmonary hypertension. *Am. J. Pathol.* 1994;144:275-285.

[10] Humbert M, Monti G, Brenot F, Sitbon O, Portier A, Grangeot-Keros L, Duroux P, Galanaud P, Simonneau G, Emilie D. Increased interleukin-1 and interleukin-6 serum concentrations in severe primary pulmonary hypertension. *Am. J. Respir Crit Care Med.* 1995;151:1628-1631.

[11] Rubin LJ. Primary pulmonary hypertension. *Chest* 1993;104:236-250.

[12] Fishman AP. Changing concepts of the pulmonary plexiform lesion. *Physiol. Res.* 2000;49:485-492.

[13] Deng Z, Morse JH, Slager SL, Cuervo N, Moore KJ, Venetos G, Kalachikov S, Cayanis E, Fischer SG, Barst RJ, et al. Familial primary pulmonary hypertension (gene pph1) is caused by mutations in the bone morphogenetic protein receptor-ii gene. *Am. J. Hum. Genet.* 2000;67:737-744.

[14] Lane KB, Machado RD, Pauciulo MW, Thomson JR, Phillips JA, 3rd, Loyd JE, Nichols WC, Trembath RC. Heterozygous germline mutations in bmpr2, encoding a tgf-beta receptor, cause familial primary pulmonary hypertension. The international pph consortium. *Nat. Genet.* 2000;26:81-84.

[15] Thomson JR, Machado RD, Pauciulo MW, Morgan NV, Humbert M, Elliott GC, Ward K, Yacoub M, Mikhail G, Rogers P, et al. Sporadic primary pulmonary hypertension is associated with germline mutations of the gene encoding bmpr-ii, a receptor member of the tgf-beta family. *J. Med. Genet.* 2000;37:741-745.

[16] McCaffrey TA. Tgf-betas and tgf-beta receptors in atherosclerosis. *Cytokine Growth Factor Rev.* 2000;11:103-114.

[17] Pepper MS. Transforming growth factor-beta: Vasculogenesis, angiogenesis, and vessel wall integrity. *Cytokine Growth Factor Rev.* 1997;8:21-43.

[18] Thannickal VJ, Fanburg BL. Reactive oxygen species in cell signaling. *Am. J. Physiol. Lung Cell Mol. Physiol.* 2000;279:L1005-1028.

[19] Comhair SA, Erzurum SC. Antioxidant responses to oxidant-mediated lung diseases. *Am J Physiol Lung Cell Mol Physiol* 2002;283:L246-255.

[20] Vinogradov AD, Grivennikova VG. Generation of superoxide-radical by the nadh:Ubiquinone oxidoreductase of heart mitochondria. *Biochemistry (Mosc)* 2005;70:120-127.

[21] Nathan C, Xie QW. Nitric oxide synthases: Roles, tolls, and controls. *Cell* 1994;78:915-918.

[22] Stuehr DJ, Griffith OW. Mammalian nitric oxide synthases. *Adv. Enzymol Relat. Areas Mol. Biol.* 1992;65:287-346.

[23] Gaston B, Drazen JM, Loscalzo J, Stamler JS. The biology of nitrogen oxides in the airways. *Am. J. Respir. Crit. Care Med.* 1994;149:538-551.

[24] Nathan C. Nitric oxide as a secretory product of mammalian cells. *Faseb J.* 1992;6:3051-3064.

[25] Guo FH, Erzurum SC. Characterization of inducible nitric oxide synthase expression in human airway epithelium. *Environ. Health Perspect.* 1998;106 Suppl 5:1119-1124.

[26] Guo FH, Uetani K, Haque SJ, Williams BR, Dweik RA, Thunnissen FB, Calhoun W, Erzurum SC. Interferon gamma and interleukin 4 stimulate prolonged expression of inducible nitric oxide synthase in human airway epithelium through synthesis of soluble mediators. *J. Clin. Invest.* 1997;100:829-838.

[27] Schmidt HH, Walter U. No at work. *Cell* 1994;78:919-925.

[28] Guo FH, Comhair SA, Zheng S, Dweik RA, Eissa NT, Thomassen MJ, Calhoun W, Erzurum SC. Molecular mechanisms of increased nitric oxide (no) in asthma: Evidence for transcriptional and post-translational regulation of no synthesis. *J. Immunol.* 2000;164:5970-5980.

[29] Stuehr DJ. Mammalian nitric oxide synthases. *Biochim. Biophys. Acta* 1999;1411:217-230.

[30] Aulak KS, Miyagi M, Yan L, West KA, Massillon D, Crabb JW, Stuehr DJ. Proteomic method identifies proteins nitrated in vivo during inflammatory challenge. *Proc. Natl. Acad. Sci. USA* 2001;98:12056-12061.

[31] Aulak KS, Koeck T, Crabb JW, Stuehr DJ. Dynamics of protein nitration in cells and mitochondria. *Am. J. Physiol. Heart Circ. Physiol.* 2004;286:H30-38.

[32] Gow AJ, Farkouh CR, Munson DA, Posencheg MA, Ischiropoulos H. Biological significance of nitric oxide-mediated protein modifications. *Am. J. Physiol. Lung Cell Mol. Physiol.* 2004;287:L262-268.

[33] Akaike T, Suga M, Maeda H. Free radicals in viral pathogenesis: Molecular mechanisms involving superoxide and no. *Proc. Soc. Exp. Biol. Med.* 1998;217:64-73.

[34] Valyi-Nagy T, Dermody TS. Role of oxidative damage in the pathogenesis of viral infections of the nervous system. *Histol. Histopathol.* 2005;20:957-967.

[35] Brennan ML, Wu W, Fu X, Shen Z, Song W, Frost H, Vadseth C, Narine L, Lenkiewicz E, Borchers MT, et al. A tale of two controversies: Defining both the role of peroxidases in nitrotyrosine formation in vivo using eosinophil peroxidase and myeloperoxidase-deficient mice, and the nature of peroxidase-generated reactive nitrogen species. *J. Biol. Chem.* 2002;277:17415-17427.

[36] Kato M, Hayashi Y, Kimura H. Oxygen radicals in inflammation and allergy related to viral infections. *Curr. Drug Targets Inflamm Allergy* 2005;4:497-501.

[37] Miller RA, Britigan BE. Role of oxidants in microbial pathophysiology. *Clin. Microbiol. Rev.* 1997;10:1-18.

[38] Sies H. Strategies of antioxidant defense. *Eur. J. Biochem.* 1993;215:213-219.

[39] Wassmann S, Wassmann K, Nickenig G. Modulation of oxidant and antioxidant enzyme expression and function in vascular cells. *Hypertension* 2004;44:381-386.

[40] Barbera JA, Peinado VI, Santos S. Pulmonary hypertension in chronic obstructive pulmonary disease. *Eur. Respir. J.* 2003;21:892-905.

[41] Dorfmuller P, Perros F, Balabanian K, Humbert M. Inflammation in pulmonary arterial hypertension. *Eur. Respir. J.* 2003;22:358-363.

[42] Joppa P, Petrasova D, Stancak B, Tkacova R. Systemic inflammation in patients with copd and pulmonary hypertension. *Chest* 2006;130:326-333.
[43] Peinado VI, Barbera JA, Abate P, Ramirez J, Roca J, Santos S, Rodriguez-Roisin R. Inflammatory reaction in pulmonary muscular arteries of patients with mild chronic obstructive pulmonary disease. *Am. J. Respir. Crit Care Med.* 1999;159:1605-1611.
[44] Dorfmuller P, Zarka V, Durand-Gasselin I, Monti G, Balabanian K, Garcia G, Capron F, Coulomb-Lhermine A, Marfaing-Koka A, Simonneau G, et al. Chemokine rantes in severe pulmonary arterial hypertension. *Am. J. Respir. Crit. Care Med.* 2002;165:534-539.
[45] Fujita M, Shannon JM, Irvin CG, Fagan KA, Cool C, Augustin A, Mason RJ. Overexpression of tumor necrosis factor-alpha produces an increase in lung volumes and pulmonary hypertension. *Am. J. Physiol. Lung Cell Mol. Physiol.* 2001;280:L39-49.
[46] Valerio A, Cardile A, Cozzi V, Bracale R, Tedesco L, Pisconti A, Palomba L, Cantoni O, Clementi E, Moncada S, et al. Tnf-alpha downregulates enos expression and mitochondrial biogenesis in fat and muscle of obese rodents. *J Clin Invest* 2006.
[47] Miyata M, Sakuma F, Yoshimura A, Ishikawa H, Nishimaki T, Kasukawa R. Pulmonary hypertension in rats. 2. Role of interleukin-6. *Int. Arch. Allergy Immunol.* 1995;108:287-291.
[48] Kimura H, Kasahara Y, Kurosu K, Sugito K, Takiguchi Y, Terai M, Mikata A, Natsume M, Mukaida N, Matsushima K, et al. Alleviation of monocrotaline-induced pulmonary hypertension by antibodies to monocyte chemotactic and activating factor/monocyte chemoattractant protein-1. *Lab. Invest* 1998;78:571-581.
[49] Stenmark KR, Morganroth ML, Remigio LK, Voelkel NF, Murphy RC, Henson PM, Mathias MM, Reeves JT. Alveolar inflammation and arachidonate metabolism in monocrotaline-induced pulmonary hypertension. *Am. J. Physiol.* 1985;248:H859-866.
[50] Chen MJ, Chiang LY, Lai YL. Reactive oxygen species and substance p in monocrotaline-induced pulmonary hypertension. *Toxicol Appl. Pharmacol.* 2001;171:165-173.
[51] Tang Z, Iqbal M, Cawthon D, Bottje WG. Heart and breast muscle mitochondrial dysfunction in pulmonary hypertension syndrome in broilers (gallus domesticus). *Comp Biochem. Physiol. A Mol. Integr. Physiol.* 2002;132:527-540.
[52] Iqbal M, Cawthon D, Wideman RF, Jr., Bottje WG. Lung mitochondrial dysfunction in pulmonary hypertension syndrome. I. Site-specific defects in the electron transport chain. *Poult. Sci.* 2001;80:485-495.
[53] Cawthon D, Iqbal M, Brand J, McNew R, Bottje WG. Investigation of proton conductance in liver mitochondria of broilers with pulmonary hypertension syndrome. *Poult Sci* 2004;83:259-265.
[54] Michelakis ED, McMurtry MS, Wu XC, Dyck JR, Moudgil R, Hopkins TA, Lopaschuk GD, Puttagunta L, Waite R, Archer SL. Dichloroacetate, a metabolic modulator, prevents and reverses chronic hypoxic pulmonary hypertension in rats: Role of increased expression and activity of voltage-gated potassium channels. *Circulation* 2002;105:244-250.

[55] Bonnet S, Michelakis ED, Porter CJ, Andrade-Navarro MA, Thebaud B, Haromy A, Harry G, Moudgil R, McMurtry MS, Weir EK, et al. An abnormal mitochondrial-hypoxia inducible factor-1alpha-kv channel pathway disrupts oxygen sensing and triggers pulmonary arterial hypertension in fawn hooded rats: Similarities to human pulmonary arterial hypertension. *Circulation* 2006;113:2630-2641.

[56] Meldrum DR. Tumor necrosis factor in the heart. *Am. J. Physiol.* 1998;274:R577-595.

[57] Fujita M, Mason RJ, Cool C, Shannon JM, Hara N, Fagan KA. Pulmonary hypertension in tnf-alpha-overexpressing mice is associated with decreased vegf gene expression. *J. Appl. Physiol.* 2002;93:2162-2170.

[58] Stevens T, Janssen PL, Tucker A. Acute and long-term tnf-alpha administration increases pulmonary vascular reactivity in isolated rat lungs. *J. Appl. Physiol.* 1992;73:708-712.

[59] Ortiz LA, Lasky J, Hamilton RF, Jr., Holian A, Hoyle GW, Banks W, Peschon JJ, Brody AR, Lungarella G, Friedman M. Expression of tnf and the necessity of tnf receptors in bleomycin-induced lung injury in mice. *Exp. Lung Res.* 1998;24:721-743.

[60] Belik J, Jankov RP, Pan J, Tanswell AK. Peroxynitrite inhibits relaxation and induces pulmonary artery muscle contraction in the newborn rat. *Free Radic. Biol. Med.* 2004;37:1384-1392.

[61] Bowers R, Cool C, Murphy RC, Tuder RM, Hopken MW, Flores SC, Voelkel NF. Oxidative stress in severe pulmonary hypertension. *Am. J. Respir. Crit. Care Med.* 2004;169:764-769.

[62] Cracowski JL, Cracowski C, Bessard G, Pepin JL, Bessard J, Schwebel C, Stanke-Labesque F, Pison C. Increased lipid peroxidation in patients with pulmonary hypertension. *Am. J. Respir. Crit. Care Med.* 2001;164:1038-1042.

[63] Lakshminrusimha S, Russell JA, Wedgwood S, Gugino SF, Kazzaz JA, Davis JM, Steinhorn RH. Superoxide dismutase improves oxygenation and reduces oxidation in neonatal pulmonary hypertension. *Am. J. Respir. Crit. Care Med.* 2006.

[64] Langen RC, Korn SH, Wouters EF. Ros in the local and systemic pathogenesis of copd. *Free Radic. Biol. Med.* 2003;35:226-235.

[65] Sin DD, Man SF. Is systemic inflammation responsible for pulmonary hypertension in copd? *Chest* 2006;130:310-312.

[66] Dinh-Xuan AT, Higenbottam TW, Clelland CA, Pepke-Zaba J, Cremona G, Butt AY, Large SR, Wells FC, Wallwork J. Impairment of endothelium-dependent pulmonary-artery relaxation in chronic obstructive lung disease. *N. Engl. J. Med.* 1991;324:1539-1547.

[67] Peinado VI, Barbera JA, Ramirez J, Gomez FP, Roca J, Jover L, Gimferrer JM, Rodriguez-Roisin R. Endothelial dysfunction in pulmonary arteries of patients with mild copd. *Am. J. Physiol.* 1998;274:L908-913.

[68] Barbera JA, Riverola A, Roca J, Ramirez J, Wagner PD, Ros D, Wiggs BR, Rodriguez-Roisin R. Pulmonary vascular abnormalities and ventilation-perfusion relationships in mild chronic obstructive pulmonary disease. *Am. J. Respir. Crit. Care Med.* 1994;149:423-429.

[69] de Godoy I, Donahoe M, Calhoun WJ, Mancino J, Rogers RM. Elevated tnf-alpha production by peripheral blood monocytes of weight-losing copd patients. *Am. J. Respir. Crit Care Med.* 1996;153:633-637.
[70] Di Francia M, Barbier D, Mege JL, Orehek J. Tumor necrosis factor-alpha levels and weight loss in chronic obstructive pulmonary disease. *Am. J. Respir. Crit. Care Med.* 1994;150:1453-1455.
[71] Janssen LJ. Isoprostanes: An overview and putative roles in pulmonary pathophysiology. *Am. J. Physiol. Lung Cell Mol. Physiol.* 2001;280:L1067-1082.
[72] Humbert M, Nunes H, Sitbon O, Parent F, Herve P, Simonneau G. Risk factors for pulmonary arterial hypertension. *Clin. Chest Med.* 2001;22:459-475.
[73] Isern RA, Yaneva M, Weiner E, Parke A, Rothfield N, Dantzker D, Rich S, Arnett FC. Autoantibodies in patients with primary pulmonary hypertension: Association with anti-ku. *Am. J. Med.* 1992;93:307-312.
[74] Balabanian K, Foussat A, Dorfmuller P, Durand-Gasselin I, Capel F, Bouchet-Delbos L, Portier A, Marfaing-Koka A, Krzysiek R, Rimaniol AC, et al. Cx(3)c chemokine fractalkine in pulmonary arterial hypertension. *Am. J. Respir. Crit. Care Med.* 2002;165:1419-1425.
[75] Xu W, Koeck T, Lara AR, Neumann D, DiFilippo FP, Koo M, Janocha AJ, Masri FA, Arroliga AC, Jennings C, et al. Alterations of cellular bioenergetics in pulmonary artery endothelial cells *Proc. Natl. Acad. Sci. USA* 2007;104:1342-1347.
[76] Langleben D, Fox RB, Jones RC, Reid LM. Effects of dimethylthiourea on chronic hypoxia-induced pulmonary arterial remodelling and ventricular hypertrophy in rats. *Clin. Invest Med.* 1989;12:235-240.
[77] Lai YL, Wu HD, Chen CF. Antioxidants attenuate chronic hypoxic pulmonary hypertension. *J Cardiovasc Pharmacol* 1998;32:714-720.
[78] Herget J, Wilhelm J, Novotna J, Eckhardt A, Vytasek R, Mrazkova L, Ostadal M. A possible role of the oxidant tissue injury in the development of hypoxic pulmonary hypertension. *Physiol Res* 2000;49:493-501.
[79] Bottje W, Enkvetchakul B, Moore R, McNew R. Effect of alpha-tocopherol on antioxidants, lipid peroxidation, and the incidence of pulmonary hypertension syndrome (ascites) in broilers. *Poult. Sci.* 1995;74:1356-1369.
[80] Hoshikawa Y, Ono S, Suzuki S, Tanita T, Chida M, Song C, Noda M, Tabata T, Voelkel NF, Fujimura S. Generation of oxidative stress contributes to the development of pulmonary hypertension induced by hypoxia. *J. Appl. Physiol.* 2001;90:1299-1306.
[81] Beckman JS, Crow JP. Pathological implications of nitric oxide, superoxide and peroxynitrite formation. *Biochem. Soc. Trans* 1993;21:330-334.
[82] Haddad IY, Pataki G, Hu P, Galliani C, Beckman JS, Matalon S. Quantitation of nitrotyrosine levels in lung sections of patients and animals with acute lung injury. *J. Clin. Invest.* 1994;94:2407-2413.
[83] Hurst JK. Whence nitrotyrosine? *J. Clin. Invest.* 2002;109:1287-1289.
[84] Fagan KA, McMurtry I, Rodman DM. Nitric oxide synthase in pulmonary hypertension: Lessons from knockout mice. *Physiol. Res.* 2000;49:539-548.

[85] Fagan KA, Tyler RC, Sato K, Fouty BW, Morris KG, Jr., Huang PL, McMurtry IF, Rodman DM. Relative contributions of endothelial, inducible, and neuronal nos to tone in the murine pulmonary circulation. *Am. J. Physiol.* 1999;277:L472-478.

[86] Steudel W, Ichinose F, Huang PL, Hurford WE, Jones RC, Bevan JA, Fishman MC, Zapol WM. Pulmonary vasoconstriction and hypertension in mice with targeted disruption of the endothelial nitric oxide synthase (nos 3) gene. *Circ. Res.* 1997;81:34-41.

[87] Champion HC, Bivalacqua TJ, Greenberg SS, Giles TD, Hyman AL, Kadowitz PJ. Adenoviral gene transfer of endothelial nitric-oxide synthase (enos) partially restores normal pulmonary arterial pressure in enos-deficient mice. *Proc. Natl. Acad. Sci. U S A* 2002;99:13248-13253.

[88] Fagan KA, Fouty BW, Tyler RC, Morris KG, Jr., Hepler LK, Sato K, LeCras TD, Abman SH, Weinberger HD, Huang PL, et al. The pulmonary circulation of homozygous or heterozygous enos-null mice is hyperresponsive to mild hypoxia. *J. Clin. Invest.* 1999;103:291-299.

[89] Xu W, Kaneko FT, Zheng S, Comhair SA, Janocha AJ, Goggans T, Thunnissen FB, Farver C, Hazen SL, Jennings C, et al. Increased arginase ii and decreased no synthesis in endothelial cells of patients with pulmonary arterial hypertension. *Faseb J.* 2004;18:1746-1748.

[90] Ozkan M, Dweik RA, Laskowski D, Arroliga AC, Erzurum SC. High levels of nitric oxide in individuals with pulmonary hypertension receiving epoprostenol therapy. *Lung* 2001;179:233-243.

[91] Girgis RE, Champion HC, Diette GB, Johns RA, Permutt S, Sylvester JT. Decreased exhaled nitric oxide in pulmonary arterial hypertension: Response to bosentan therapy. *Am. J. Respir. Crit. Care Med.* 2005;172:352-357.

[92] Cella G, Bellotto F, Tona F, Sbarai A, Mazzaro G, Motta G, Fareed J. Plasma markers of endothelial dysfunction in pulmonary hypertension. *Chest* 2001;120:1226-1230.

[93] Luzzi SD, Marletta MA. L-arginine analogs as alternate substrates for nitric oxide synthase. *Bioorg. Med. Chem. Lett.* 2005;15:3934-3941.

[94] Dweik RA, Laskowski D, Abu-Soud HM, Kaneko F, Hutte R, Stuehr DJ, Erzurum SC. Nitric oxide synthesis in the lung. Regulation by oxygen through a kinetic mechanism. *J. Clin. Invest.* 1998;101:660-666.

[95] Mason NA, Springall DR, Burke M, Pollock J, Mikhail G, Yacoub MH, Polak JM. High expression of endothelial nitric oxide synthase in plexiform lesions of pulmonary hypertension. *J Pathol* 1998;185:313-318.

[96] Veyssier-Belot C, Cacoub P. Role of endothelial and smooth muscle cells in the physiopathology and treatment management of pulmonary hypertension. *Cardiovasc. Res.* 1999;44:274-282.

[97] Guo FH, De Raeve HR, Rice TW, Stuehr DJ, Thunnissen FB, Erzurum SC. Continuous nitric oxide synthesis by inducible nitric oxide synthase in normal human airway epithelium in vivo. *Proc. Natl. Acad. Sci. U S A* 1995;92:7809-7813.

[98] Khoo JP, Zhao L, Alp NJ, Bendall JK, Nicoli T, Rockett K, Wilkins MR, Channon KM. Pivotal role for endothelial tetrahydrobiopterin in pulmonary hypertension. *Circulation* 2005;111:2126-2133.

[99] Cosentino F, Barker JE, Brand MP, Heales SJ, Werner ER, Tippins JR, West N, Channon KM, Volpe M, Luscher TF. Reactive oxygen species mediate endothelium-dependent relaxations in tetrahydrobiopterin-deficient mice. *Arterioscler. Thromb. Vasc. Biol.* 2001;21:496-502.

[100] Katusic ZS. Vascular endothelial dysfunction: Does tetrahydrobiopterin play a role? *Am J. Physiol. Heart Circ. Physiol.* 2001;281:H981-986.

[101] Alp NJ, Channon KM. Regulation of endothelial nitric oxide synthase by tetrahydrobiopterin in vascular disease. *Arterioscler Thromb. Vasc. Biol.* 2004;24:413-420.

[102] Nandi M, Miller A, Stidwill R, Jacques TS, Lam AA, Haworth S, Heales S, Vallance P. Pulmonary hypertension in a gtp-cyclohydrolase 1-deficient mouse. *Circulation* 2005;111:2086-2090.

[103] Pearson DL, Dawling S, Walsh WF, Haines JL, Christman BW, Bazyk A, Scott N, Summar ML. Neonatal pulmonary hypertension--urea-cycle intermediates, nitric oxide production, and carbamoyl-phosphate synthetase function. *N. Engl. J. Med.* 2001;344:1832-1838.

[104] Leiper J, Vallance P. Biological significance of endogenous methylarginines that inhibit nitric oxide synthases. *Cardiovasc. Res.* 1999;43:542-548.

[105] Leiper JM, Santa Maria J, Chubb A, MacAllister RJ, Charles IG, Whitley GS, Vallance P. Identification of two human dimethylarginine dimethylaminohydrolases with distinct tissue distributions and homology with microbial arginine deiminases. *Biochem J* 1999;343 Pt 1:209-214.

[106] Gotoh T, Araki M, Mori M. Chromosomal localization of the human arginase ii gene and tissue distribution of its mrna. *Biochem. Biophys. Res. Commun.* 1997;233:487-491.

[107] Mori M, Gotoh T. Regulation of nitric oxide production by arginine metabolic enzymes. *Biochem. Biophys. Res. Commun.* 2000;275:715-719.

[108] Wei LH, Jacobs AT, Morris SM, Jr., Ignarro LJ. Il-4 and il-13 upregulate arginase i expression by camp and jak/stat6 pathways in vascular smooth muscle cells. *Am. J. Physiol. Cell Physiol.* 2000;279:C248-256.

[109] Wei LH, Wu G, Morris SM, Jr., Ignarro LJ. Elevated arginase i expression in rat aortic smooth muscle cells increases cell proliferation. *Proc. Natl. Acad. Sci. USA* 2001;98:9260-9264.

[110] Zimmermann N, King NE, Laporte J, Yang M, Mishra A, Pope SM, Muntel EE, Witte DP, Pegg AA, Foster PS, et al. Dissection of experimental asthma with DNA microarray analysis identifies arginase in asthma pathogenesis. *J. Clin. Invest.* 2003;111:1863-1874.

[111] Morris CR, Morris SM, Jr., Hagar W, Van Warmerdam J, Claster S, Kepka-Lenhart D, Machado L, Kuypers FA, Vichinsky EP. Arginine therapy: A new treatment for pulmonary hypertension in sickle cell disease? *Am. J. Respir. Crit. Care Med.* 2003;168:63-69.

[112] Drexler H, Zeiher AM, Meinzer K, Just H. Correction of endothelial dysfunction in coronary microcirculation of hypercholesterolaemic patients by l-arginine. *Lancet* 1991;338:1546-1550.

[113] Clarkson P, Adams MR, Powe AJ, Donald AE, McCredie R, Robinson J, McCarthy SN, Keech A, Celermajer DS, Deanfield JE. Oral l-arginine improves endothelium-dependent dilation in hypercholesterolemic young adults. *J. Clin. Invest.* 1996;97:1989-1994.

[114] Harrison DG. Cellular and molecular mechanisms of endothelial cell dysfunction. *J. Clin. Invest.* 1997;100:2153-2157.

[115] Granger DL, Hibbs JB, Jr., Perfect JR, Durack DT. Metabolic fate of l-arginine in relation to microbiostatic capability of murine macrophages. *J. Clin. Invest.* 1990;85:264-273.

[116] Mehta S, Stewart DJ, Langleben D, Levy RD. Short-term pulmonary vasodilation with l-arginine in pulmonary hypertension. *Circulation* 1995;92:1539-1545.

[117] Nagaya N, Uematsu M, Oya H, Sato N, Sakamaki F, Kyotani S, Ueno K, Nakanishi N, Yamagishi M, Miyatake K. Short-term oral administration of l-arginine improves hemodynamics and exercise capacity in patients with precapillary pulmonary hypertension. *Am. J. Respir. Crit. Care Med.* 2001;163:887-891.

[118] Stamler JS, Singel DJ, Loscalzo J. Biochemistry of nitric oxide and its redox-activated forms. *Science* 1992;258:1898-1902.

[119] Wink DA, Hanbauer I, Grisham MB, Laval F, Nims RW, Laval J, Cook J, Pacelli R, Liebmann J, Krishna M, et al. Chemical biology of nitric oxide: Regulation and protective and toxic mechanisms. *Curr Top Cell Regul* 1996;34:159-187.

[120] Moya MP, Gow AJ, McMahon TJ, Toone EJ, Cheifetz IM, Goldberg RN, Stamler JS. S-nitrosothiol repletion by an inhaled gas regulates pulmonary function. *Proc. Natl. Acad. Sci. USA* 2001;98:5792-5797.

[121] Gaston B, Singel D, Doctor A, Stamler JS. S-nitrosothiol signaling in respiratory biology. *Am. J. Respir. Crit. Care Med.* 2006;173:1186-1193.

[122] McMahon TJ, Ahearn GS, Moya MP, Gow AJ, Huang YC, Luchsinger BP, Nudelman R, Yan Y, Krichman AD, Bashore TM, et al. A nitric oxide processing defect of red blood cells created by hypoxia: Deficiency of s-nitrosohemoglobin in pulmonary hypertension. *Proc. Natl. Acad. Sci. USA* 2005;102:14801-14806.

[123] Comhair SA, Xu W, Ghosh S, Thunnissen FB, Almasan A, Calhoun WJ, Janocha AJ, Zheng L, Hazen SL, Erzurum SC. Superoxide dismutase inactivation in pathophysiology of asthmatic airway remodeling and reactivity. *Am. J. Pathol.* 2005;166:663-674.

[124] Ghosh S, Janocha AJ, Aronica MA, Swaidani S, Comhair SA, Xu W, Zheng L, Kaveti S, Kinter M, Hazen SL, et al. Nitrotyrosine proteome survey in asthma identifies oxidative mechanism of catalase inactivation. *J. Immunol.* 2006;176:5587-5597.

[125] Fukai T, Siegfried MR, Ushio-Fukai M, Cheng Y, Kojda G, Harrison DG. Regulation of the vascular extracellular superoxide dismutase by nitric oxide and exercise training. *J. Clin. Invest.* 2000;105:1631-1639.

In: Oxidative Stress: Clinical and Biomedical Implications ISBN: 978-1-60021-888-0
Editors: B. M. Matata, M. M. Elahi, pp. 277-290

Chapter XVII

Peroxynitrite and Nitrosative Stress in Heart Failure: Role of Matrix Metalloproteinases (MMPS) and Poly(ADP)-Ribose Polymerase (PARP)

***Pál Pacher*[*1] *and Csaba Szabó*[2]**
[1]Section on Oxidative Stress Tissue Injury,
Laboratory of Physiological Studies, National Institutes of Health,
NIAAA, 5625 Fishers Lane MSC 9413,
Room 2N17, Bethesda MD 20892-9413, USA
[2]Department of Surgery, UMD NJ-New Jersey Medical School,
Newark, New Jersey 07103, USA

Abstract

Heart failure is the major cause of hospitalization, morbidity and mortality worldwide. There is considerable evidence suggesting that the increased production of reactive oxygen and nitrogen species play a crucial role in the pathogenesis of various forms of myocardial injury and heart failure. Peroxynitrite, a reactive nitrogen species produced from the diffusion-controlled reaction between nitric oxide and superoxide anion, has been shown to impair cardiac function via multiple mechanisms including activation of matrix metalloproteinases (MMPs) and nuclear enzyme poly(ADP-ribose) polymerase (PARP). Recent studies have demonstrated that pharmacological neutralization of peroxynitrite or inhibition of PARP provides significant benefits in various forms of heart failure. This chapter discusses the role of peroxynitrite and nitrosative stress and downstream pathways in various forms of heart failure.

---

[*] Correspondence to: Pál Pacher M.D., Ph.D., F.A.P.S., F.A.H.A. Section on Oxidative Stress and Tissue Injury, Laboratory of Physiologic Studies, National Institutes of Health/NIAAA, 5625 Fishers Lane, MSC-9413, Bethesda, Maryland 20892-9413, USA. Phone: 301 443-4830; Fax: 301 480-0257; E-mail: pacher@mail.nih.gov

Introduction

Acute and chronic heart failure are major causes of hospitalization, morbidity and mortality worldwide. The mechanisms leading to cardiac failure are of multiple origins and involve activation of various secondary pathways (e.g. cytokines, neurohormones, oxidative and nitrosative stress) resulting in abnormalities in cardiac receptors and signaling processes, calcium homeostasis, contractile protein function and endothelial function. The resultant alterations ultimately lead to cardiac and vascular remodeling with hypertrophy, fibrosis, cardiac dilation and myocardial apoptosis or necrosis. Accumulating evidence supports the crucial role of peroxynitrite formation and consequent poly(ADP-ribose) polymerase-1 (PARP-1) and MMP activation in the development and progression of endothelial and cardiac dysfunction associated with heart failure (reviewed in (Pacher et al., 2007; Pacher et al., 2005b; Ungvari et al., 2005)). This chapter focuses on the role of peroxynitrite-nitrosative stress-MMP-PARP pathway in various forms of cardiomyopathies and heart failure.

Mechanisms of Increased Oxidative Stress in Heart Failure

Increased production of reactive oxygen species (ROS: superoxide, hydrogen peroxide, hydroxyl radical) has been demonstrated in various experimental and clinical models of heart failure (reviewed in Ungvari et al., 2005). Repetitive episodes of myocardial ischemia/reperfusion, increased level of inflammatory cytokines (e.g. TNF-alpha, IL-6), impaired antioxidant defense mechanisms and catecholamine auto-oxidation may trigger increased myocardial ROS generation. Multiple sources, including NAD(P)H oxidases, xanthine oxidase, cyclooxygenases, the mitochondrial electron transport chain, activated neutrophils, nitric oxide synthases, and auto oxidation of certain tissue metabolites, may contribute to the increased production of superoxide, hydrogen peroxide and hydroxyl radical in failing myocardium (reviewed in (Pacher et al., 2007; Pacher et al., 2006b; Sorescu and Griendling, 2002; Ungvari et al., 2005)).

Evidence for Impaired Regulation of Nitric Oxide Synthases in Heart Failure

Nitric oxide (NO), produced by NO synthases (NOS) in cardiomyocytes, endothelial cells and cardiac nerves, serves a number of important physiological roles in the regulation of cardiovascular function including coronary vasodilation, inhibition of platelet and neutrophil adhesion and activation, modulation of cardiac contractile function, and inhibition of cardiac oxygen consumption (Massion et al., 2003; Paulus and Bronzwaer, 2002). NO is necessary for normal cardiac physiology and plays a protective role in the ischemic heart by several mechanisms. These mechanisms include stimulation of soluble guanylate cyclase (sGC) and thus decrease of intracellular Ca^{2+} (Pacher et al., 2007). NO can also terminate chain-propagating lipid radical reactions caused by oxidative stress (Jones and Bolli, 2006; Pacher et al., 2007; Shah and MacCarthy, 2000). NO, through its interactions with components of the mitochondrial respiratory chain, may function not only as a physiological regulator of cell respiration, but also as a modulator of the generation of reactive oxygen species by mitochondria, thereby affecting mechanisms of cell survival or death (Moncada and Erusalimsky, 2002; Pacher et al., 2007). In higher levels, or in the presence of reactive oxygen species, NO can also exert cytotoxic effects (Pacher et al., 2007; Shah and MacCarthy, 2000). However, most of the toxic actions of NO initially attributed to NO are in fact mediated via the production of peroxynitrite (Pacher et al., 2007; Szabo, 2003).

Despite the accumulating evidence derived from experimental and human studies demonstrating overexpression and increased activity of iNOS in the myocardium of animals and patients with various forms of heart failure and beneficial effect of iNOS inhibition on cardiac function (Chen et al., 2002; Feng et al., 2001; Ferreiro et al., 2004; Funakoshi et al., 2002; Haywood et al., 1996; Kobayashi et al., 2002; Mihm et al., 2001a; Vanderheyden et al., 2004; Yang et al., 2004; Ziolo et al., 2001), the role of NO and iNOS in the development and progression of the heart failure is a subject of intensive debate (Mungrue et al., 2003).

Increased myocardial iNOS activity appears to be responsible for beta-adrenergic hypo responsiveness and depressed myocardial contractility and in a rat model of volume-overload HF (Gealekman et al., 2002), but iNOS deficiency did not attenuate congestive heart failure in mice (Jones et al., 2005). Furthermore, while one mouse study found positive correlation between the chronic overexpression of iNOS and peroxynitrite generation with cardiac enlargement, conduction defects, sudden cardiac death, and less commonly heart failure in mice (Mungrue et al., 2002), this was not confirmed by an other one (Heger et al., 2002). Neuronal (Damy et al., 2004) and endothelial NO synthases (Dai et al., 2004) may also contribute to the increased NO production in the failing myocardium. Importantly, impaired production of NO or its excessive degradation or neutralization with oxidants such as superoxide, also impairs the soluble guanylate cyclase (sGC)-cGMP signaling pathway (Evgenov et al., 2006; Pacher et al., 2005b).

Mechanisms of Peroxynitrite-Induced Cardiovascular Dysfunction in Heart Failure

Peroxynitrite generation has been demonstrated in various forms of myocardial I/R and myocarditis both in experimental animals and biopsies obtained from human subjects (reviewed in (Pacher et al., 2007; Pacher et al., 2005b; Turko and Murad, 2002)). Increased myocardial iNOS expression, nitrotyrosine formation and/or matrix metalloproteinas-2 (MMP-2) activation have been reported in heart failure induced by permanent left anterior coronary artery ligation in mice (Feng et al., 2001) and rats (Mihm et al., 2001a; Pacher et al., 2002c)) or by pacing in dogs (Cesselli et al., 2001), and in acute and chronic mouse models of doxorubicin-induced cardiomyopathy (Bai et al., 2004; Mihm et al., 2002; Pacher et al., 2003; Weinstein et al., 2000). Notably, increased myocardial iNOS expression and nitrotyrosine formation correlated with deterioration of cardiac function in mice (Mihm et al., 2002; Weinstein et al., 2000). A correlation between increased NT formation with myofibrillar creatine kinase inhibition was also found in human atrial appendages from patient with fibrillation (Mihm et al., 2001b). Increased iNOS protein expression also correlated with increased nitrotyrosine formation in myocardial specimens of patients with heart failure. Although iNOS-positive patients were generally characterized by larger LV volume and depressed function, the preserved NO generation appeared to be associated with higher cardiac work due to the preserved Frank-Starling relationship in end-stage heart failure (Vanderheyden et al., 2004). In idiopathic dilated cardiomyopathic human hearts increased nitration of SERCA2a was observed which correlated with the time to half relaxation in myocytes (Lokuta et al., 2005).

There are numerous additional targets of peroxynitrite-induced cytotoxicity which is a subjects of numerous recent detailed overviews (Pacher et al., 2007; Szabo, 2003). Peroxynitrite-induced tyrosine nitration may lead to dysfunction of large number of critical cellular proteins involved in the regulation of cardiovascular function (Pacher et al., 2007; Szabo, 2003; Turko and Murad, 2002). Furthermore, peroxynitrite is a major trigger of apoptosis in cardiomyocytes (Arstall et al., 1999; Grishko et al., 2003; Levrand et al., 2006), endothelial (Dickhout et al., 2005; Mihm et al., 2000) and vascular smooth muscle (Li et al., 2004; Li et al., 2003) cells, induces decreased spontaneous contractions of cardiomyocytes (Ishida et al., 1996), nitration and inhibition of myofibrillar creatine kinase (Mihm and Bauer, 2002; Mihm et al., 2003), and causes irreversible inhibition of the mitochondrial respiratory chain (Brown and Borutaite, 2001; Wolin et al., 1999). Peroxynitrite also activates extracellular signal-related kinase (ERK; a MAP kinase which has been linked with hypertrophic and anti-apoptotic responses in the heart) and modulates NF-kappaB activation triggered by inflammatory stimuli in cardiac and endothelial cell lines (Levrand et al., 2005; Pesse et al., 2005). Additionally, it induces upregulation of adhesion molecules in endothelial cells, disruption of endothelial glycocalyx, and may enhance the adhesion of neutrophils to the endothelium, through complex interactions with various cell signaling pathways (Galkina et al., 2004; Kurzelewski et al., 2005; Sohn et al., 2003; Zhao et al., 2004; Zouki et al., 2001), and depending on the environment it stimulates or inhibits platelet aggregation (Moro et al., 1994; Nowak and Wachowicz, 2002).

Peroxynitrite also activates matrix metalloproteinases (MMPs) and nuclear enzyme poly(ADP-ribose) polymerase (PARP) which contribute to impaired cardiovascular function in most cardiovascular pathologies and inflammatory disorders discussed below (reviewed in Jagtap et al., 2005; Pacher et al., 2007; Pacher et al., 2005b; Virag and Szabo, 2002). Peroxynitrite decomposition catalysts exert beneficial effects on cardiac and endothelial function in various models of myocardial injury (e.g. heart failure, myocardial I/R, myocarditis, cardiac allograft rejection, and diabetic cardiomyopathy) by multiple mechanisms involving protection against vascular and myocardial tyrosine nitration, lipid peroxidation, and inactivation of contractile proteins, MMP and PARP activation, and multiple other mechanisms reviewed in (Pacher et al., 2007; Pacher et al., 2005b; Szabo, 2003).

Role of Peroxynitrite-Induced Matrix Metalloproteinase (MMP) Activation in Heart Failure

Matrix metalloproteinases, a family of proteolytic enzymes involved in the degradation and remodeling of the extracellular matrix, play important roles both under physiological (e.g.: angiogenesis, remodeling of the endometrium) and pathophysiological (e.g.: neutrophil infiltration, cancer cell invasion, connective tissue remodeling) conditions (Schulz, 2007). There is emerging evidence suggesting that MMPs have a variety of novel actions as proteases on substrates apart from extracellular matrix (Schulz, 2007). MMPs are thought to be secreted in an inactive form from the cells and activated at the cell membrane via a proteolytic mechanism. However, this paradigm of only extracellular activation and action of MMPs, particularly that for MMP-2, has been challenged by recent evidence (Schulz, 2007). In the heart, collagens are the major extracellular matrix proteins surrounding and interconnecting myocytes and muscular fibers to provide a mechanical support, loss of which leads to myocytes slippage, ventricular dilation, and progressive contractile dysfunction. The balance between matrix synthesis and degradation plays a pivotal role in the process of ventricular remodeling during heart failure (reviewed in Spinale, 2002; Schulz, 2007).

Recent studies have suggested that the increased oxidative and nitrative stress play an important role in the activation of MMPs both in vitro and in vivo (reviewed in Schulz, 2007 #8481). Oxidative stress causes tissue injury through activation of the proMMPs. The activation of proMMPs is triggered by the reaction of peroxynitrite and cytosolic glutathione by S-glutathiolation of a critical cysteine residue in the autoinhibitory domain, resulting in a catalytically active proMMP (Okamoto et al, 2001). Infusion of peroxynitrite into isolated rat hearts resulted in MMP-2 activation, which preceded the onset of mechanical dysfunction, and could be abolished by MMP inhibitors (Wang et al., 2002). MMP-2 was activated and released from the reperfused heart following ischemia and inhibition of MMP-2 activity functionally protected the heart (Cheung et al., 2000). MMP-2 was found to be localized to the sarcomeric thin myofilaments upon ischemia/reperfusion injury and was responsible for the proteolytic cleavage of troponin I (Wang et al., 2002).

MMP inhibitors ameliorate ventricular dysfunction in a porcine model induced by rapid ventricular pacing (King et al., 2003). MMP-2 activity is increased in the myocardium of spontaneously hypertensive rats with CHF and correlates with ventricular dilation and age-dependent dysfunction. Inhibition of MMP-2 activity ameliorates these changes (Peterson et al., 2001). Targeted deletion of MMP-9 attenuates left ventricular remodeling after experimental myocardial infarction in mice (Ducharme et al., 2000). Peroxynitrite-dependent early activation of MMP-2 also occurs in doxorubicin-induced murine heart failure model Bai, 2004 and in rat hearts exposed to pro-inflammatory cytokines (Gao et al., 2003). Contractile failure of the heart was reduced either by peroxynitrite decomposition catalyst (Pacher et al., 2003) or MMP inhibitors (Gao et al., 2003). The activities of MMP-2 and MMP-9 are increased in atrial biopsies of patients undergoing cardiopulmonary bypass (Lalu et al., 2005) and in the ventricles of patients suffering from dilated cardiomyopathy (Spinale et al., 2000). Furthermore, the concentration of plasma MMP-9 may predict adverse cardiovascular events and a particular polymorphism of MMP-9 (R279Q) is highly associated with the occurrence of stable angina in humans (Blankenberg et al., 2003).

Role of Poly(ADP-Ribose) Polymerase (PARP) Activation in Heart Failure

Poly(ADP-ribose) polymerase-1 (PARP-1) is a nuclear enzymes, the major isoform belonging to a family of nuclear enzymes, with multiple regulatory functions. PARP-1 cleaves NAD+ to nicotinamide and ADP-ribose to form long branches (ADP-ribose) polymers on glutamic acid residues of a number of target proteins including histones and PARP-1 (automodification domain) itself. Poly(ADP-ribosyl)ation is involved in the regulation of many cellular processes such as DNA repair, gene transcription, cell cycle progression, cell death, chromatin function and genomic stability (reviewed in (Jagtap and Szabo, 2005; Szabo et al., 2006; Virag and Szabo, 2002)). Initially, it was assumed that the regulation of PARP occurs primarily at the level of DNA breakage: recognition of DNA breaks was considered to be the primary regulator (activator) or the catalytic activity of PARP (Szabo et al., 2006; Virag and Szabo, 2002). Recent studies have provided evidence that PARP-1 activity can also be modulated by several endogenous factors, including various kinases, purines and caffeine metabolites (Szabo et al., 2006). There is a gender difference in the contribution of PARP-1 to stroke and inflammatory responses, which is due, at least in part, to endogenous estrogen levels (Szabo et al., 2006).

When PARP is activated by oxygen- and nitrogen-derived radical-induced DNA single-strand breaks, catalyzes the cleavage of NAD+ into nicotinamide and ADP-ribose and uses the latter to synthesize branched nucleic acid-like polymers poly(ADP-ribose) covalently attached to nuclear acceptor proteins. As a result, in oxidatively or nitrosatively stressed cells NAD becomes depleted, impairing glycolysis, Krebs cycle, mitochondrial electron transport and eventually resulting in ATP depletion (Jagtap and Szabo, 2005; Szabo et al., 2006; Virag and Szabo, 2002). PARP-1 also regulates the expression of various proteins at the transcriptional level including iNOS and ICAM-1 and various cytokines and chemokines. Pharmacological inhibition or genetic deletion of PARP attenuates the production of a variety of inflammatory mediators in vivo (Jagtap and Szabo, 2005; Szabo, 2005; Szabo et al., 2006; Virag and Szabo, 2002). The triggers of PARP activation, in general, are DNA strand breakage and increased intracellular calcium levels (Virag and Szabo, 2002): both of these triggers have been demonstrated, both in acute and chronic heart failure.

Genetic deletion of PARP-1 or pharmacological inhibition or markedly improves the outcome of myocardial ischemia-reperfusion injury (Szabo and Bahrle, 2005), heart transplantation (Fiorillo et al., 2002; Fiorillo et al., 2003; Szabo et al., 2002b) and and various forms of cardiomyopathy and chronic heart failure (Bai et al., 2004; Pacher et al., 2002a; Pacher et al., 2002b; Pacher et al., 2002c; Pacher et al., 2006a; Pacher et al., 2002d; Pacher et al., 2002e; Pacher et al., 2002f; Pacher et al., 2005b; Pacher et al., 2004; Szabo et al., 2002a; Szabo et al., 2004; Szenczi et al., 2005). In chronic heart failure induced by chronic coronary artery ligation and associated with advanced aging in rats, increased nitrosative stress and poly(ADP-ribosyl)ation were found both in myocytes and endothelial cells. Pharmacological inhibition of PARP attenuated the myocardial hypertrophy and improved endothelial function (Pacher et al., 2002c; Pacher et al., 2006a; Pacher et al., 2004). By using PARP-1 knockout mice or pharmacological inhibitors the role of oxidative and nitrosative stress-induced PARP activation in the development of cardiac dysfunction induced by doxorubicin was also demonstrated (Bai et al., 2004; Pacher et al., 2003; Pacher et al., 2002b; Pacher et al., 2006a). Likewise, in diabetic cardiomyopathy model the depression of myocardial contractile function was associated with a significant increase in protein nitration and poly(ADP-ribosyl)ation in the cardiac myocytes (Pacher et al., 2002d; Pacher et al., 2005a; Pacher and Szabo, 2006). Recent studies have also demonstrated overexpression of the PARP-1 and increased poly(ADP-ribosyl)ation in myocardium of mice with aortic banding-induced CHF (Pillai et al., 2005; Xiao et al., 2005). Furthermore, genetic deletion of PARP-1 or pharmacological inhibition protected against hypertrophy response, heart failure and endothelial dysfunction induced by aortic banding or angiotensin II infusion, and prevented the mitochondrial-to nuclear translocation of the cell death factor apoptosis-inducing factor (Pillai et al., 2006; Pillai et al., 2005; Szabo et al., 2004; Xiao et al., 2005). Additional studies have demonsrated overexpression of PARP-1 or increased activity in biopsies from human subjects with chronic heart failure (Molnar et al., 2006) or sepsis (Soriano et al., 2006).

Conclusion

The evidence outlined above strongly support the crucial pathogenetic role of peroxynitrite-nitrosative stress-MMP-PARP pathway in cardiac and endothelial dysfunction associated with various forms of cardiovascular injury and heart failure. It would be interesting to see if various drugs with known cardiac and vascular protective effects used to treat human heart failure (antioxidant therapies, statins, β-blockers, ACE-inhibitors, etc.) are able to suppress the over activation of peroxynitrite-MMP-PARP pathway in the cardiovascular system. Several novel pharmacological inhibitors of PARP have entered clinical testing as cytoprotective agents in cardiovascular diseases and as adjunct anti-tumor therapeutics (Jagtap and Szabo, 2005; Szabo et al., 2006), and several tetracycline antibiotics turned out to be potent PARP-1 inhibitors (Szabo et al., 2006). Therefore, there is considerable hope and promise that these new drugs may become efficient therapies in the near future against various cardiovascular disorders.

References

Arstall MA, Sawyer DB, Fukazawa R and Kelly RA (1999) Cytokine-mediated apoptosis in cardiac myocytes: the role of inducible nitric oxide synthase induction and peroxynitrite generation. *Circ. Res.* 85:829-840.

Bai P, Mabley JG, Liaudet L, Virag L, Szabo C and Pacher P (2004) Matrix metalloproteinase activation is an early event in doxorubicin-induced cardiotoxicity. *Oncol. Rep.* 11:505-508.

Blankenberg S, Rupprecht HJ, Poirier O, Bickel C, Smieja M, Hafner G, Meyer J, Cambien F and Tiret L (2003) Plasma concentrations and genetic variation of matrix metalloproteinase 9 and prognosis of patients with cardiovascular disease. *Circulation* 107:1579-1585.

Brown GC and Borutaite V (2001) Nitric oxide, mitochondria, and cell death. *IUBMB Life* 52:189-195.

Cesselli D, Jakoniuk I, Barlucchi L, Beltrami AP, Hintze TH, Nadal-Ginard B, Kajstura J, Leri A and Anversa P (2001) Oxidative stress-mediated cardiac cell death is a major determinant of ventricular dysfunction and failure in dog dilated cardiomyopathy. *Circ. Res*. 89:279-286.

Chen Y, Traverse JH, Du R, Hou M and Bache RJ (2002) Nitric oxide modulates myocardial oxygen consumption in the failing heart. *Circulation* 106:273-279.

Cheung PY, Sawicki G, Wozniak M, Wang W, Radomski MW and Schulz R (2000) Matrix metalloproteinase-2 contributes to ischemia-reperfusion injury in the heart. *Circulation* 101:1833-1839.

Dai ZK, Tan MS, Chai CY, Yeh JL, Chou SH, Chiu CC, Jeng AY, Chen IJ and Wu JR (2004) Upregulation of endothelial nitric oxide synthase and endothelin-1 in pulmonary hypertension secondary to heart failure in aorta-banded rats. *Pediatr. Pulmonol.* 37:249-256.

Damy T, Ratajczak P, Shah AM, Camors E, Marty I, Hasenfuss G, Marotte F, Samuel JL and Heymes C (2004) Increased neuronal nitric oxide synthase-derived NO production in the failing human heart. *Lancet* 363:1365-1367.

Dickhout JG, Hossain GS, Pozza LM, Zhou J, Lhotak S and Austin RC (2005) Peroxynitrite causes endoplasmic reticulum stress and apoptosis in human vascular endothelium: implications in atherogenesis. *Arterioscler Thromb Vasc. Biol*. 25:2623-2629.

Ducharme A, Frantz S, Aikawa M, Rabkin E, Lindsey M, Rohde LE, Schoen FJ, Kelly RA, Werb Z, Libby P and Lee RT (2000) Targeted deletion of matrix metalloproteinase-9 attenuates left ventricular enlargement and collagen accumulation after experimental myocardial infarction. *J. Clin. Invest*. 106:55-62.

Evgenov OV, Pacher P, Schmidt PM, Hasko G, Schmidt HHW and J-P S (2006) Nitric oxide-independent stimulators and activators of soluble guanylate cyclase. *Nature Reviews Drug Discovery* 5:755-768.

Feng Q, Lu X, Jones DL, Shen J and Arnold JM (2001) Increased inducible nitric oxide synthase expression contributes to myocardial dysfunction and higher mortality after myocardial infarction in mice. *Circulation* 104:700-704.

Ferreiro CR, Chagas AC, Carvalho MH, Dantas AP, Scavone C, Souza LC, Buffolo E and da Luz PL (2004) Expression of inducible nitric oxide synthase is increased in patients with heart failure due to ischemic disease. *Braz. J. Med. Biol. Res*. 37:1313-1320.

Fiorillo C, Pace S, Ponziani V, Nediani C, Perna AM, Liguori P, Cecchi C, Nassi N, Donzelli GP, Formigli L and Nassi P (2002) Poly(ADP-ribose) polymerase activation and cell injury in the course of rat heart heterotopic transplantation. *Free Rad Res*. 36:79-87.

Fiorillo C, Ponziani V, Giannini L, Cecchi C, Celli A, Nediani C, Perna AM, Liguori P, Nassi N, Formigli L, Tani A and Nassi P (2003) Beneficial effects of poly (ADP-ribose) polymerase inhibition against the reperfusion injury in heart transplantation. *Free Rad Res.* 37:331-339.

Funakoshi H, Kubota T, Kawamura N, Machida Y, Feldman AM, Tsutsui H, Shimokawa H and Takeshita A (2002) Disruption of inducible nitric oxide synthase improves beta-adrenergic inotropic responsiveness but not the survival of mice with cytokine-induced cardiomyopathy. *Circ. Res.* 90:959-965.

Galkina SI, Dormeneva EV, Bachschmid M, Pushkareva MA, Sud'ina GF and Ullrich V (2004) Endothelium-leukocyte interactions under the influence of the superoxide-nitrogen monoxide system. *Med. Sci. Monit.* 10:BR307-316.

Gao CQ, Sawicki G, Suarez-Pinzon WL, Csont T, Wozniak M, Ferdinandy P and Schulz R (2003) Matrix metalloproteinase-2 mediates cytokine-induced myocardial contractile dysfunction. *Cardiovasc. Res.* 57:426-433.

Gealekman O, Abassi Z, Rubinstein I, Winaver J and Binah O (2002) Role of myocardial inducible nitric oxide synthase in contractile dysfunction and beta-adrenergic hyporesponsiveness in rats with experimental volume-overload heart failure. *Circulation* 105:236-243.

Grishko V, Pastukh V, Solodushko V, Gillespie M, Azuma J and Schaffer S (2003) Apoptotic cascade initiated by angiotensin II in neonatal cardiomyocytes: role of DNA damage. *Am. J. Physiol.* 285:H2364-2372.

Haywood GA, Tsao PS, von der Leyen HE, Mann MJ, Keeling PJ, Trindade PT, Lewis NP, Byrne CD, Rickenbacher PR, Bishopric NH, Cooke JP, McKenna WJ and Fowler MB (1996) Expression of inducible nitric oxide synthase in human heart failure. *Circulation* 93:1087-1094.

Heger J, Godecke A, Flogel U, Merx MW, Molojavyi A, Kuhn-Velten WN and Schrader J (2002) Cardiac-specific overexpression of inducible nitric oxide synthase does not result in severe cardiac dysfunction. *Circulation research* 90:93-99.

Ishida H, Ichimori K, Hirota Y, Fukahori M and Nakazawa H (1996) Peroxynitrite-induced cardiac myocyte injury. *Free Radic. Biol. Med.* 20:343-350.

Jagtap P and Szabo C (2005) Poly(ADP-ribose) polymerase and the therapeutic effects of its inhibitors. *Nature Reviews Drug Discovery* 4:421-440.

Jagtap PG, Baloglu E, Southan GJ, Mabley JG, Li H, Zhou J, van Duzer J, Salzman AL and Szabo C (2005) Discovery of potent poly(ADP-ribose) polymerase-1 inhibitors from the modification of indeno[1,2-c]isoquinolinone. *J. Med. Chem.* 48:5100-5103.

Jones SP and Bolli R (2006) The ubiquitous role of nitric oxide in cardioprotection. *J. Mol. Cell Cardiol.* 40:16-23.

Jones SP, Greer JJ, Ware PD, Yang J, Walsh K and Lefer DJ (2005) Deficiency of iNOS does not attenuate severe congestive heart failure in mice. *Am. J. Physiol.* 288:H365-370.

King MK, Coker ML, Goldberg A, McElmurray JH, 3rd, Gunasinghe HR, Mukherjee R, Zile MR, O'Neill TP and Spinale FG (2003) Selective matrix metalloproteinase inhibition with developing heart failure: effects on left ventricular function and structure. *Circ. Res.* 92:177-185.

Kobayashi N, Horinaka S, Mita S, Yoshida K, Honda T, Kobayashi T, Hara K, Nishikimi T and Matsuoka H (2002) Aminoguanidine inhibits mitogen-activated protein kinase and improves cardiac performance and cardiovascular remodeling in failing hearts of salt-sensitive hypertensive rats. *J. Hypertens* 20:2475-2485.

Kurzelewski M, Czarnowska E and Beresewicz A (2005) Superoxide- and nitric oxide-derived species mediate endothelial dysfunction, endothelial glycocalyx disruption, and enhanced neutrophil adhesion in the post-ischemic guinea-pig heart. *J. Physiol. Pharmacol.* 56:163-178.

Lalu MM, Pasini E, Schulze CJ, Ferrari-Vivaldi M, Ferrari-Vivaldi G, Bachetti T and Schulz R (2005) Ischaemia-reperfusion injury activates matrix metalloproteinases in the human heart. *Eur. Heart J.* 26:27-35.

Levrand S, Pesse B, Feihl F, Waeber B, Pacher P, Rolli J, Schaller MD and Liaudet L (2007) Peroxynitrite is a potent inhibitor of NF-kappa B activation triggered by inflammatory stimuli in cardiac and endothelial cell lines. *J. Biol. Chem.*, in press.

Levrand S, Vannay-Bouchiche C, Pesse B, Pacher P, Feihl F, Waeber B and Liaudet L (2006) Peroxynitrite is a major trigger of cardiomyocyte apoptosis in vitro and in vivo. *Free Radic. Biol. Med.* 41:886-895.

Li J, Li W, Su J, Liu W, Altura BT and Altura BM (2004) Peroxynitrite induces apoptosis in rat aortic smooth muscle cells: possible relation to vascular diseases. *Exp. Biol. Med.* (Maywood) 229:264-269.

Li J, Su J, Li W, Liu W, Altura BT and Altura BM (2003) Peroxynitrite induces apoptosis in canine cerebral vascular muscle cells: possible relation to neurodegenerative diseases and strokes. Neurosci Lett 350:173-177.

Lokuta AJ, Maertz NA, Meethal SV, Potter KT, Kamp TJ, Valdivia HH and Haworth RA (2005) Increased nitration of sarcoplasmic reticulum Ca2+-ATPase in human heart failure. *Circulation* 111:988-995.

Massion PB, Feron O, Dessy C and Balligand JL (2003) Nitric oxide and cardiac function: ten years after, and continuing. *Circ. Res.* 93:388-398.

Mihm MJ and Bauer JA (2002) Peroxynitrite-induced inhibition and nitration of cardiac myofibrillar creatine kinase. *Biochimie* 84:1013-1019.

Mihm MJ, Coyle CM, Schanbacher BL, Weinstein DM and Bauer JA (2001a) Peroxynitrite induced nitration and inactivation of myofibrillar creatine kinase in experimental heart failure. *Cardiovasc. Res.* 49:798-807.

Mihm MJ, Jing L and Bauer JA (2000) Nitrotyrosine causes selective vascular endothelial dysfunction and DNA damage. *J. Cardiovasc. Pharmacol.* 36:182-187.

Mihm MJ, Yu F, Carnes CA, Reiser PJ, McCarthy PM, Van Wagoner DR and Bauer JA (2001b) Impaired myofibrillar energetics and oxidative injury during human atrial fibrillation. *Circulation* 104:174-180.

Mihm MJ, Yu F, Reiser PJ and Bauer JA (2003) Effects of peroxynitrite on isolated cardiac trabeculae: selective impact on myofibrillar energetic controllers. *Biochimie* 85:587-596.

Mihm MJ, Yu F, Weinstein DM, Reiser PJ and Bauer JA (2002) Intracellular distribution of peroxynitrite during doxorubicin cardiomyopathy: evidence for selective impairment of myofibrillar creatine kinase. *Br. J. Pharmacol.* 135:581-588.

Molnar A, Toth A, Bagi Z, Papp Z, Edes I, Vaszily M, Galajda Z, Papp JG, Varro A, Szuts V, Lacza Z, Gero D and Szabo C (2006) Activation of the poly(ADP-ribose) polymerase pathway in human heart failure. *Mol. Med.* 12:143-152.

Moncada S and Erusalimsky JD (2002) Does nitric oxide modulate mitochondrial energy generation and apoptosis? *Nat. Rev. Mol. Cell Biol.* 3:214-220.

Moro MA, Darley-Usmar VM, Goodwin DA, Read NG, Zamora-Pino R, Feelisch M, Radomski MW and Moncada S (1994) Paradoxical fate and biological action of peroxynitrite on human platelets. *Proc. Natl. Acad. Sci. USA* 91:6702-6706.

Mungrue IN, Gros R, You X, Pirani A, Azad A, Csont T, Schulz R, Butany J, Stewart DJ and Husain M (2002) Cardiomyocyte overexpression of iNOS in mice results in peroxynitrite generation, heart block, and sudden death. *J. Clin. Invest* 109:735-743.

Mungrue IN, Stewart DJ and Husain M (2003) The Janus faces of iNOS. *Circ. Res*. 93:e74.

Nowak P and Wachowicz B (2002) Peroxynitrite-mediated modification of fibrinogen affects platelet aggregation and adhesion. *Platelets* 13:293-299.

Pacher P, Beckman JS and Liaudet L (2007) Nitric oxide and peroxynitrite in health and disease. *Physiol. Rev*. 87:315-424.

Pacher P, Cziraki A, Mabley JG, Liaudet L, Papp L and Szabo C (2002a) Role of poly(ADP-ribose) polymerase activation in endotoxin-induced cardiac collapse in rodents. *Biochem. Pharmacol.* 64:1785-1791.

Pacher P, Liaudet L, Bai P, Mabley JG, Kaminski PM, Virag L, Deb A, Szabo E, Ungvari Z, Wolin MS, Groves JT and Szabo C (2003) Potent metalloporphyrin peroxynitrite decomposition catalyst protects against the development of doxorubicin-induced cardiac dysfunction. *Circulation* 107:896-904.

Pacher P, Liaudet L, Bai P, Virag L, Mabley JG, Hasko G and Szabo C (2002b) Activation of poly(ADP-ribose) polymerase contributes to development of doxorubicin-induced heart failure. *J. Pharmacol. Exp. Ther.* 300:862-867.

Pacher P, Liaudet L, Mabley J, Komjati K and Szabo C (2002c) Pharmacologic inhibition of poly(adenosine diphosphate-ribose) polymerase may represent a novel therapeutic approach in chronic heart failure. *J. Am. Coll. Cardiol.* 40:1006-1016.

Pacher P, Liaudet L, Mabley JG, Cziraki A, Hasko G and Szabo C (2006a) Beneficial effects of a novel ultrapotent poly(ADP-ribose) polymerase inhibitor in murine models of heart failure. *Int. J. Mol. Med.* 17:369-375.

Pacher P, Liaudet L, Soriano FG, Mabley JG, Szabo E and Szabo C (2002d) The role of poly(ADP-ribose) polymerase activation in the development of myocardial and endothelial dysfunction in diabetes. *Diabetes* 51:514-521.

Pacher P, Mabley JG, Soriano FG, Liaudet L, Komjati K and Szabo C (2002e) Endothelial dysfunction in aging animals: the role of poly(ADP-ribose) polymerase activation. *Br. J. Pharmacol.* 135:1347-1350.

Pacher P, Mabley JG, Soriano FG, Liaudet L and Szabo C (2002f) Activation of poly(ADP-ribose) polymerase contributes to the endothelial dysfunction associated with hypertension and aging. *Int. J. Mol. Med.* 9:659-664.

Pacher P, Nivorozhkin A and Szabo C (2006b) Therapeutic effects of xanthine oxidase inhibitors: renaissance half a century after the discovery of allopurinol. *Pharmacol. Rev.* 58:87-114.

Pacher P, Obrosova IG, Mabley JG and Szabo C (2005a) Role of nitrosative stress and peroxynitrite in the pathogenesis of diabetic complications. Emerging new therapeutical strategies. *Curr. Med. Chem*. 12:267-275.

Pacher P, Schulz R, Liaudet L and Szabo C (2005b) Nitrosative stress and pharmacological modulation of heart failure. *Trends Pharmacol. Sci.* 26:302-310.

Pacher P and Szabo C (2006) Role of peroxynitrite in the pathogenesis of cardiovascular complications of diabetes. *Curr. Opin. Pharmacol.* 6:136-141.

Pacher P, Vaslin A, Benko R, Mabley JG, Liaudet L, Hasko G, Marton A, Batkai S, Kollai M and Szabo C (2004) A new, potent poly(ADP-ribose) polymerase inhibitor improves cardiac and vascular dysfunction associated with advanced aging. *J. Pharmacol. Exp. Ther.* 311:485-491.

Paulus WJ and Bronzwaer JG (2002) Myocardial contractile effects of nitric oxide. *Heart Fail Rev.* 7:371-383.

Pesse B, Levrand S, Feihl F, Waeber B, Gavillet B, Pacher P and Liaudet L (2005) Peroxynitrite activates ERK via Raf-1 and MEK, independently from EGF receptor and p21Ras in H9C2 cardiomyocytes. *J. Mol. Cell Cardiol.* 38:765-775.

Peterson JT, Hallak H, Johnson L, Li H, O'Brien PM, Sliskovic DR, Bocan TM, Coker ML, Etoh T and Spinale FG (2001) Matrix metalloproteinase inhibition attenuates left ventricular remodeling and dysfunction in a rat model of progressive heart failure. *Circulation* 103:2303-2309.

Pillai JB, Gupta M, Rajamohan SB, Lang R, Raman J and Gupta MP (2006) Poly(ADP-ribose) polymerase-1-deficient mice are protected from angiotensin II-induced cardiac hypertrophy. *Am. J. Physiol.* 291:H1545-1553.

Pillai JB, Russell HM, Raman J, Jeevanandam V and Gupta MP (2005) Increased expression of poly(ADP-ribose) polymerase-1 contributes to caspase-independent myocyte cell death during heart failure. *Am. J. Physiol.* 288:H486-496.

Schulz R (2007) Intracellular targets of matrix metalloproteinase-2 in cardiac disease: rationale and therapeutic approaches. *Ann. Rev. Pharmacol. Toxicol.* 47:211-242.

Shah AM and MacCarthy PA (2000) Paracrine and autocrine effects of nitric oxide on myocardial function. *Pharmacol. Ther.* 86:49-86.

Sohn HY, Krotz F, Zahler S, Gloe T, Keller M, Theisen K, Schiele TM, Klauss V and Pohl U (2003) Crucial role of local peroxynitrite formation in neutrophil-induced endothelial cell activation. *Cardiovasc. Res.* 57:804-815.

Sorescu D and Griendling KK (2002) Reactive oxygen species, mitochondria, and NAD(P)H oxidases in the development and progression of heart failure. *Congest Heart Fail* 8:132-140.

Soriano FG, Nogueira AC, Caldini EG, Lins MH, Teixeira AC, Cappi SB, Lotufo PA, Bernik MM, Zsengeller Z, Chen M and Szabo C (2006) Potential role of poly(adenosine 5'-diphosphate-ribose) polymerase activation in the pathogenesis of myocardial contractile dysfunction associated with human septic shock. *Crit. Care Med.* 34:1073-1079.

Spinale FG, Coker ML, Heung LJ, Bond BR, Gunasinghe HR, Etoh T, Goldberg AT, Zellner JL and Crumbley AJ (2000) A matrix metalloproteinase induction/activation system exists in the human left ventricular myocardium and is upregulated in heart failure. *Circulation* 102:1944-1949.

Szabo C (2003) Multiple pathways of peroxynitrite cytotoxicity. *Toxicol. Lett.* 140-141:105-112.

Szabo C (2005) Poly(ADP-ribose) polymerase activation by reactive nitrogen species-Relevance for the pathogenesis of inflammation. *Nitric Oxide.*

Szabo C, Mabley JG, Moeller SM, Shimanovich R, Pacher P, Virag L, Soriano FG, Van Duzer JH, Williams W, Salzman AL and Groves JT (2002a) Part I: pathogenetic role of peroxynitrite in the development of diabetes and diabetic vascular complications: studies with FP15, a novel potent peroxynitrite decomposition catalyst. *Mol. Med.* 8:571-580.

Szabo C, Pacher P and Swanson RA (2006) Novel modulators of poly(ADP-ribose) polymerase. *Trends Pharmacol. Sci*. 27:626-630.

Szabo C, Pacher P, Zsengeller Z, Vaslin A, Komjati K, Benko R, Chen M, Mabley JG and Kollai M (2004) Angiotensin II-mediated endothelial dysfunction: role of poly(ADP-ribose) polymerase activation. *Mol. Med.* 10:28-35.

Szabo G and Bahrle S (2005) Role of nitrosative stress and poly(ADP-ribose) polymerase activation in myocardial reperfusion injury. *Curr. Vasc. Pharmacol.* 3:215-220.

Szabo G, Bahrle S, Stumpf N, Sonnenberg K, Szabo EE, Pacher P, Csont T, Schulz R, Dengler TJ, Liaudet L, Jagtap PG, Southan GJ, Vahl CF, Hagl S and Szabo C (2002b) Poly(ADP-Ribose) polymerase inhibition reduces reperfusion injury after heart transplantation. *Circ. Res*. 90:100-106.

Szenczi O, Kemecsei P, Holthuijsen MF, van Riel NA, van der Vusse GJ, Pacher P, Szabo C, Kollai M, Ligeti L and Ivanics T (2005) Poly(ADP-ribose) polymerase regulates myocardial calcium handling in doxorubicin-induced heart failure. *Biochem. Pharmacol.* 69:725-732.

Turko IV and Murad F (2002) Protein nitration in cardiovascular diseases. *Pharmacol. Rev.* 54:619-634.

Ungvari Z, Gupte SA, Recchia FA, Batkai S and Pacher P (2005) Role of oxidative-nitrosative stress and downstream pathways in various forms of cardiomyopathy and heart failure. *Curr. Vasc. Pharmacol.* 3:221-229.

Vanderheyden M, Bartunek J, Knaapen M, Kockx M, De Bruyne B and Goethals M (2004) Hemodynamic effects of inducible nitric oxide synthase and nitrotyrosine generation in heart failure. *J. Heart Lung Transplant.* 23:723-728.

Virag L and Szabo C (2002) The therapeutic potential of poly(ADP-ribose) polymerase inhibitors. *Pharmacol. Rev*. 54:375-429.

Wang W, Sawicki G and Schulz R (2002) Peroxynitrite-induced myocardial injury is mediated through matrix metalloproteinase-2. *Cardiovasc. Res*. 53:165-174.

Weinstein DM, Mihm MJ and Bauer JA (2000) Cardiac peroxynitrite formation and left ventricular dysfunction following doxorubicin treatment in mice. J. Pharmacol. Exp. Ther. 294:396-401.

Wolin MS, Xie YW and Hintze TH (1999) Nitric oxide as a regulator of tissue oxygen consumption. *Curr. Opin. Nephrol. Hypertens.* 8:97-103.

Xiao CY, Chen M, Zsengeller Z, Li H, Kiss L, Kollai M and Szabo C (2005) Poly(ADP-Ribose) polymerase promotes cardiac remodeling, contractile failure, and translocation of apoptosis-inducing factor in a murine experimental model of aortic banding and heart failure. *J. Pharmacol. Exp. Ther.* 312:891-898.

Yang B, Larson DF and Watson RR (2004) Modulation of iNOS activity in age-related cardiac dysfunction. *Life Sci.* 75:655-667.

Zhao S, Zhang Y, Gu Y, Lewis DF and Wang Y (2004) Heme oxygenase-1 mediates up-regulation of adhesion molecule expression induced by peroxynitrite in endothelial cells. *J. Soc. Gynecol. Investig.* 11:465-471.

Ziolo MT, Katoh H and Bers DM (2001) Expression of inducible nitric oxide synthase depresses beta-adrenergic-stimulated calcium release from the sarcoplasmic reticulum in intact ventricular myocytes. *Circulation* 104:2961-2966.

Zouki C, Zhang SL, Chan JS and Filep JG (2001) Peroxynitrite induces integrin-dependent adhesion of human neutrophils to endothelial cells via activation of the Raf-1/MEK/Erk pathway. *Faseb J.* 15:25-27.

In: Oxidative Stress: Clinical and Biomedical Implications ISBN: 978-1-60021-888-0
Editors: B. M. Matata, M. M. Elahi, pp. 291-306

Chapter XVIII

Reactive Oxidant Species and Emerging Therapies in Clinical Medicine

Farzaneh Daghigh*
Department of Biochemistry and Molecular Biology,
Philadelphia College of Osteopathic Medicine,
Philadelphia, PA 19131, USA

Abstract

Nitric oxide (NO) is a dynamic and reactive molecule that impacts a wide range of disease. It reversibly alters the expression of specific genes, activity of proteins and signaling pathways, regulating both physiologic processes and disease states. NO has a basic role in regulating vascular tone of the pulmonary circulation, and recent animal models have suggested a key role on peritoneal lung development. Inhaled nitric oxide (inhaled NO) is a well tolerated and effective treatment for newborns with pulmonary hypertension and hypoxemic respiratory failure and bronchopulmonary dysplasia (BPD). The success of inhaled NO therapy in infants has led to active investigations of its potential role in treating adults. The effect of inhaled NO in human infectious diseases is being evaluated; pulmonary tuberculosis is a likely candidate in which inhaled NO may be of therapeutic benefits. Red blood cells (RBCs) both scavenge NO and carry it in form of S-nitrosothiols (SNOs). It has been postulated that lack of vasodilatation in diabetes and sickle cell disease (SCD) may be a consequence of hypoxia-induced defects in NO processing by RBCs.

In management of patients with airway diseases who are treated with inhaled corticosteroids, exhaled NO (FE_{NO}) and exhaled breath condensate (EBC) are widely utilized to determine the appropriate dose of the drug. Also, low concentrations of S-nitrosoglutathione (GSNO) compared to normal in patients with cystic Fibrosis (CF) or asthma is associated with decreased pulmonary function. Treatments with GSNO have shown to stimulate chloride ion secretion in CF or enhance bronchodilation in asthmatic patients. There is growing evidence that arginine is beneficial for conditions of

* Tel: (215) 871-6449; Fax (215) 871-6865; E-mail: farzanehd@pcom.edu

endothelial dysfunction such as CF or SCD. Inhaled arginine therapy improved pulmonary functions in CF, and combined treatments of inhaled arginine and hydroxyurea proved effective for the vaso-occlusive crisis. Although many advances have been made in vascular gene therapy, translation of preclinical research to the clinic has been slower than expected. The key barriers to the translation of this research are suitable vectors and adequate delivery devices. The optimal vector should be safe, target-specific, regulatable, and result in a suitable temporal pattern of therapeutic transgene expression. Currently available vectors meet a number of criteria for specific interventions such as improving endothelial function and treating hypertension, but significant advances are still required. The identification of the biological properties of NO is one of the scientific discoveries of the century, and its fate in clinical medicine is yet to be fully realized.

Introduction

The medical significance of reactive oxygen and nitrogen species (ROS and RNS, respectively) has become increasingly recognized as a component of virtually every disease process. Oxidative research is moving in the direction of transferring the knowledge of oxidative pathology from basic science into clinical medicine and the pharmacy. In management of ischemia-reperfusion injury, chronic inflammation, neurodegenerative diseases, and other diseases, new pharmacologic strategies have aimed at supplementing antioxidant defense systems as well as antagonizing redox-sensitive signal transduction. However, there is no routine use of antioxidants in a variety of clinical settings.

Selective pharmacological inhibition of inducible nitric oxide synthase (iNOS) in shock and inflammation is expected to be of significant therapeutic benefit, inhibiting the cytotoxic concentration of NO and maintaining its physiological concentrations. In light of the critical roles of superoxide in disease and cellular signaling, the superoxide dismutase (SOD) "synzymes" can serve as powerful tools in the treatment of numerous diseases ranging from acute and chronic inflammation to shock and ischemia-reperfusion injury [1]. In treatment of wounds, novel therapies have emerged involving the use of NO, such as monochromatic near-infrared treatment [2] and topical oxygen therapy [3].

The role of NO in human growth and metastasis is complicated and remains unclear [4, 5]. Diverse inter-tumor and intra-tumor heterogeneity and complex genetic background and functional status of both tumor and stromal cells are critical variables in iNOS expression [6]. Researchers should clearly understand the mechanism of iNOS expression and NO production before designing an effective strategy targeting tumor associated iNOS/NO.

A noninvasive device has been designed to potentially elevate the NO release from vascular endothelium in response to increased pulsatile shear stress, by whole-body acceleration [7]. This method of passive exercise may be beneficial to patients with diseases of endothelial dysfunction who are incapable of exercising [8].

Prevention of cardiopulmonary bypass (CPB) significantly reduced myocardial ischemia-reperfusion injury and other adverse postoperative responses. CPB induces oxidative stress and complex systemic inflammatory response [9]. The off-pump coronary artery bypass grafting approach has shown some promising result in reducing the ischemia-reperfusion injury in some case. Further investigation of this approach has been recommended [10].

Several enzymes have been implicated as possible sources of elevated ROS/RNS in diseases, including NADPH oxidase, xanthine oxidase, NO synthase (NOS), cyclooxygenase, and mitochondrial transport enzymes. Also, reductions of enzymes that scavenge ROS/RNS, such as SOD, catalase, and glutathione peroxidase, may play a role. Therefore, inhibiting oxidative producing enzymes and restoring antioxidant enzymes have been targeted as useful therapies. Recent technologies have allowed scientists to develop treatments using NOS inhibitors and NO donors; the literature on these pharmacological advancements is colossal and well beyond the scope of this chapter. Herein, I have discussed the diverse utilization of inhaled NO therapy, exhaled NO and breath condensate, arginine and gene therapies.

Effects of Inhaled NO Therapy in Newborns

Chronic lung disease associated with prolonged hospitalization is the most important long-term pulmonary complication in preterm infants. Drug therapies such as calcium-channel blockers, adenosine, and hydralazine have shown only modest improvement in the pulmonary arterial vascular resistance with concurrent side effects like systemic hypertension. The insignificant effectiveness of these therapies has led to the use of inhaled NO, which improves arterial oxygenation and reduces the need for extracorporeal life support with decreased oxygen requirements [11]. NO can be safely inhaled when delivered by face mask, by nasal cannula, or via an endotracheal tube. Those infants who respond to inhaled NO therapy were more likely to survive to hospital discharge [11]. Nitrosative and oxidative stress may be related to the severity of pulmonary injury and, consequently, the overall outcome in prematurely born infants with severe BPD [12]. Formation of 3-nitrotyrosine (3-NT) and 3-chlorotyrosine (3-CT), generated from nitrating and chlorinating the aromatic ring of tyrosine residues, respectively, has been indicated in asthma, acute respiratory syndrome, and cystic fibrosis [13-15]. Lorch et al. determined that after a 72 hr treatment with inhaled NO of 22 infants with severe BPD, the infants whose plasma levels of 3-NT decreased were more likely to be weaned from mechanical ventilation ($P=0.03$). There was no significant association between 3-CT and improvement in BPD [12]. Inhaled NO may play a role in reducing oxidant stress.

It has been known that inhaled NO is an effective treatment for pulmonary hypertension in infants, is largely due to its properties as a selective pulmonary vasodilator [16,17]. There is increasing evidence that endogenous NO plays an essential role in pulmonary vascular and parenchymal function in the developing immature lung, and low dose inhaled NO therapy may have beneficial effects on both the acute and chronic perturbations that are associated with the pathogenesis of BPD in the premature newborn [18].

In a recent report of the largest trials of inhaled NO therapy it has been shown that this therapy also improves the pulmonary outcome for premature infants who are at high risk for BPD [19,20]. In this randomized, double-blinded, placebo-controlled study, treatments were initiated in preterm infants between the age of 7 and 21 days undergoing mechanical ventilation, to study the efficacy of inhaled NO treatments [19, 20]. Administration of inhaled

NO to infants significantly increased survival in those without BPD and had no apparent short term effects, as compared with infants who received placebo gas. Infants who were treated with inhaled NO were hospitalized for fewer days and needed supplemental oxygen for a shorter period [19, 20]. In this multi-center trial, unlike the previously reported studies of infants with pulmonary hypertension being treated shortly after birth, the infants were enrolled at 7 days of age due to a concern related to possible brain injury resulting from interaction with NO [19, 21]. The infants were exposed to longer (24 days) and greater dose (20 ppm and lower) than did infants in other trials with controversial results [20, 22, 23]. Kinsella et al. and Van Meurs et al. found administration of low dose (5 ppm) of inhaled NO to be beneficial only among infants with a birth weight of more than 1 Kg [22,23]. Ongoing long-term follow-up developmental studies of premature infants will determine future pulmonary and cognitive outcomes of inhaled NO therapy [19, 20].

The role of endothelin-1 as either a marker or a mediator for neonatal pulmonary disorders in response to the inhaled NO therapy was investigated [24]. Endothelin-1, a potent vasoconstrictor and proliferating inducing substance, has been found to be elevated in infants with persistent pulmonary hypertension [25]. Thirty three infants with chronic lung disease, a birth weight of less than 1.25 Kg and between 10 and 30 days, as well as twenty one infants with persistent pulmonary hypertension between 36 and 72 h of age, were treated with 20 ppm of inhaled NO. The data reconfirmed the presence of immunoreactive endothelin-1 in two different neonatal pulmonary disorders. A significant relationship between decreased plasma levels of endothelin-1 and magnitude of improvement in arterial PO_2 in response to inhaled NO was detected [24]. Therefore, Truog et al. suggested that inhibitors of endothekin-1 synthesis or receptor antagonists may be helpful as an intervention therapy.

Effects of Inhaled NO Therapy in Adults

Although inhaled NO has been used for treatment of acute respiratory failure and pulmonary hypertension since 1991 in adult patients in the preoperative setting and in critical care, the potential therapeutic role of inhaled NO in adults remains uncertain; licensed indications are restricted to pediatric practice. The potential benefits of inhaled NO therapy have been reported in patients with the acute respiratory distress syndrome (ARDS), pulmonary arterial hypertension, pulmonary thromboembolism, and immediately following cardiopulmonary bypass and cardiac and lung transplantation. Multiple investigators have reported that: (i) inhaled NO may vary in dose dependency over time and between patients for immediate selective pulmonary vasodilatation (the optimum dose ranging from 0.1-10 ppm). Since only the vasculature associated with ventilated lung units are within the reach, inhaled NO diffuse across the alveolar-capillary membrane, improving ventilation/perfusion mismatch and subsequent oxygenation in acute lung injury in adults [26] and that (ii) administration of small doses of inhaled NO to patients with pulmonary hypertension and/or hypoxemia can rapidly decrease pulmonary artery pressure and improve the PaO_2, without obvious systemic effects or toxicity.

An interdisciplinary panel of experts in the field of inhaled NO has made several evidence-based recommendations on the safe use of inhaled NO therapy in adults [27]. The expert panel identified the use of inhaled NO as a rescue treatment in patients with severe acute pulmonary hypertension and/or severe refractory arterial hypoxemia. Moreover, inhaled NO is a useful drug for testing pulmonary vasoreactivity in patients undergoing heart transplantation or in patients with pulmonary arterial hypertension. None of the clinically relevant outcome of inhaled NO therapy has provided a long term effect, or an improvement in length of stay in the intensive care unit, survival, or duration of mechanical ventilation [28].

Expert panels' recommendations on safety of inhaled NO [27] include considering toxicity at higher doses (i.e. 20 ppm and higher), monitoring of inhaled NO therapy, measurement of met-hemoglobin, minimizing the generation of NO_2 and conducting long-term follow-up studies to ascertain adverse effects following inhaled NO therapy. Responses to inhaled NO treatments may identify patients suitable for heart/lung transplantation or intervention in patients with congenital heart disease [27]. Despite many favorable studies [29-39], the panel of experts found insufficient data to treat patients with heart failure or long-term treatments of pulmonary arterial hypertension, thromboembolism, sickle cell disease, prevention of ischemia/ reperfusion injury after lung transplantation or thromboendarterectomy, chronic obstructive pulmonary disease, and prevention of one-lung ventilation in adults [27].

Effects of Inhaled NO on Infections

The selectivity of inhaled NO therapy has been attributed to the rapid scavenging and inactivation of NO by hemoglobin (Hb) through reactions with the heme iron. Jia et al. demonstrated that Hb also reacts with NO to form S-nitrosothiol on a conserved cysteine-β93 [40]. Stamler et al. has proposed that S-nitrosohemoglobin (SNO-Hb), a relatively stable protein, could carry NO bioactively within the vasculature to distal organs outside of lung [41]. Immune cells in the vasculature of distal organs may be one of the targets of inhaled NO therapy. NO plays an important role in innate immune response to intracellular organism; it either kills pathogens (bacteria, viruses, parasites, and fungi) or inhibits their replication [42]. The effect of inhaled NO on infections in humans is just beginning to be evaluated. NO may play an essential role in maintaining pathogen latency. Several organisms, such as *M. tuberculosis*, *Leishmania major*, *Toxoplasma gondii*, and Epstein-Barr virus, persist in host cells in a non-replicative dormant state [43]. Pulmonary tuberculosis is an infectious disease in which inhaled NO is likely to be of therapeutic benefits since it reaches *M. tuberculosis* in the lung. Long et al. showed 90 ppm NO gas kills *M. tuberculosis* in culture [44]. Also in animal models, *M. tuberculosis* infection was exacerbated by either inhibition or targeted deletion of iNOS [45, 46]. In a randomized study, however, in patients with acute lung injury who received 5 ppm inhaled NO, higher rates of infection were observed when compared to the placebo group [47]. The precise effect of inhaled NO on the immune response is likely to depend on the concentration of inhaled NO. In another study, patients with tuberculosis tolerated their treatment of 80 ppm very well, but didn't improve the bacteria load after the 3 days of inhaled NO therapy [48]. Therefore, additional studies using longer courses or increased concentration of inhaled NO therapy are needed to assess its therapeutic efficacy in tuberculosis.

The Extrapulmonary Effects of Inhaled NO

The mechanism of hypoxia regulation of pulmonary arterial pressure is not completely understood. Sustained hypoxemia is a result of lung disease and a cause of pulmonary hypertension. Recently, an active role for RBCs has been postulated as the carriers of NO rather than its consumers [49,50]. Mc Mahan et al. have demonstrated that (i) hypoxia depletes RBCs of SNO-Hb, lowering oxygenation and impairing their ability to counteract pulmonary vasoconstriction, (ii) RBCs from patients with elevated pulmonary arterial pressure exhibit depletion of SNO-Hb and consequent impairment in pO_2 regulated vasodilatation, and (iii) *in situ* repletion of SNO-Hb corrects these physiological deficits in both animals and humans [50]. These data point to a physiological role for blood-borne S-nitrosothiol bioactivity in V/Q matching. Hypoxia-induced defects in NO processing by RBCs may contribute to pulmonary hypertension. In two separate studies, RBCs from patients with diabetes [51] and sickle cell disease [52] show impaired NO vasodilatation activity. The lack of vasodilatation in diabetes is attributed to Hb glycosylation and mutation in sickle cell disease. Hypoxia-induced depletion of SNO-Hb leads to the lack of vasodilatation and may underlie a range of impairments, including systemic and pulmonary hypertension, and vascular diatheses associated with hemoglobinopathies [50].

Exhaled NO (FE_{NO})

FE_{NO} is widely accepted as a non-invasive marker of eosinophilic airway inflammation for research and an alternative for adjustment of inhaled-corticosteroid treatment in patients with chronic asthma, offering practitioners an accurate, easy, and reproducible technique for diagnosis and management of asthma. In a randomized, placebo controlled study by Smith et al., dose adjustments of inhaled corticosteroid were evaluated comparing the conventional guidelines versus FE_{NO} measurements in patients with chronic asthma [53]. The data showed a 40% reduction in the required dose of inhaled corticosteroid without compromising any major clinical outcomes, including exacerbation rates and prednisone use. The published normal range of FE_{NO} cutoff rate of 15 ppb at flow rate of 250 mL/S was used [54,55]. The use of FE_{NO} measurements may help to minimize the long term side effects related to corticosteroids at higher doses. A joint task force of European Respiratory Society (ERS) and American Thoracic Society (ATS) published their recommendations on standardized procedures for measurement of FE_{NO} [56].

Exhaled Breath Condensate (EBC)

EBC is a noninvasive test for monitoring airway diseases such as asthma and chronic obstructive pulmonary disease. EBC contains a large number of mediators including adenosine, ammonia, hydrogen peroxide, isoprostanes, leukotrienes, nitrogen oxides, peptides and cytokines. Concentration and pH of these mediators are influenced by airway diseases. Analysis of EBC could identify coexisting pathogens. Collection of the EBC biomarkers is still in its infancy with methodological issues surrounding EBC collection and assay. The ERS/ATS task force has published their recommendations for researchers interested in the field, as many improvements to EBC technique are required before it can be used in clinical practice [57].

S-Nitrosoglutathione (GSNO) Inhalation Therapy

There is substantial evidence that ROS/RNS, generated by environmental pollutant and activated inflammatory or airway cells, may alter the properties of a number of ion channels either by direct modification or through signal transduction pathways [58]. The biological effects of NO on transport proteins have been associated with formation of SNOs [58]. GSNO is an endogenous nitrogen oxide bronchodilator present in high concentrations (~ 0.5-10 μM) in the extracellular fluids of the lung and brain [59,60] which relaxes airway smooth muscle [59]. NO generated from GSNO can stimulate cGMP or react with superoxide to form peroxynitrite. Levels of epithelial NOS and GSNO are low in the airways of patients with CF [61] and asthma [62]. In CF, GSNO stimulates chloride secretion in patients with the delF508 mutation [63] or in Calu-3 cells [64], even though its effect on cystic fibrosis transmembrane conductance is controversial [63-65]. Zaman et al. have shown that GSNO levels of >5 μM inhibit 5-lipooxygenase expression and activity in asthmatic airway, thus promoting bronchodilation [66]. Hence, GSNO may be of value as a candidate to correct the basic defect in CF and asthma.

Arginine Therapy

Morris et al. have proposed that chronic hemolysis puts patients with diseases of hereditary hemolytic anemias at risk for pulmonary hypertension (PHT). Hemolysis induces the release of hemoglobin and red blood cell arginase enzyme, limiting the availability of L-arginine to NOS, leading to diminished L-arginine and hence, NO [67]. Impaired NO bioavailability is the central dogma of endothelial dysfunction and a key factor in the pathophysiology of SCD [67], thalassemia [68], and asthma [69]. In a large (228 patients) cohort study of patients with SCD, significant elevations of plasma arginase activity were directly correlated with increased intravascular hemolytic rate and, to a lesser extent, with markers of inflammation and soluble adhesion molecule levels [70]. Investigators followed up the patients for survival up to 49 months and have shown an association between morbidity resulting from PHT due to dysregulated arginine metabolism [70].

There is growing evidence that arginine is beneficial for conditions of endothelial dysfunction [71]. Administration of oral or parenteral arginine at 6-30 g/day has shown no adverse effect [72,73]. Inhaled arginine improved pulmonary functions in patients with cystic fibrosis [74]. In SCD patients experiencing vaso-occlusive crisis, combined arginine and hydroxyurea treatment maximized both arginine and NO bioavailibilties in plasma [73]. As mentioned in this chapter, inhaled NO therapy has demonstrated benefits in children with SCD [75] as well as treating the complications of PHT [76]. However, future controlled clinical investigations of arginine supplementation in a larger number of patients may prove it to be a more affective alternative to inhaled NO.

Gene Therapy

In preventing islet cell destruction, gene transfer to islet cells may be useful in prolonging islet survival after transplantation in patients with type 1 diabetes mellitus (TIDM) [77]. Currently, successful application of gene therapy to islet modification is limited by the availability of a single, ideal gene delivery [77] and viral vectors have offered the most efficient means of gene delivery to non-dividing cells. However, currently available expression vectors suffer from significant limitations, and further study is required to determine the optimal vector for clinical islet cell gene transfer. Transplanted islets are susceptible to destruction caused by inflammation in the graft microenvironment [78]. Cytokine induced islet cell destruction occurs in part through generation of intracellular toxic oxygen radicals such as superoxide radical [79,80] and NO [81]. ROS and NO cause activation of redox sensitive NF-κB, culminating in a diverse series of events ranging from basic functional disturbances to beta cell apoptosis and/or necrosis [82]. Antioxidant gene therapy studies have led to a body of conflicting literature with respect to the protection afforded by antioxidant gene transfer. McCabe et al. have summarized the outcome of several studies into two tables signifying cytoprotectivity or lack of it by antioxidant gene transfer in insulinoma and intact islet cells [77]. These inconsistent outcomes could contribute to variability in the choice of cell line, cytokine concentration, combination of exposure, and period of exposure. Strategies aimed at targeting the inhibition iNOS, as opposed to ROS, may harbor therapeutic potential in preserving β-islet cells survival when exposed to proinflammatory cytokines [83].

Vascular gene transfer has numerous potential therapeutic applications including but not limited to the following: atherosclerosis, cerebral vasospasm, diabetes mellitus, hypercholesterolemia, hypertension, hypoxia, and sub-arachnoid hemorrhage [84-88]. Post-angioplasty restenosis, a process which results from dysregulated smooth muscle cell proliferation and migration, leads to endothelial dysfunction and ultimately to reduced production of NO or ROS. The therapeutic effect of eNOS and iNOS gene delivery in a model of endothelial denudation demonstrated the ability of adenoviral-mediated gene transfer of both NOS isoforms to inhibit intimal hyperplasia in the injured rabbit carotid artery [89]. Over expression of eNOS enhanced endothelial regeneration, while iNOS had an inhibitory effect compared to control, highlighting the potential of eNOS gene therapy as a potential therapeutic strategy for the prevention of restenosis after vascular injury [90]. Other groups have confirmed these results and extended them to the neuronal NOS isoform [90]. Vascular reactivity to adenoviral gene transfer of iNOS had a deleterious effect on atherosclerotic plaque progression [91]. In contrast, short-term, high level expression of iNOS inhibits intimal hyperplasia after mechanical injury and transplantation. In hypercholesterolemic rabbit, iNOS gene transfer resulted in increased superoxide production and reduced relaxation. Whereas, in chow fed animals, vascular relaxation or superoxide production are not affected [92]. Therefore, individual NOS isoforms may have unique characteristics when used for vascular gene modification. Post-angioplasty neointimal hyperplasia, allograft arteriosclerosis, impaired wound healing, metastatic disease, and erectile dysfunction are likely to be among the first clinical targets of iNOS gene transfer [92].

An alterative gene therapy approach to vascular dysfunction would aim to increase the antioxidant capacity of the blood vessel wall by overexpressing genes such as SOD. Endothelial dysfunction in hypertensive rat model improved with adenovirus mediated expression of human extracellular SOD [93]. In some cases, hypercholesterolemia and diabetes mellitus, endothelial dysfunction can be reversed by such strategy [94,95]. The future application of gene transfer technology to diseases of the vasculature holds significant promise.

Conclusion

Elucidation of the multiple biological roles of NO in homeostasis has led to the development of many therapies in clinical medicine. Although public has become familiar with NO through the discovery of Viagra, more medically significant areas are pulmonary hypertension in premature babies, wound healing, endothelial dysfunction with atherosclerosis, hypertension, or diabetes. Useful clinical therapies are evolving for the specific site of action of a ROS or RNS without deleterious effects on other signaling pathways.

Acknowledgement

I thank Dr. Ruth Thornton for her editorial help and Mr. Douglas Koch for his assistance with literature search.

References

[1] Cuzzocrea, S., Riley, D.P., Caputi, A.P. and Salvemini, D. (2001) Antioxidant therapy: a new pharmacological approach in shock, inflammation, and ischemia/reperfusion injury. *Pharmacol. Rev.* 53, 135-59.

[2] Leonard, D.R., Farooqi, M.H. and Myers, S. (2004) Restoration of sensation, reduced pain, and improved balance in subjects with diabetic peripheral neuropathy: a double-blind, randomized, placebo-controlled study with monochromatic near-infrared treatment. *Diabetes Care* 27, 168-72.

[3] Sen, C.K., Khanna, S., Gordillo, G., Bagchi, D., Bagchi, M. and Roy, S. (2002) Oxygen, oxidants, and antioxidants in wound healing: an emerging paradigm. *Ann. NY Acad. Sci.* 957, 239-49.

[4] Xie, K. and Huang, S. (2003) Contribution of nitric oxide-mediated apoptosis to cancer metastasis inefficiency. *Free Radic. Biol. Med.* 34, 969-86.

[5] Lala, P.K. and Orucevic, A. (1998) Role of nitric oxide in tumor progression: lessons from experimental tumors. *Cancer Metastasis Rev*. 17, 91-106.

[6] Lancaster, J.R., Jr. and Xie, K. (2006) Tumors face NO problems? *Cancer Res.* 66, 6459-62.

[7] Abraham, W.M., Ahmed, A., Serebriakov, I., Lauredo, I.T., Bassuk, J., Adams, J.A. and Sackner, M.A. (2006) Whole-body periodic acceleration modifies experimental asthma in sheep. *Am. J. Respir. Crit. Care Med*. 174, 743-52.

[8] Sackner, M.A., Gummels, E. and Adams, J.A. (2005) Effect of moderate-intensity exercise, whole-body periodic acceleration, and passive cycling on nitric oxide release into circulation. *Chest* 128, 2794-803.

[9] Matata, B.M. and Galinanes, M. (2000) Cardiopulmonary bypass exacerbates oxidative stress but does not increase proinflammatory cytokine release in patients with diabetes compared with patients without diabetes: regulatory effects of exogenous nitric oxide. *J. Thorac. Cardiovasc Surg*. 120, 1-11.

[10] Elahi, M.M., Khan, J.S. and Matata, B.M. (2006) Deleterious effects of cardiopulmonary bypass in coronary artery surgery and scientific interpretation of off-pump's logic. *Acute Card Care* 8, 196-209.

[11] Banks, B.A., Seri, I., Ischiropoulos, H., Merrill, J., Rychik, J. and Ballard, R.A. (1999) Changes in oxygenation with inhaled nitric oxide in severe bronchopulmonary dysplasia. *Pediatrics* 103, 610-8.

[12] Lorch, S.A., Banks, B.A., Christie, J., Merrill, J.D., Althaus, J., Schmidt, K., Ballard, P.L., Ischiropoulos, H. and Ballard, R.A. (2003) Plasma 3-nitrotyrosine and outcome in neonates with severe bronchopulmonary dysplasia after inhaled nitric oxide. *Free Radic. Biol. Med.* 34, 1146-52.

[13] Sittipunt, C., Steinberg, K.P., Ruzinski, J.T., Myles, C., Zhu, S., Goodman, R.B., Hudson, L.D., Matalon, S. and Martin, T.R. (2001) Nitric oxide and nitrotyrosine in the lungs of patients with acute respiratory distress syndrome. *Am. J. Respir. Crit. Care Med.* 163, 503-10.

[14] Saleh, D., Ernst, P., Lim, S., Barnes, P.J. and Giaid, A. (1998) Increased formation of the potent oxidant peroxynitrite in the airways of asthmatic patients is associated with induction of nitric oxide synthase: effect of inhaled glucocorticoid. *Faseb J.* 12, 929-37.

[15] Jones, K.L., Hegab, A.H., Hillman, B.C., Simpson, K.L., Jinkins, P.A., Grisham, M.B., Owens, M.W., Sato, E. and Robbins, R.A. (2000) Elevation of nitrotyrosine and nitrate concentrations in cystic fibrosis sputum. *Pediatr. Pulmonol*. 30, 79-85.

[16] Roberts, J.D., Jr., Fineman, J.R., Morin, F.C., 3rd, Shaul, P.W., Rimar, S., Schreiber, M.D., Polin, R.A., Zwass, M.S., Zayek, M.M., Gross, I., Heymann, M.A. and Zapol, W.M. (1997) Inhaled nitric oxide and persistent pulmonary hypertension of the newborn. The Inhaled Nitric Oxide Study Group. *N. Engl. J. Med.* 336, 605-10.

[17] The Neonatal Inhaled Nitric Oxide Study Group. (1997) Inhaled nitric oxide in full-term and nearly full-term infants with hypoxic respiratory failure. *N. Engl. J. Med.* 336, 597-604.

[18] Kinsella, J.P. (2006) Inhaled nitric oxide therapy in premature newborns. *Curr. Opin. Pediatr.* 18, 107-11.

[19] Ballard, R.A., Truog, W.E., Cnaan, A., Martin, R.J., Ballard, P.L., Merrill, J.D., Walsh, M.C., Durand, D.J., Mayock, D.E., Eichenwald, E.C., Null, D.R., Hudak, M.L., Puri, A.R., Golombek, S.G., Courtney, S.E., Stewart, D.L., Welty, S.E., Phibbs, R.H., Hibbs, A.M., Luan, X., Wadlinger, S.R., Asselin, J.M. and Coburn, C.E. (2006) Inhaled nitric oxide in preterm infants undergoing mechanical ventilation. *N. Engl. J. Med.* 355, 343-53.

[20] Kinsella, J.P., Cutter, G.R., Walsh, W.F., Gerstmann, D.R., Bose, C.L., Hart, C., Sekar, K.C., Auten, R.L., Bhutani, V.K., Gerdes, J.S., George, T.N., Southgate, W.M., Carriedo, H., Couser, R.J., Mammel, M.C., Hall, D.C., Pappagallo, M., Sardesai, S., Strain, J.D., Baier, M. and Abman, S.H. (2006) Early inhaled nitric oxide therapy in premature newborns with respiratory failure. *N. Engl. J. Med.* 355, 354-64.

[21] Schreiber, M.D., Gin-Mestan, K., Marks, J.D., Huo, D., Lee, G. and Srisuparp, P. (2003) Inhaled nitric oxide in premature infants with the respiratory distress syndrome. *N. Engl. J. Med.* 349, 2099-107.

[22] Van Meurs, K.P., Wright, L.L., Ehrenkranz, R.A., Lemons, J.A., Ball, M.B., Poole, W.K., Perritt, R., Higgins, R.D., Oh, W., Hudak, M.L., Laptook, A.R., Shankaran, S., Finer, N.N., Carlo, W.A., Kennedy, K.A., Fridriksson, J.H., Steinhorn, R.H., Sokol, G.M., Konduri, G.G., Aschner, J.L., Stoll, B.J., D'Angio, C.T. and Stevenson, D.K. (2005) Inhaled nitric oxide for premature infants with severe respiratory failure. N Engl *J. Med.* 353, 13-22.

[23] Kinsella, J.P., Walsh, W.F., Bose, C.L., Gerstmann, D.R., Labella, J.J., Sardesai, S., Walsh-Sukys, M.C., McCaffrey, M.J., Cornfield, D.N., Bhutani, V.K., Cutter, G.R., Baier, M. and Abman, S.H. (1999) Inhaled nitric oxide in premature neonates with severe hypoxaemic respiratory failure: a randomised controlled trial. *Lancet* 354, 1061-5.

[24] Truog, W.E., Pallotto, E., Clark, P., Banks, B., Kaftan, H.A., Ekekezie, II, Norberg, M. and Ballard, R.A. (2002) Interaction of endogenous endothelin-1 and inhaled nitric oxide in term and preterm infants. *Clin. Sci. (Lond) 103 Suppl.* 48, 294S-297S.

[25] Rosenberg, A.A., Kennaugh, J., Koppenhafer, S.L., Loomis, M., Chatfield, B.A. and Abman, S.H. (1993) Elevated immunoreactive endothelin-1 levels in newborn infants with persistent pulmonary hypertension. *J. Pediatr*. 123, 109-14.

[26] Frostell, C.G., Blomqvist, H., Hedenstierna, G., Lundberg, J. and Zapol, W.M. (1993) Inhaled nitric oxide selectively reverses human hypoxic pulmonary vasoconstriction without causing systemic vasodilation. *Anesthesiology* 78, 427-35.

[27] Germann, P., Braschi, A., Della Rocca, G., Dinh-Xuan, A.T., Falke, K., Frostell, C., Gustafsson, L.E., Herve, P., Jolliet, P., Kaisers, U., Litvan, H., Macrae, D.J., Maggiorini, M., Marczin, N., Mueller, B., Payen, D., Ranucci, M., Schranz, D., Zimmermann, R. and Ullrich, R. (2005) Inhaled nitric oxide therapy in adults: European expert recommendations. *Intensive Care Med.* 31, 1029-41.

[28] Griffiths, M.J. and Evans, T.W. (2005) Inhaled nitric oxide therapy in adults. *N. Engl. J. Med.* 353, 2683-95.

[29] Hare, J.M., Shernan, S.K., Body, S.C., Graydon, E., Colucci, W.S. and Couper, G.S. (1997) Influence of inhaled nitric oxide on systemic flow and ventricular filling pressure in patients receiving mechanical circulatory assistance. *Circulation* 95, 2250-3.

[30] Morales-Blanhir, J., Santos, S., de Jover, L., Sala, E., Pare, C., Roca, J., Rodriguez-Roisin, R. and Barbera, J.A. (2004) Clinical value of vasodilator test with inhaled nitric oxide for predicting long-term response to oral vasodilators in pulmonary hypertension. *Respir. Med.* 98, 225-34.

[31] Bottiger, B.W., Motsch, J., Dorsam, J., Mieck, U., Gries, A., Weimann, J. and Martin, E. (1996) Inhaled nitric oxide selectively decreases pulmonary artery pressure and pulmonary vascular resistance following acute massive pulmonary microembolism in piglets. *Chest* 110, 1041-7.

[32] Estagnasie, P., Le Bourdelles, G., Mier, L., Coste, F. and Dreyfuss, D. (1994) Use of inhaled nitric oxide to reverse flow through a patent foramen ovale during pulmonary embolism. *Ann. Intern. Med.* 120, 757-9.

[33] Atz, A.M. and Wessel, D.L. (1997) Inhaled nitric oxide in sickle cell disease with acute chest syndrome. *Anesthesiology* 87, 988-90.

[34] Weiner, D.L., Hibberd, P.L., Betit, P., Cooper, A.B., Botelho, C.A. and Brugnara, C. (2003) Preliminary assessment of inhaled nitric oxide for acute vaso-occlusive crisis in pediatric patients with sickle cell disease. *Jama* 289, 1136-42.

[35] Vonbank, K., Ziesche, R., Higenbottam, T.W., Stiebellehner, L., Petkov, V., Schenk, P., Germann, P. and Block, L.H. (2003) Controlled prospective randomised trial on the effects on pulmonary haemodynamics of the ambulatory long term use of nitric oxide and oxygen in patients with severe COPD. *Thorax* 58, 289-93.

[36] Gianetti, J., Del Sarto, P., Bevilacqua, S., Vassalle, C., De Filippis, R., Kacila, M., Farneti, P.A., Clerico, A., Glauber, M. and Biagini, A. (2004) Supplemental nitric oxide and its effect on myocardial injury and function in patients undergoing cardiac surgery with extracorporeal circulation. *J. Thorac. Cardiovasc. Surg.* 127, 44-50.

[37] Macdonald, P.S., Keogh, A., Mundy, J., Rogers, P., Nicholson, A., Harrison, G., Jansz, P., Kaan, A.M. and Spratt, P. (1998) Adjunctive use of inhaled nitric oxide during implantation of a left ventricular assist device. *J. Heart Lung Transplant* 17, 312-6.

[38] Argenziano, M., Choudhri, A.F., Moazami, N., Rose, E.A., Smith, C.R., Levin, H.R., Smerling, A.J. and Oz, M.C. (1998) Randomized, double-blind trial of inhaled nitric oxide in LVAD recipients with pulmonary hypertension. *Ann. Thorac. Surg.* 65, 340-5.

[39] Rabkin, D.G., Sladen, R.N., DeMango, A., Steinglass, K.M. and Goldstein, D.J. (2001) Nitric oxide for the treatment of postpneumonectomy pulmonary edema. *Ann. Thorac. Surg.* 72, 272-4.

[40] Jia, L., Bonaventura, C., Bonaventura, J. and Stamler, J.S. (1996) S-nitrosohaemoglobin: a dynamic activity of blood involved in vascular control. *Nature* 380, 221-6.

[41] Stamler, J.S., Jaraki, O., Osborne, J., Simon, D.I., Keaney, J., Vita, J., Singel, D., Valeri, C.R. and Loscalzo, J. (1992) Nitric oxide circulates in mammalian plasma

primarily as an S-nitroso adduct of serum albumin. *Proc. Natl. Acad. Sci. USA* 89, 7674-7.

[42] Fang, F.C. (2004) Antimicrobial reactive oxygen and nitrogen species: concepts and controversies. *Nat. Rev. Microbiol.* 2, 820-32.

[43] Mannick, J.B. (2006) Immunoregulatory and antimicrobial effects of nitrogen oxides. *Proc. Am. Thorac. Soc.* 3, 161-5.

[44] Long, R., Light, B. and Talbot, J.A. (1999) Mycobacteriocidal action of exogenous nitric oxide. *Antimicrob. Agents Chemother* 43, 403-5.

[45] Flynn, J.L., Scanga, C.A., Tanaka, K.E. and Chan, J. (1998) Effects of aminoguanidine on latent murine tuberculosis. *J. Immunol.* 160, 1796-803.

[46] MacMicking, J.D., North, R.J., LaCourse, R., Mudgett, J.S., Shah, S.K. and Nathan, C.F. (1997) Identification of nitric oxide synthase as a protective locus against tuberculosis. *Proc. Natl. Acad. Sci. USA* 94, 5243-8.

[47] Taylor, R.W., Zimmerman, J.L., Dellinger, R.P., Straube, R.C., Criner, G.J., Davis, K., Jr., Kelly, K.M., Smith, T.C. and Small, R.J. (2004) Low-dose inhaled nitric oxide in patients with acute lung injury: a randomized controlled trial. *Jama* 291, 1603-9.

[48] Long, R., Jones, R., Talbot, J., Mayers, I., Barrie, J., Hoskinson, M. and Light, B. (2005) Inhaled nitric oxide treatment of patients with pulmonary tuberculosis evidenced by positive sputum smears. *Antimicrob Agents Chemother* 49, 1209-12.

[49] Gow, A.J. (2005) Nitric oxide, hemoglobin, and hypoxic vasodilation. *Am. J. Respir. Cell Mol. Biol.* 32, 479-82.

[50] McMahon, T.J., Ahearn, G.S., Moya, M.P., Gow, A.J., Huang, Y.C., Luchsinger, B.P., Nudelman, R., Yan, Y., Krichman, A.D., Bashore, T.M., Califf, R.M., Singel, D.J., Piantadosi, C.A., Tapson, V.F. and Stamler, J.S. (2005) A nitric oxide processing defect of red blood cells created by hypoxia: deficiency of S-nitrosohemoglobin in pulmonary hypertension. *Proc. Natl. Acad. Sci. USA* 102, 14801-6.

[51] James, P.E., Lang, D., Tufnell-Barret, T., Milsom, A.B. and Frenneaux, M.P. (2004) Vasorelaxation by red blood cells and impairment in diabetes: reduced nitric oxide and oxygen delivery by glycated hemoglobin. *Circ. Res.* 94, 976-83.

[52] Pawloski, J.R., Hess, D.T. and Stamler, J.S. (2005) Impaired vasodilation by red blood cells in sickle cell disease. *Proc. Natl. Acad. Sci. USA* 102, 2531-6.

[53] Smith, A.D., Cowan, J.O., Brassett, K.P., Herbison, G.P. and Taylor, D.R. (2005) Use of exhaled nitric oxide measurements to guide treatment in chronic asthma. *N. Engl. J. Med.* 352, 2163-73.

[54] Kharitonov, S.A., Gonio, F., Kelly, C., Meah, S. and Barnes, P.J. (2003) Reproducibility of exhaled nitric oxide measurements in healthy and asthmatic adults and children. *Eur. Respir. J.* 21, 433-8.

[55] Jones, S.L., Kittelson, J., Cowan, J.O., Flannery, E.M., Hancox, R.J., McLachlan, C.R. and Taylor, D.R. (2001) The predictive value of exhaled nitric oxide measurements in assessing changes in asthma control. *Am. J. Respir. Crit. Care Med.* 164, 738-43.

[56] (2005) ATS/ERS recommendations for standardized procedures for the online and offline measurement of exhaled lower respiratory nitric oxide and nasal nitric oxide, 2005. *Am. J. Respir. Crit. Care Med.* 171, 912-30.

[57] Horvath, I., Hunt, J., Barnes, P.J., Alving, K., Antczak, A., Baraldi, E., Becher, G., van Beurden, W.J., Corradi, M., Dekhuijzen, R., Dweik, R.A., Dwyer, T., Effros, R., Erzurum, S., Gaston, B., Gessner, C., Greening, A., Ho, L.P., Hohlfeld, J., Jobsis, Q., Laskowski, D., Loukides, S., Marlin, D., Montuschi, P., Olin, A.C., Redington, A.E., Reinhold, P., van Rensen, E.L., Rubinstein, I., Silkoff, P., Toren, K., Vass, G., Vogelberg, C. and Wirtz, H. (2005) Exhaled breath condensate: methodological recommendations and unresolved questions. *Eur. Respir. J.* 26, 523-48.

[58] Matalon, S., Hardiman, K.M., Jain, L., Eaton, D.C., Kotlikoff, M., Eu, J.P., Sun, J., Meissner, G. and Stamler, J.S. (2003) Regulation of ion channel structure and function by reactive oxygen-nitrogen species. *Am. J. Physiol. Lung Cell Mol. Physiol.* 285, L1184-9.

[59] Gaston, B., Reilly, J., Drazen, J.M., Fackler, J., Ramdev, P., Arnelle, D., Mullins, M.E., Sugarbaker, D.J., Chee, C., Singel, D.J. and et al. (1993) Endogenous nitrogen oxides and bronchodilator S-nitrosothiols in human airways. *Proc. Natl. Acad. Sci. USA* 90, 10957-61.

[60] Kluge, I., Gutteck-Amsler, U., Zollinger, M. and Do, K.Q. (1997) S-nitrosoglutathione in rat cerebellum: identification and quantification by liquid chromatography-mass spectrometry. *J. Neurochem.* 69, 2599-607.

[61] Grasemann, H., Gaston, B., Fang, K., Paul, K. and Ratjen, F. (1999) Decreased levels of nitrosothiols in the lower airways of patients with cystic fibrosis and normal pulmonary function. *J. Pediatr.* 135, 770-2.

[62] Gaston, B., Sears, S., Woods, J., Hunt, J., Ponaman, M., McMahon, T. and Stamler, J.S. (1998) Bronchodilator S-nitrosothiol deficiency in asthmatic respiratory failure. *Lancet* 351, 1317-9.

[63] Servetnyk, Z., Krjukova, J., Gaston, B., Zaman, K., Hjelte, L., Roomans, G.M. and Dragomir, A. (2006) Activation of chloride transport in CF airway epithelial cell lines and primary CF nasal epithelial cells by S-nitrosoglutathione. *Respir. Res.* 7, 124.

[64] Chen, L., Patel, R.P., Teng, X., Bosworth, C.A., Lancaster, J.R., Jr. and Matalon, S. (2006) Mechanisms of cystic fibrosis transmembrane conductance regulator activation by S-nitrosoglutathione. *J. Biol. Chem.* 281, 9190-9.

[65] Zaman, K., Carraro, S., Doherty, J., Henderson, E.M., Lendermon, E., Liu, L., Verghese, G., Zigler, M., Ross, M., Park, E., Palmer, L.A., Doctor, A., Stamler, J.S. and Gaston, B. (2006) S-nitrosylating agents: a novel class of compounds that increase cystic fibrosis transmembrane conductance regulator expression and maturation in epithelial cells. *Mol. Pharmacol.* 70, 1435-42.

[66] Zaman, K., Hanigan, M.H., Smith, A., Vaughan, J., Macdonald, T., Jones, D.R., Hunt, J.F. and Gaston, B. (2006) Endogenous S-nitrosoglutathione modifies 5-lipoxygenase expression in airway epithelial cells. *Am. J. Respir. Cell Mol. Biol.* 34, 387-93.

[67] Morris, C.R. (2006) New Strategies for the Treatment of Pulmonary Hypertension in Sickle Cell Disease : The Rationale for Arginine Therapy. Treat Respir Med 5, 31-45.

[68] Morris, C.R., Kuypers, F.A., Kato, G.J., Lavrisha, L., Larkin, S., Singer, T. and Vichinsky, E.P. (2005) Hemolysis-associated pulmonary hypertension in thalassemia. *Ann. NY Acad. Sci.* 1054, 481-5.

[69] Morris, C.R., Poljakovic, M., Lavrisha, L., Machado, L., Kuypers, F.A. and Morris, S.M., Jr. (2004) Decreased arginine bioavailability and increased serum arginase activity in asthma. *Am. J. Respir. Crit. Care Med.* 170, 148-53.
[70] Morris, C.R., Kato, G.J., Poljakovic, M., Wang, X., Blackwelder, W.C., Sachdev, V., Hazen, S.L., Vichinsky, E.P., Morris, S.M., Jr. and Gladwin, M.T. (2005) Dysregulated arginine metabolism, hemolysis-associated pulmonary hypertension, and mortality in sickle cell disease. *Jama* 294, 81-90.
[71] Gornik, H.L. and Creager, M.A. (2004) Arginine and endothelial and vascular health. *J. Nutr.* 134, 2880S-2887S; discussion 2895S.
[72] Maxwell, A.J. and Cooke, J.P. (1998) Cardiovascular effects of L-arginine. *Curr. Opin. Nephrol. Hypertens* 7, 63-70.
[73] Morris, C.R., Vichinsky, E.P., van Warmerdam, J., Machado, L., Kepka-Lenhart, D., Morris, S.M., Jr. and Kuypers, F.A. (2003) Hydroxyurea and arginine therapy: impact on nitric oxide production in sickle cell disease. *J. Pediatr. Hematol. Oncol.* 25, 629-34.
[74] Solomons, C.C., Cotton, E.K. and Dubois, R. (1971) The use of buffered L-arginine in the treatment of cystic fibrosis. *Pediatrics* 47, 384-90.
[75] Surdacki, A., Zmudka, K., Bieron, K., Kostka-Trabka, E., Dubiel, J.S. and Gryglewski, R.J. (1994) *Lack of beneficial effects of L-arginine infusion in primary pulmonary hypertension.* Wien Klin Wochenschr 106, 521-6.
[76] Morris, C.R., Morris, S.M., Jr., Hagar, W., Van Warmerdam, J., Claster, S., Kepka-Lenhart, D., Machado, L., Kuypers, F.A. and Vichinsky, E.P. (2003) Arginine therapy: a new treatment for pulmonary hypertension in sickle cell disease? *Am. J. Respir. Crit .Care Med.* 168, 63-9.
[77] McCabe, C., Samali, A. and O'Brien, T. (2006) Cytoprotection of beta cells: rational gene transfer strategies. *Diabetes Metab. Res. Rev.* 22, 241-52.
[78] Giannoukakis, N., Rudert, W.A., Robbins, P.D. and Trucco, M. (1999) Targeting autoimmune diabetes with gene therapy. *Diabetes* 48, 2107-21.
[79] Eizirik, D.L. and Mandrup-Poulsen, T. (2001) A choice of death--the signal-transduction of immune-mediated beta-cell apoptosis. *Diabetologia* 44, 2115-33.
[80] Rabinovitch, A. (1998) Animal models of type 1 diabetes are relevant to human IDDM--use caution. *Diabetes Metab. Rev.*14, 189-90.
[81] Suarez-Pinzon, W.L., Strynadka, K., Schulz, R. and Rabinovitch, A. (1994) Mechanisms of cytokine-induced destruction of rat insulinoma cells: the role of nitric oxide. *Endocrinology* 134, 1006-10.
[82] Mandrup-Poulsen, T. (2003) Apoptotic signal transduction pathways in diabetes. *Biochem. Pharmacol.* 66, 1433-40.
[83] McCabe, C., Samali, A. and O'Brien, T. (2006) beta cell cytoprotective strategies: establishing the relative roles for iNOS and ROS. *Biochem. Biophys. Res. Commun.* 342, 1240-8.
[84] Khurana, V.G., Smith, L.A., Baker, T.A., Eguchi, D., O'Brien, T. and Katusic, Z.S. (2002) Protective vasomotor effects of in vivo recombinant endothelial nitric oxide synthase gene expression in a canine model of cerebral vasospasm. *Stroke* 33, 782-9.

[85] Zanetti, M., Sato, J., Katusic, Z.S. and O'Brien, T. (2000) Gene transfer of endothelial nitric oxide synthase alters endothelium-dependent relaxations in aortas from diabetic rabbits. *Diabetologia* 43, 340-7.

[86] Mozes, G., Kullo, I.J., Mohacsi, T.G., Cable, D.G., Spector, D.J., Crotty, T.B., Gloviczki, P., Katusic, Z.S. and O'Brien, T. (1998) Ex vivo gene transfer of endothelial nitric oxide synthase to atherosclerotic rabbit aortic rings improves relaxations to acetylcholine. *Atherosclerosis* 141, 265-71.

[87] Cable, D.G., Pompili, V.J., O'Brien, T. and Schaff, H.V. (1999) Recombinant gene transfer of endothelial nitric oxide synthase augments coronary artery relaxations during hypoxia. *Circulation* 100, II335-9.

[88] Onoue, H., Tsutsui, M., Smith, L., Stelter, A., O'Brien, T. and Katusic, Z.S. (1998) Expression and function of recombinant endothelial nitric oxide synthase gene in canine basilar artery after experimental subarachnoid hemorrhage. *Stroke* 29, 1959-65; discussion 1965-6.

[89] Cooney, R., Hynes, S.O., Sharif, F., Howard, L. and O'Brien, T. (2006) Effect of gene delivery of NOS isoforms on intimal hyperplasia and endothelial regeneration after balloon injury. *Gene Ther*.

[90] Channon, K.M., Qian, H., Neplioueva, V., Blazing, M.A., Olmez, E., Shetty, G.A., Youngblood, S.A., Pawloski, J., McMahon, T., Stamler, J.S. and George, S.E. (1998) In vivo gene transfer of nitric oxide synthase enhances vasomotor function in carotid arteries from normal and cholesterol-Fed rabbits. *Circulation* 98, 1905-11.

[91] Behr-Roussel, D., Rupin, A., Simonet, S., Bonhomme, E., Coumailleau, S., Cordi, A., Serkiz, B., Fabiani, J.N. and Verbeuren, T.J. (2000) Effect of chronic treatment with the inducible nitric oxide synthase inhibitor N-iminoethyl-L-lysine or with L-arginine on progression of coronary and aortic atherosclerosis in hypercholesterolemic rabbits. *Circulation* 102, 1033-8.

[92] Zanetti, M., d'Uscio, L.V., Kovesdi, I., Katusic, Z.S. and O'Brien, T. (2003) In vivo gene transfer of inducible nitric oxide synthase to carotid arteries from hypercholesterolemic rabbits. *Stroke* 34, 1293-8.

[93] Fennell, J.P., Brosnan, M.J., Frater, A.J., Hamilton, C.A., Alexander, M.Y., Nicklin, S.A., Heistad, D.D., Baker, A.H. and Dominiczak, A.F. (2002) Adenovirus-mediated overexpression of extracellular superoxide dismutase improves endothelial dysfunction in a rat model of hypertension. *Gene Ther* 9, 110-7.

[94] Zanetti, M., Sato, J., Katusic, Z.S. and O'Brien, T. (2001) Gene transfer of superoxide dismutase isoforms reverses endothelial dysfunction in diabetic rabbit aorta. *Am. J. Physiol. Heart Circ. Physiol*. 280, H2516-23.

[95] Zanetti, M., Sato, J., Jost, C.J., Gloviczki, P., Katusic, Z.S. and O'Brien, T. (2001) Gene transfer of manganese superoxide dismutase reverses vascular dysfunction in the absence but not in the presence of atherosclerotic plaque. *Hum. Gene Ther*. 12, 1407-16.

Index

A

B

C

D

E

F

G

I

J

K

L

M

P

Q

R

S

T

U

V

W

X

Y

Z